AF838082

Lecture Notes in Networks and Systems

Volume 1895

Series Editor

Janusz Kacprzyk, Systems Research Institute, Polish Academy of Sciences, Warsaw, Poland

Advisory Editors

Fernando Gomide, Department of Computer Engineering and Automation—DCA, School of Electrical and Computer Engineering—FEEC, University of Campinas—UNICAMP, São Paulo, Brazil

Okyay Kaynak, Department of Electrical and Electronics Engineering, Bogazici University, Istanbul, Türkiye

Derong Liu, Department of Electrical and Computer Engineering, University of Illinois at Chicago, Chicago, USA
Institute of Automation, Chinese Academy of Sciences, Beijing, China

Witold Pedrycz, Department of Electrical and Computer Engineering, University of Alberta, Edmonton, Alberta, Canada
Systems Research Institute, Polish Academy of Sciences, Warsaw, Poland

Marios M. Polycarpou, Department of Electrical and Computer Engineering, KIOS Research Center for Intelligent Systems and Networks, University of Cyprus, Nicosia, Cyprus

Imre J. Rudas, Óbuda University, Budapest, Hungary

Jun Wang, Department of Computer Science, City University of Hong Kong, Kowloon, Hong Kong

The series "Lecture Notes in Networks and Systems" publishes the latest developments in Networks and Systems—quickly, informally and with high quality. Original research reported in proceedings and post-proceedings represents the core of LNNS.

Volumes published in LNNS embrace all aspects and subfields of, as well as new challenges in, Networks and Systems.

The series contains proceedings and edited volumes in systems and networks, spanning the areas of Cyber-Physical Systems, Autonomous Systems, Sensor Networks, Control Systems, Energy Systems, Automotive Systems, Biological Systems, Vehicular Networking and Connected Vehicles, Aerospace Systems, Automation, Manufacturing, Smart Grids, Nonlinear Systems, Power Systems, Robotics, Social Systems, Economic Systems and other. Of particular value to both the contributors and the readership are the short publication timeframe and the world-wide distribution and exposure which enable both a wide and rapid dissemination of research output.

The series covers the theory, applications, and perspectives on the state of the art and future developments relevant to systems and networks, decision making, control, complex processes and related areas, as embedded in the fields of interdisciplinary and applied sciences, engineering, computer science, physics, economics, social, and life sciences, as well as the paradigms and methodologies behind them.

Indexed by SCOPUS, EI Compendex, INSPEC, WTI Frankfurt eG, zbMATH, SCImago.

All books published in the series are submitted for consideration in Web of Science.

For proposals from Asia please contact Aninda Bose (aninda.bose@springer.com).

Anastasia Kulachinskaya · Mustika Sari ·
Andrea Bencsik · Zuzana Dvořáková ·
Artur Kuchumov
Editors

Digital Transformation: What Kind of Smart Life do we have today?

Volume 1

Springer

Editors
Anastasia Kulachinskaya
Peter the Great St. Petersburg Polytechnic University
St. Petersburg, Russia

Andrea Bencsik
Wekerle International University
Budapest, Hungary

Artur Kuchumov
St. Petersburg State University of Economics
St. Petersburg, Russia

Mustika Sari
Universitas Indonesia
Jawa Barat, Indonesia

Zuzana Dvořáková
Masaryk Institute of Advanced Studies
Czech Technical University in Prague
Prague, Czech Republic

ISSN 2367-3370 ISSN 2367-3389 (electronic)
Lecture Notes in Networks and Systems
ISBN 978-3-032-20581-0 ISBN 978-3-032-20582-7 (eBook)
https://doi.org/10.1007/978-3-032-20582-7

© The Editor(s) (if applicable) and The Author(s), under exclusive license to Springer Nature Switzerland AG 2026

This work is subject to copyright. All rights are solely and exclusively licensed by the Publisher, whether the whole or part of the material is concerned, specifically the rights of translation, reprinting, reuse of illustrations, recitation, broadcasting, reproduction on microfilms or in any other physical way, and transmission or information storage and retrieval, electronic adaptation, computer software, or by similar or dissimilar methodology now known or hereafter developed.
The use of general descriptive names, registered names, trademarks, service marks, etc. in this publication does not imply, even in the absence of a specific statement, that such names are exempt from the relevant protective laws and regulations and therefore free for general use.
The publisher, the authors and the editors are safe to assume that the advice and information in this book are believed to be true and accurate at the date of publication. Neither the publisher nor the authors or the editors give a warranty, expressed or implied, with respect to the material contained herein or for any errors or omissions that may have been made. The publisher remains neutral with regard to jurisdictional claims in published maps and institutional affiliations.

This Springer imprint is published by the registered company Springer Nature Switzerland AG
The registered company address is: Gewerbestrasse 11, 6330 Cham, Switzerland

If disposing of this product, please recycle the paper.

Contents

Smart Destinations, Sport, and Tourism

Contributors

Matluba Nematovna Abdullayeva Tashkent State University of Economics, Tashkent, Uzbekistan

Khasanboy Abdusamatov ISFT Institute, Tashkent, Uzbekistan

Sanobar Ikrom Kizi Akhmedova University of Science and Technologies, Tashkent, Uzbekistan

Iroda Abdushukurovna Bakiyeva University of science and technologies, Tashkent, Uzbekistan

Andrey A. Bogatyrev University of Science and Technology MISIS, Moscow, Russia

Olga P. Bogatyreva Tver State University, Tver, Russia

Tatyana Boriskina Volgograd State Technisch University, Volgograd, Russia

Irina Aleksandrovna Brusakova Saint-Petersburg State Electrotechnical University "LETI", Saint-Petersburg, Russia

K. E. Chelyshev University of Tyumen, Tyumen, Russia

T. Y. Chernysheva University of Tyumen, Tyumen, Russia

Ivan Chusov Volgograd Cooperative Institute (Branch), Russian University of Cooperation, Volgograd, Russia

S. E. Demidova Financial University Under the Government of the Russian Federation, Moscow, Russia

A. I. Demina Higher School of Economics, National Research University, St Petersburg, Russia

Alissa Dubgorn Peter the Great St.Petersburg Polytechnic University, Saint-Petersburg, Russia

Anora Norboy kizi Eshtemirova Tashkent State Transport University, Tashkent, Uzbekistan

M. Gabibova Herzen State Pedagogical University of Russia, St Petersburg, Russia

Sardorkhon Sunnatillaevich Gaibnazarov Tashkent State Medical University, Tashkent, Uzbekistan

Sunatilla Baxodirjanovich Gaibnazarov Tashkent State Technical University, Tashkent, Uzbekistan

Zumrat Talatovna Gaibnazarova University of Science and Technologies, Tashkent, Uzbekistan

Olga Vasilevna Gasheva Euphrosyne Polotskaya State University of Polotsk, Novopolotsk, Belarus

Leysan Gatina Kazan National Research Technological University, Kazan, Russia

Valeriya Glazkova National Research Moscow State University of Civil Engineering (NRU MGSU), Moscow, Russia

Anait Vardanovna Grigorian Department of Business Economics, Entrepreneurship and Innovation, St. Petersburg State University, St. Petersburg, Russian Federation

Nadezhda Guseva Saint Petersburg State University, Saint Petersburg, Russia

Elmira Huzieva Kazan State Power Engineering University, Kazan, Russia

Feruza Axtamovna Ibragimova University of science and technologies, Tashkent, Uzbekistan

Natalia Ketko Volgograd State Technisch University, Volgograd, Russia

Osman Khalov Plekhanov Russian University of Economics, Moscow, Russia

Radmila Khizbullina Kazan State Power Engineering University, Kazan, Russia

Oleg Kolyshev Volgograd State Socio-Pedagogical University, Volgograd, Russia

Yuri Kolyshev Volgograd State Technisch University, Volgograd, Russia

Arthur Kuchumov Saint Petersburg State University of Economics, Saint Petersburg, Russia

Sevara Marat kizi Kurbanbaeva Tashkent State University of Economics, Tashkent, Uzbekistan

L. V. Larchenko Herzen State Pedagogical University of Russia, St Petersburg, Russia

Anastasia Levina Peter the Great St.Petersburg Polytechnic University, Saint-Petersburg, Russia

Alena Lisichonak Euphrosyne Polotskaya State University of Polotsk, Novopolotsk, Belarus

Elena A. Litash-Sorokina Russian Presidential Academy of National Economy and Public Administration (RANEPA), Moscow, Russia

Elena A. Litvintseva Russian Presidential Academy of National Economy and Public Administration, Institute of Public Administration and Civil Service, Moscow, Russia

Alena Borisovna Maley Euphrosyne Polotskaya State University of Polotsk, Novopolotsk, Belarus

Elena Marusinina Volgograd State University, Volgograd, Russia

Elena Maslennikova Moscow Metropolitan Governance Yury Luzhkov University, Moscow, Russia

Elena G. Milyugina Tver State University, Tver, Russia

Nikolay Nikolaevich Molchanov Department of Business Economics, Entrepreneurship and Innovation, St. Petersburg State University, St. Petersburg, Russian Federation

Oksana Sergeevna Muraveva Department of Business Economics, Entrepreneurship and Innovation, St. Petersburg State University, St. Petersburg, Russian Federation

Natalia Mushketova Volgograd State University, Volgograd, Russia

A. N. Paranina Herzen State Pedagogical University of Russia, St Petersburg, Russia

M. A. Plakhotnikova University of Tyumen, Tyumen, Russia

Nadezhda N. Pokrovskaia Saint-Petersburg State Electrotechnical University "LETI", Saint-Petersburg, Russia

Irina Pozdnyakova Euphrosyne Polotskaya State University of Polotsk, Novopolotsk, Belarus

Viktoriya Razletovskaia MGIMO University, Moscow, Russian Federation

Oksana Nikolaevna Rozhko Plekhanov Russian University of Economics, Moscow, Russia

Polina Rukina Peter the Great St.Petersburg Polytechnic University, Saint-Petersburg, Russia

Nodira Rustamova ISFT Institute, Tashkent, Uzbekistan

Tatiana A. Savich Russian Presidential Academy of National Economy and Public Administration, Institute of Public Administration and Civil Service, Moscow, Russia

Gulnora Mardievna Shadieva Samarkand Institute of Economics and Service, Samarkand, Uzbekistan

Elena Sharafanova International Banking Institute Named After Anatoliy Sobchak, Saint Petersburg, Russia

Adel Shokhnekh Volgograd State Technisch University, Volgograd, Russia

Anna Shokhnekh Volgograd State Socio-Pedagogical University, Volgograd, Russia

Natalya Skiter Volgograd State Technisch University, Volgograd, Russia

Arsenii Sluzhaev Peter the Great St.Petersburg Polytechnic University, Saint-Petersburg, Russia

Svetlana Soldatova FSBI 'Directorate of Scientific and Technical Programs', Moscow, Russia

Igor Stepnov MGIMO University, Moscow, Russian Federation

Lyudmila A. Stepnova Peoples' Friendship University of Russia (RUDN), Moscow, Russia;
Russian Presidential Academy of National Economy and Public Administration, Institute of Public Administration and Civil Service, Moscow, Russia;
Moscow Institute of Psychoanalysis, Moscow, Russia

Mukhtor Nematovich Sunnatov Tashkent State University of Economics, Tashkent, Uzbekistan

Anzhelika Syrbu The Russian Presidential Academy of National Economy and Public Administration, Volgograd, Russia

Victoria Telyatnikova Volgograd State Socio-Pedagogical University, Volgograd, Russia

Yana Testina Saint Petersburg State University, Saint Petersburg, Russia

Elizaveth Tetkina Volgograd State Socio-Pedagogical University, Volgograd, Russia

Artem Turenko FSBI 'Directorate of Scientific and Technical Programs', Moscow, Russia

Madina Gafarovna Tuychiyeva Tashkent State Transport University, Tashkent, Uzbekistan

Liudmila A. Vasilenko Russian Presidential Academy of National Economy and Public Administration (RANEPA), Moscow, Russia

Aliaksandr Yemialyanau Euphrosyne Polotskaya State University of Polotsk, Novopolotsk, Belarus

Flor Zakharchenko Volgograd State University, Volgograd, Russia

O. V. Zakharova University of Tyumen, Tyumen, Russia

Irina A. Zhuravleva Financial University under the Government of the Russian Federation, Moscow, Russia

Inha Vladimirovna Ziankova Euphrosyne Polotskaya State University of Polotsk, Novopolotsk, Belarus

Elena Zudina Volgograd State Socio-Pedagogical University, Volgograd, Russia

Smart Road Services and Traffic Management

From Lines to Networks: How Modern Digital Technologies Are Transforming Supply Chains into Intelligent Distribution Ecosystems

Oksana Nikolaevna Rozhko and Osman Khalov

Abstract The traditional supply chain model, which represents a linear sequence of actions from the supplier to the end consumer, has ceased to meet the demands of the modern economy. In the context of globalization, rapid market changes, uncertainty, and high requirements for supply chain efficiency, this approach has become insufficient. Modern technologies enable the creation of more flexible and resilient systems, transforming simple supply chains into complex intellectual ecosystems. This chapter examines the transformation of distribution channels under the influence of the convergence of key digital technologies: artificial intelligence (AI), machine learning, the Internet of Things (IoT), blockchain, digital twins, and cloud computing. The analysis focuses on how their integration creates a synergistic effect, leading to the emergence of autonomous, self-optimizing, and customer-centric distribution networks. These advanced networks are capable of predicting demand, adapting with flexibility, and ensuring an unprecedented level of transparency and operational efficiency. The study posits that this paradigm shift from sequential chains to dynamic, intelligent networks represents a fundamental change in the logistics and distribution landscape, driven by the pervasive adoption of disruptive digital tools and data-centric decision-making processes.

Keywords Supply networks · Digital transformation · Artificial intelligence · Distribution channels · Cognitive infocommunication system

1 Introduction

Distribution logistics constitutes the final link of the supply chain, directly connecting the company with the customer. Its efficiency determines customer satisfaction, operational costs, and overall profitability. Today, the organization of supply chains stands

O. N. Rozhko (✉) · O. Khalov
Plekhanov Russian University of Economics, Moscow, Russia
e-mail: oxana.rozhcko@yandex.ru

© The Author(s), under exclusive license to Springer Nature Switzerland AG 2026

A. Kulachinskaya et al. (eds.), *Digital Transformation: What Kind of Smart Life do we have today?*, Lecture Notes in Networks and Systems 1895,
https://doi.org/10.1007/978-3-032-20582-7_1

on the brink of a fundamental transformation. The combination of artificial intelligence and modern technologies does not merely optimize existing processes but creates a fundamentally new paradigm for their organization. The subject of this research is modern supply chains, and the object is the process of their transformation from a linear to a network structure of organization based on innovative management approaches.

Supply chains have traditionally been built on the "supplier-to-consumer" principle, with each link representing an independent segment, most often connected by centralized management and limited information about the state of the entire system. This model was effective in a stable world with predictable demand; however, in the modern reality, characterized by market volatility, growing consumer expectations, and global crises, its shortcomings have become most evident [1]. Its fragility has been revealed, where the disruption of a single link, for instance, a failure in the supply of components to a manufacturing plant, paralyzes the entire chain [2], and its inflexibility, where responding to changes in demand takes weeks, leading to stockouts or surplus inventory [3].

Digital technologies are radically altering this paradigm, transforming linear supply chains into dynamic, adaptive, and intelligent supply networks [3, 4]. If a Supply Chain is a predetermined, hierarchical route focused on efficiency and cost minimization at each individual segment, then a Supply Network is a decentralized ecosystem where every participant (supplier, logistics provider, manufacturer, distributor) can be directly connected to any other [5]. The network's focus shifts towards resilience, speed, and overall value for the end customer. A network is a web of interconnections, where each node interacts not only with the preceding and subsequent participants but also with other nodes horizontally, creating extensive information communications among the participants of the distribution system [6].

This model is more flexible, adaptive, and efficient than the linear one [6] because, firstly, it ensures flexibility. Network models allow for a faster response to changes in customer needs, demand fluctuations, and external shocks. This makes systems more resilient and adaptive. Secondly, it enhances transparency [2]. In a network structure, data on the movement of goods, production processes, inventory status, and other parameters are accessible to all participants in the chain, which facilitates better planning and real-time decision-making [2, 6]. Thirdly, it reduces costs. In the network model, the optimization of inventory and processes occurs based on real-time data, enabling the reduction of excess inventory and storage costs [7]. Digital technologies act as the catalyst and the "nervous system" for the transformation of Supply Chains into Supply Networks [8].

2 Literature Review

The strategy and practice of implementing key digital technologies that help transform traditional chains into dynamic, self-learning networks are being studied by numerous researchers who propose various economic mechanisms for applying the available modern digital toolkit. Accumulated scientific and practical experience demonstrates that the digitalization of supply chains facilitates their transition into intellectual ecosystems [9–11]. The technological drivers of this transition that have seen the most widespread adoption currently include artificial intelligence and machine learning, the Internet of Things, blockchain, digital twins, and cloud computing. Let us examine them in more detail.

IoT and Sensors. Connected devices and sensors at all levels (supply, production, transport, warehouse) provide a stream of real-time data. This data is the foundation of the network: it enables the measurement of status, tracking of movements, forecasting of loads, and prompt response. Reference [12] notes that IoT technologies are one of the dominant driving forces behind the digital transformation of supply chains.

Big Data and Analytics. Collecting data is not enough; the crucial step is its analysis. Reference [3] explores the concept of transitioning from descriptive analytics to predictive and prescriptive analytics. According to the authors, this allows for the identification of demand patterns, optimization of loads, and prediction of disruptions. In the network model, it becomes possible to consolidate data from different participants (e.g., manufacturer + logistics operator + distributor) and process it within a unified information system, which, in turn, enhances the competitive advantage of the network's actors.

Cloud Technologies serve as the foundation. They enable the creation of a unified digital space for all network participants, integrating disparate ERP, WMS, and TMS systems within the environment of a single information space of a cloud platform (SaaS, KaaS), accessible in real-time to all partners. The primary technological effect is the creation of a so-called "transparent network" or a "single source of truth" for the entire network, meaning that informational "gaps" are eliminated [2]. A manufacturer sees data on a retailer's inventory, and a logistics company gains access to production plans to optimize the transportation schedule [11].

Research [13] notes that strategic digital transformation based on artificial intelligence and machine learning technologies imparts flexibility to networks. AI algorithms can analyze complex patterns of connections between network nodes, analyze and create graphical structures, predict supplier behavior, and optimize logistics. The modern transition from reactive to proactive logistics is made possible precisely by AI.

The results of the scientific work [14] clearly demonstrate how decentralized identifiers and a hybrid blockchain approach can improve data management in supply chains. In a network model involving numerous participants (suppliers, logistics providers, warehouses, partners), trust, traceability, and data immutability are particularly important. Blockchain and distributed ledgers make it possible to ensure the transparency and integrity of the chain/network data.

A Digital Twin is a virtual model of a real system as a whole or one of its functional units: a production site, a logistics flow, a transport network. In network logistics, such models allow for scenario simulation, real-time monitoring, change management, and thus, support for the network-ecosystem. A number of studies emphasize that enhancing such a model provides opportunities for adaptation and resilience [13–15].

Research [16] shows that the digital transformation of a chain into a network enhances information processing and necessitates new relationships between firms and their partners. An integrated digital infrastructure of IT systems is created: ERP, WMS, TMS, MES, not only within a company but also with external partners, as the network model requires data standards, APIs, and exchange platforms. This implies an architecture of a "thick" data layer and a "thin" application layer, where all nodes are connected and integrated into a unified big data processing information system.

The presented review allows for the conclusion that the transition from a linear supply chain model to a network one facilitates the chain's rapid response to disruptions, demand variability, and route changes by introducing scenario planning and adaptability into management processes [13]. Thus, digital twins, data analytics, and AI enable the simulation of "what-if" situations, switching of routes, and redistribution of flows.

3 Research Methodology

The study employed analytical methods based on the mental division of the object into its constituent elements to identify its structure, properties, and interrelationships. The research required a combination of various types of analysis to examine the problem from different perspectives and obtain a more comprehensive picture.

To identify the stable connections between the elements of the network distribution system that determine its integrity, structural analysis was applied. To investigate the role of individual components in ensuring the functioning of the system as a whole, functional analysis was used. To consider the object of research as part of a complex of interconnected elements, taking into account internal and external interactions, systems analysis was applied.

In the application of these methods, the principles of objectivity, determinism, and comprehensiveness were observed.

4 Results

While managing a business in the modern complex (rapidly changing and intricately organized) competitive environment, supply chain actors strive to create dynamic and simultaneously resilient distribution channels to respond promptly to volatility in the freight services market and customer needs. Consequently, flexible IT solutions are becoming an increasingly critical component in designing the functional logistics cycle.

The network model of supply chains, despite its service orientation and segmentation of services, performs the traditional functions of their planning and controlling, ensuring the execution of interactions agreed upon among the supply chain actors. At the same time, digital transformation leads to an architectural restructuring of management processes within the chain of physical goods movement in distribution channels, shifting them from a system of vertical interaction to a horizontal one.

In a traditional linear chain, interaction is predominantly vertical: top-down levels. In the network model, horizontal interaction between homogeneous nodes (e.g., competing suppliers, distribution centers, logistics platforms) is added. In essence, the network transforms the management system into a combination of vertical and horizontal linkages. For instance, it is not merely a "production → warehouse" connection, but a "production ↔ distribution center ↔ third-party logistics operator ↔ customer" nexus with the capability for exchange and interaction among warehouse nodes, transport providers, and service partners.

Thus, management transitions from a centralized form of organization to a distributed one; that is, the centralized planning of the focal company in the supply chain is replaced by planning and decision-making distributed among the participants of the ecosystem. This necessitates new management models that consolidate the data of all supply chain participants within a unified big data information space system.

We propose the use of a cognitive info-communication environment model for the supply network as a convergence of all services provided by it, based on a complex of software and hardware for processing and storing large volumes of information. Today, cognitive info-communication systems have become virtually an integral part of most IT developments in the design and modeling of self-learning artificial neural networks, which, through computational decision-making processes, form new rules of system behavior, including those of intelligent transport systems [17, 18].

An artificial cognitive system is organized by analogy with the abilities of a biological one, which perceives its environment and then adapts to it or changes it through knowledge accumulated during operation and acquired skills. It includes [19]:

- A communication domain, which integrates the functions of monitoring and information gathering, sensors and actuators, communication, navigation, and location systems.
- An information domain, which is aimed at solving tasks related to identifying significant correlations and trends in database volumes.
- A cognitive domain, which performs the intellectual analysis of the distribution network system's databases, enabling the solution of tasks related to forecasting the behavior of its elements.

The cognitive domain functions through actuators equipped with database processing software, which includes the following stages [19]:

- Detection of patterns in the behavior of information system elements through free search.
- Predictive modeling, during which the identified patterns are used to predict unknown values or potential behaviors of the system and its elements.
- Identification, interpretation, and analysis of exceptions in the found patterns and the making of computational optimization decisions.

The modern functionality of cognitive info-communication systems is capable of creating, based on learning, self-organizing systems that change internally and possess built-in mechanisms for adapting to environmental conditions. The most promising concept is Knowledge as a Service (KaaS), which is a cloud service model where knowledge, information, and expertise are provided to a user or system on-demand, via an API or web interface, as a standardized service. Instead of owning data and processing it independently, the client receives a ready-made answer, analysis, or solution.

The key characteristics of the KaaS model are [20]:

1. On-Demand Availability—the user receives knowledge at the right moment, without the need for its preliminary collection or storage.
2. Centralization and Aggregation—the KaaS provider consolidates data from multiple disparate sources (databases, sensors, news feeds, scientific articles, etc.) into a single, logically organized system.
3. Analytics and Contextualization—disparate "raw" data is transformed into valuable information through analysis, machine learning, and semantic technologies; as a result, the system understands the context of the request.
4. Automated Delivery—the processing results ("knowledge") are delivered automatically—via chatbots, API integrations, push notifications—that is, to where they are needed for decision-making.
5. Subscription Model—the service is typically provided on a monthly or annual subscription basis, or on a pay-as-you-go principle.

The architecture and components of a typical KaaS platform include several layers [20]:

1. Data Layer (Data Sources): Diverse internal and external sources of information (SQL/NoSQL databases, CRM, ERP, cloud storage, public APIs, IoT devices).
2. Integration and Processing Layer:
   - ETL/ELT processes—for extracting, transforming, and loading data.
   - Data Lakes—for storing raw data in any format.
   - Data Warehouses—a structured repository for analysis.
3. Knowledge Layer (the most important component):
   - Ontologies and Knowledge Bases—define the connections and meanings between data.
   - Knowledge Graphs—visualize and store complex interrelationships.
   - Machine Learning and NLP—these are models for classification, forecasting, entity extraction, and natural language understanding.
4. API & Delivery Layer—these are interfaces (REST API, GraphQL) through which knowledge is consumed by client applications, chatbots, and other systems.
5. Presentation Layer—web portals, dashboards, mobile applications where users interact with the knowledge.

Figure 1 presents our proposed model for organizing network-based horizontal supply chain management within a cognitive information system environment.

The implementation of the model optimizes:

- The management processes of material flows in supply chains, taking into account the congestion of networks, logistics centers, and terminals;
- The management processes of logistics infrastructure areas;
- The management processes of transportation and interaction with transport systems (by mode of transport);

Cognitive subsystem

•decision-making based on accumulated knowledge;
•contextual processing of information on demand;
•problem solving by generating and issuing a vector of control actions for changing the state of actuators and devices;
•planning, scheduling, and coordinating the system's operating mode;
•control of actuator actions

Information subsystem

•communication system data acquisition subsystem;
•large information database storage subsystem;
•large database correlation and structuring subsystem;
•input information analysis subsystem (data organization);
•image generation subsystem for transmitting and processing video signals for information exchange;
•semantic information processing subsystem (memorization, encoding, structure)

Communication subsystem

•subsystems for interactive response of satellite navigation and geolocation systems;
•electronic maps;
•subsystems for all types of wireless communications;
•subsystems for transmitting information for rolling stock monitoring;
•subsystems for transmitting information for transport route monitoring;
•subsystems for transmitting information for cargo monitoring

Subsystem of initial data for key access of system participants

•ITS networks and access databases for transport modes located on rolling stock;
•touchscreen identification tags on containers, packaging, and packaging;
•sensors of automated and robotic devices supporting logistics operations at infrastructure facilities and logistics centers;
•onboard ITS of unmanned vehicles (cars, drones, locomotives, boats, etc.)
•automated systems for monitoring and video recording of transport routes;
•data from personal devices on documentary support for rolling stock and delivered cargo (including certificates, insurance, permits, and customs clearance), certified with an electronic signature;
•data from personal devices on financial, regulatory, and legal support for delivery, certified with an electronic signature.

Fig. 1 A model of a multi-level management architecture for a network ecosystem within a cognitive info-communication environment of supply chain actors

- The management processes of routes, accounting for the dynamics of information changes;
- The processes of transportation control;
- The management processes of security and logistics risks;
- The processes of cargo safety and integrity control;
- The processes of financial and legal support.

5 Discussion

The conducted research, as well as the accumulated scientific and practical experience, demonstrates that the digitalization of supply chains facilitates their transition into intellectual ecosystems. According to research, an intelligent distribution network ecosystem can be understood as a system in which:

- Nodes (suppliers, manufacturers, logistics operators, customers) are in constant digital interconnection;
- Data is collected, processed, and transmitted in real-time (or near-real-time);
- Decision-making employs analytics, AI, machine learning, and automation;
- The system does not merely react but forecasts, adapts, and learns;
- The chain transforms into a network where horizontal and vertical linkages are equivalent, and participants can transform their roles, interacting as equal members of the ecosystem.

The key parameters determining the system's operational efficiency are:

- Transparency across all segments of the network;
- Flexibility and speed of operational management response;
- Resilience and adaptability to disruptions;
- Compatibility and integration of participant processes within the distribution channels.

The projected outcomes of managing the supply chain network model based on the cognitive network model are presented in Table 1.

What does network-centric distribution look like in practice? Let us consider an example.

- Scenario in a Linear Chain:

A ship is delayed at port → Production halts at the factory → Deliveries to stores are disrupted → Management spends a week in meetings assigning blame.

- Scenario in an Intelligent Network:

1. Sensors on the ship and in the port register the delay. Data is fed into a cloud platform in real-time.

Table 1 Projected outcomes of implementing the cognitive distribution network management model

Consolidation of logistical activities	Integration of information exchange
Reduction of transport costs through the organization of terminal cargo transportation technologies and shortening of the delivery period	Coordinated planning and risk reduction in supply chains through the creation of a unified information system with a publicly accessible database
Reduction of administrative and management costs and elimination of duplicate functions through the operation of shared infrastructure facilities	Ensuring the necessary speed and reliability of information to reduce decision-making time in logistical business processes
Reduction of specific aggregate costs of warehouse logistics services and enhancement of their competitiveness through the expansion of the range and quality of services	Reduction of costs for forecasting and inventory management due to high-quality information exchange among process participants
Generation of additional income through the development of supplementary production services for cargo processing and the redistribution of consolidated production and trade inventories	Improvement in the quality of logistics services through continuous information tracking and accounting for individual consumer requirements and changes in potential demand for types of services

2. An AI algorithm, having analyzed the data, instantly assesses the impact: which orders will be disrupted, which customers will be affected, and where alternative inventories exist.
3. The system automatically proposes several scenarios:
   - Option A: Initiate backup production at another plant within the network, utilizing local components.
   - Option B: Redirect a portion of the goods from another distribution center, even if it belongs to a different legal entity but is part of the network.
   - Option C: Offer individual end-consumers an alternative product from the nearest fulfillment center of a marketplace that is part of the network.
4. A manager approves the optimal option, and smart contracts automatically activate the reallocation of resources and payments.

Result. The problem is resolved within hours, not weeks. The customer may never even learn of the disruption.

Advantages of the Network Model:

1. Resilience. The presence of multiple alternative connections and suppliers makes the network impervious to the disruption of a single channel.
2. Responsiveness. The network instantly adapts to changes in demand and market conditions.
3. Efficiency. End-to-end transparency and collaborative optimization eliminate waste and excess inventory across the entire ecosystem, not just within a single link.

4. Innovativeness. The ease of connecting new technology partners (e.g., startups with drone delivery services) enables the rapid implementation of innovations.

The conducted study, together with accumulated scientific and practical experience [2, 3, 9, 17, 18], demonstrates that the ongoing digitalization of supply chains is driving their evolution into intelligent ecosystems [1, 4, 10, 11]. The proposed model of a cognitive infocommunication environment, grounded in the "Knowledge-as-a-Service" (KaaS) paradigm, represents a logical and promising step forward in the development of networked structures [18–20]. However, its widespread adoption and operational deployment raise profound questions that extend well beyond mere technological feasibility. While these issues have been acknowledged in prior research [5, 7, 8, 12], they have not yet been addressed in sufficient depth.

Below, we explore the key challenges in greater detail:

1. From Technological Feasibility to Organizational Barriers: The Challenge of Trust and Data Sovereignty

Although technological enablers—such as IoT, AI, and blockchain—act as catalysts for transformation, the primary constraints are no longer technological but rather rooted in human and organizational capital. The networked model fundamentally redefines governance principles:

- First, its implementation demands a shift from control to trust. In a decentralized network, the traditional hierarchical control exerted by a focal firm weakens. Governance increasingly relies on trust—technologically assured through smart contracts and blockchain, and institutionally reinforced through clear service-level agreements (SLAs) and co-designed business processes. Yet building such trust among potential competitors, who are compelled to collaborate within a shared network, remains a non-trivial challenge.
- Second, the paradigm of a "single version of the truth" presupposes seamless data sharing. In practice, however, companies face a fundamental dilemma: the more data they share with partners, the greater the overall network efficiency—but also the higher the risks of exposing trade secrets, eroding competitive advantage, or violating regulatory frameworks (e.g., GDPR).

This raises a critical question: How can organizations strike an optimal balance between transparency and confidentiality? Standardized data-access tiers and advances in confidential computing may offer partial solutions, but they require substantial refinement and consensus.

2. A New Management Paradigm: From Supply Chain Management to Ecosystem Orchestration

The role of logistics or operations managers is evolving into that of ecosystem conductors, necessitating a new skill set:

- The ability to establish equitable relationships with heterogeneous network participants and negotiate shared interaction standards and protocols.

- Proficiency not only in reading reports but in actively engaging with cognitive systems—formulating complex queries and interpreting outputs from predictive and prescriptive analytics.
- Capacity to make high-stakes decisions under uncertainty. In highly automated networks, human roles shift toward resolving ill-defined, strategic challenges where AI presents multiple scenarios, yet the final choice—laden with ethical and strategic implications—remains with the manager.

3. Vulnerabilities of Cognitive Networks: Emerging Risk Categories.

As noted in [11, 13, 18, 20], while intelligent networks enhance resilience to operational disruptions, they simultaneously introduce new systemic vulnerabilities:

- Cybersecurity: The unified digital environment becomes a high-value target for cyberattacks. Compromising a single critical node—such as the KaaS platform—could paralyze not just one company but the entire ecosystem. This demands unprecedented investments in cyber defense and the design of fault-tolerant, segmented architectures.
- Complexity and the "Black Box" Effect: As cognitive systems grow more sophisticated and autonomous, understanding their decision logic becomes increasingly difficult. An AI might propose a route that is algorithmically optimal yet conflicts with a company's strategic priorities or ethical standards. Thus, Explainable AI (XAI) becomes a critical requirement.
- Technological Dependence: Complete reliance on complex IT infrastructure introduces risks stemming from technical failures or vendor lock-in, where business continuity hinges on a single technology provider.

There remains one crucial, yet underexplored, question without which no cognitive model can be practically implemented: the economics of value creation and cost allocation in digital supply networks.

Deploying our proposed model entails significant upfront investment—not only from the focal firm but from all key network participants. This raises fundamental issues of fairness: Who should bear the costs of building and maintaining the shared cognitive platform, and in what proportion? How can the collective economic benefits (e.g., reduced total costs) be accurately measured and equitably distributed among participants?

Addressing this requires further research. The KaaS digital model itself functions as an enabler of a new business paradigm: logistics service providers may evolve into vendors of "logistics intelligence," selling not just transportation but guaranteed end-to-end value-chain optimization. This transformation reshapes market structures and competitive dynamics alike.

6 Conclusions

Today, the network is becoming the new competitive paradigm. Modern technologies are altering the very architecture of distribution channels in logistics. The transformation from a chain into a network represents a shift from a paradigm of control to a paradigm of collaboration, from the competition of individual companies to the competition of ecosystems. The company of the future is not the one that owns the longest chain, but the one that is the most valuable node in the most flexible, intelligent, and reliable supply network. Success will be determined not by assets on the balance sheet, but by the speed at which a company can orchestrate the resources of its entire ecosystem to meet the needs of the end consumer.

The shift from linear supply chains to cognitive supply networks is not merely a technological upgrade, but a fundamental transformation of the underlying business paradigm. The success of this transition will depend on supply chain actors' ability to overcome not only technological barriers, but also managerial and economic ones.

Our future research will therefore focus not on whether such a model is feasible—the present work demonstrates that it is—but rather on how to effectively manage its associated risks, foster trust-based relationships among participants, cultivate new managerial competencies, and design equitable mechanisms for value distribution.

Ultimately, the network-centric logistics of the future is, first and foremost, an ecosystem built on trust and collaboration—where technology serves not as the goal, but as an enabler of this new management philosophy.

References

1. Mohsen BM, Mohsen M (2025) Intelligent supply chain networks: integrating AI, big data, and future communication technologies for enhanced decision-making. Procedia Comput Sci 265:8–16. https://doi.org/10.1016/j.procs.2025.07.150
2. Li P, Chen Y, Guo X (2025) Digital transformation and supply chain resilience. Int Rev Econ Financ 99:104033. https://doi.org/10.1016/j.iref.2025.104033
3. Farfán Chilicaus GC, Licapa-Redolfo GS, Arbulú Ballesteros MA, Corrales Otazú CD, Apaza Miranda SJ, Flores Castillo MM, Castro Ijiri GL, Guzmán Valle MDlÁ, Arbulú Castillo JC (2025) Digital transformation and sustainability in post-pandemic supply chains: a global bibliometric analysis of technological evolution and research patterns (2020–2024). Sustainability 17(7):30092025 (2025). https://doi.org/10.3390/su17073009
4. Yadykin VK, Barykin SE, Kossukhina MA (2020) Network concept of intelligent digital supply chain. Innovations 4(258):46–50. https://doi.org/10.26310/2071-3010.2020.258.4.006
5. Iskanderov Y, Pautov M (2020) Actor-network approach to self-organisation in global logistics networks. In: Kotenko I, Badica C, Desnitsky V, El Baz D, Ivanovic M (eds) Intelligent distributed computing XIII. IDC 2019 studies in computational intelligence, vol 868, pp 117–127. Springer, Cham. https://doi.org/10.1007/978-3-030-32258-8_14
6. Legenvre H, Hameri A-P, Golini R (2022) Ecosystems and supply chains: how do they differ and relate. Digit Bus 2(2):100029. https://doi.org/10.1016/j.digbus.2022.100029
7. Michou H, Richet J-L (2025) Digital transformation, emergence and complexity in supply networks. Int J Inf Manage 84:102917. https://doi.org/10.1016/j.ijinfomgt.2025.102917

8. Massari GF, Nacchiero R, Giannoccaro I (2025) Transformative supply chains: the enabling role of digital technologies. Int J Prod Econ 283:109562. https://doi.org/10.1016/j.ijpe.2025.109562
9. Tubis AA, Grzybowska K, Król B (2023) Supply chain in the digital age: a scientometric-thematic literature review. Sustainability 15:11391. https://doi.org/10.3390/su151411391
10. Najmi MH, Iqbal SMA, Khan S (2024) Aligning supply chain functions with emerging technologies: a strategic approach. Eng Proc 76(1):34. https://doi.org/10.3390/engproc2024076034
11. Zaman SAA, Qazi H, Zafar A, Jiang Y (2025) Emerging technologies and smart supply chain management: a roadmap for future success. In: Mubarik MS, Khan SA (eds) Smart supply chain management. Springer, Singapore. https://doi.org/10.1007/978-981-96-1333-5_8
12. Ning L, Yao D (2023) The impact of digital transformation on supply chain capabilities and supply chain competitive performance. Sustainability 15:10107. https://doi.org/10.3390/su151310107
13. Wu L, Huang J, Wang M, Kumar A (2024) Unleashing supply chain agility: leveraging data network effects for digital transformation. Int J Prod Econ 277:109402. https://doi.org/10.1016/j.ijpe.2024.109402
14. Herbke P et al (2024) DIDChain: advancing supply chain data management with decentralized identifiers and blockchain. In: 2024 IEEE international conference on service-oriented system engineering (SOSE). Shanghai, China, pp 54–63. https://doi.org/10.1109/SOSE62363.2024.00013
15. Aslam H, Rehman A, Iftikhar A, ul Zia M, H, Akbar, U., Kamal, M. M.: Digital transformation: unlocking supply chain resilience through adaptability and innovation. Technol Forecast Soc Change 219:124234 (2025). https://doi.org/10.1016/j.techfore.2025.124234
16. Hamann-Lohmer J, Bendig M, Lasch R (2023) Investigating the impact of digital transformation on relationship and collaboration dynamics in supply chains and manufacturing networks—a multi-case study. Int J Prod Econ 262:108932. https://doi.org/10.1016/j.ijpe.2023.108932
17. Cordoş N, Duma I, Moldovanu D, Todoruţ A, Barabás I (2025) An overview of intelligent transportation systems in Europe. World Electr Veh J 16:387. https://doi.org/10.3390/wevj16070387
18. Garg T, Kaur GA (2022) Systematic review on intelligent transport systems. J Comput Cogn Eng 2(3):175–188 (2022). https://doi.org/10.47852/bonviewJCCE2202245
19. Komashinskiy V, Malygin I, Korolev O (2020) Introduction into cognitive multimodal transportation systems. Transp Res Procedia 50:273–279. https://doi.org/10.1016/j.trpro.2020.10.033
20. Sivaiah J, Venkataramana K (2015) Cloud database storage model by using Key-as-a-Service (KaaS). Int J Eng Comput Sci 4(7):13284–13288

Digital and Logistics Infrastructure of the Arctic: Modelling Approach

Arsenii Sluzhaev, Anastasia Levina, Alissa Dubgorn, and Polina Rukina

Abstract This chapter introduces a methodology for simulation-based modelling of smart digital and logistics infrastructure elements in the Arctic, with a particular focus on transport and logistics systems operating under extreme environmental conditions. The study addresses key components of Arctic infrastructure, including data centers, communication channels, local networks and IoT infrastructures, energy supply systems, and smart logistics services supporting transportation corridors and remote facilities. A structured methodological framework is proposed for the formalisation of operational, emergency, climatic, integration, and development scenarios that reflect the specific challenges of Arctic transport and logistics, such as seasonal accessibility, limited connectivity, and high infrastructure vulnerability. The chapter defines a system of key performance indicators covering reliability, network performance, logistics efficiency, energy sustainability, economic costs, and socio-environmental impacts. These metrics enable quantitative assessment of infrastructure resilience and support scenario-based analysis of smart transport and logistics services. Various simulation paradigms, including discrete-event, agent-based, and system dynamics modelling, are considered within a hybrid approach. Methodological limitations related to data availability, model scalability, and system complexity are also discussed. The proposed methodology provides a foundation for the systematic analysis, design, and improvement of smart digital and logistics infrastructure, contributing to the sustainable and resilient development of Arctic transport systems.

Keywords Arctic · Smart transportation infrastructure · Digital infrastructure · Smart logistics · Simulation modelling

---

A. Sluzhaev (✉) · A. Levina · A. Dubgorn · P. Rukina
Peter the Great St.Petersburg Polytechnic University, Saint-Petersburg, Russia
e-mail: sluzhaevars@outlook.com

© The Author(s), under exclusive license to Springer Nature Switzerland AG 2026

A. Kulachinskaya et al. (eds.), *Digital Transformation: What Kind of Smart Life do we have today?*, Lecture Notes in Networks and Systems 1895,
https://doi.org/10.1007/978-3-032-20582-7_2

1 Introduction

The Arctic is one of the most strategically important regions of the twenty-first century. According to various estimates, the Arctic Zone of the Russian Federation (AZRF) contains up to 20% of the world's oil and gas reserves, as well as substantial deposits of non-ferrous metals and hydrocarbon resources [1]. In addition, the Northern Sea Route (NSR) is viewed as a key international transport corridor capable of providing Russia with stable competitive advantages in global trade and reshaping global logistics patterns [2–4]. The efficiency and safety of this corridor, along with the performance of auxiliary terrestrial and maritime transport networks, increasingly depend on the development of advanced digital technologies.

Government documents, including the Strategy for the Development of the Arctic Zone of the Russian Federation and Ensuring National Security until 2035 (Presidential Decree No. 645 of 26.10.2020), designate Arctic development as a national priority [5, 6]. A particular emphasis is placed on the integrated modernisation of infrastructure, especially transport corridors, port logistics systems, road networks, energy supply, and digital services that support mobility, routing, cargo monitoring, and sustainable transport operations.

At the same time, the region faces intensifying challenges: permafrost degradation, rising air temperatures, growing risks of infrastructure failure, and threats to the stability of energy and transport systems [7]. Studies indicate that up to 70% of buildings and structures in the permafrost zone may experience damage by the mid-twenty-first century, affecting not only stationary infrastructure but also roads, port facilities, logistics hubs, and seasonal transport routes. Under these conditions, digitalisation becomes an essential tool for adaptation and for maintaining the reliability of transport and logistic services.

The use of digital twins, multi-agent systems, and the Internet of Things enables more resilient management of logistics chains, optimisation of routing under climatic uncertainty, predictive maintenance of transport infrastructure, and real-time monitoring of cargo flows [8]. Scientific publications highlight that digital infrastructure is becoming a cornerstone for smart transportation services, intelligent logistics management, and sustainable mobility in territories characterised by high climatic and transport vulnerability.

Thus, the study of digital infrastructure development in the Arctic holds both academic and practical significance. It contributes to advancing global approaches to resilient logistics, strengthens the reliability and sustainability of high-latitude transport routes, and provides a technological foundation for ensuring the stable operation of critical mobility and supply chain systems under accelerating climate change.

2 Materials and Methods

The information base of the study includes statistical data from federal and regional government bodies, regulatory documents of the Russian Federation, materials from telecommunications operators and industrial companies (Rostelecom, NOVATEK, Gazprom, Rosneft), as well as scientific publications indexed in Scopus and Web of Science that address issues of digital infrastructure and simulation modelling in the Arctic.

The research methodology is based on the systematisation and generalisation of information related to the functioning and development of the digital infrastructure of the Arctic Zone of the Russian Federation. The study employs methods of structural and comparative analysis and a scenario-based approach. Architectural modelling was used as a research tool, in particular through the Archi modelling software.

3 Literature Review

The Arctic Zone of the Russian Federation (AZRF) represents a complex socio-economic, technological, and environmental system characterised by extreme climatic conditions, significant territorial fragmentation, and a high concentration of critical infrastructure. These features necessitate the development of modelling approaches capable of integrating heterogeneous processes, from the functioning of digital systems to logistics and population support.

The study identified several critically important components of digital infrastructure, listed below.

Data Centers

The development of digital infrastructure in the AZRF is impossible without the establishment of a resilient system of data centers, which serve as key elements for storing, processing, and transmitting information. In the Arctic, where distances between settlements are large and access to external networks is limited, local and regional data centers become critically important for ensuring the continuity of business processes, government services, and scientific research [9].

The following categories of data centers are distinguished:

- Regional data centers, which support government and corporate systems and function as major nodes for data storage and analytics. Their specific feature lies in the need for high energy efficiency and resilience to external factors. Due to the remoteness of the region, such data centers must meet heightened requirements for fault tolerance and cybersecurity;
- Modular (edge) data centers, which are mobile and compact solutions located closer to extraction sites or logistics hubs. Their advantages include reduced data-processing latency and higher efficiency of Internet-of-Things (IoT) services,

which is particularly relevant for extractive industries, equipment condition monitoring, and the prediction of technological risks;
- Backup and distributed data centers, which ensure the continuity of digital services in the event of accidents or cyberattacks. In the AZRF, where the likelihood of climatic disasters (storms, icing, power interruptions) is high, the presence of a distributed network of reserve nodes becomes a factor of national digital security [10].

Energy supply plays a particularly important role in designing Arctic data centers. Traditional diesel generators remain the dominant source of power, but solutions based on renewable energy (wind and solar generation), supplemented by energy storage systems, are increasingly considered [11]. Low temperatures allow the use of natural cooling, which reduces energy consumption for climate control and enables high energy-efficiency performance [12].

Thus, data centers in the Arctic zone serve not only as technological nodes but also as strategic elements ensuring the sovereignty and security of the digital environment. Their development must be based on the principles of fault tolerance, distribution, energy efficiency, and integration with other subsystems of the region's digital infrastructure.

Communication Channels

Reliable and resilient communication channels form the foundation of the AZRF's digital infrastructure, supporting government systems, industry, scientific projects, and social services. Given the remoteness of the region, harsh climate, and limited availability of traditional telecommunications solutions, there is a need for a comprehensive application of various communication technologies that ensure redundancy and fault tolerance.

The following key communication channels can be identified:

- Satellite communications (LEO/MEO/GEO, VSAT), which currently represent the primary means of data transmission in the AZRF. They typically offer wide coverage but high signal latency, however, with the development of LEO systems, data transmission speeds are increasing;
- Terrestrial backbone channels (fibre-optic lines), which integrate the Arctic territories into the national information network of the High North [13].

A comparative description of satellite communications and terrestrial backbone channels can be presented in tabular form (see Table 1) [14].

Local Networks and IoT Infrastructure

Local networks and IoT infrastructure also play an important role in the digital development of the Arctic Zone of the Russian Federation, as they enable the collection, transmission, and preliminary processing of critical information from remote facilities. In the Arctic, where traditional telecommunications solutions are constrained by harsh climatic factors and the high cost of deployment, IoT systems become a fundamental tool for monitoring and managing industrial, transport, and energy

Table 1 Comparison of satellite communications and fibre-optic channels in the Arctic Zone of the Russian Federation

Criterion	Satellite communications (LEO/MEO/GEO, VSAT)	Terrestrial backbone channels
Coverage	Global, including hard-to-reach areas, oceans, and ice fields	Localised along deployment routes (NSR, coastline, winter roads). Limited by physical infrastructure
Signal latency	High for GEO (500–700 ms), medium for MEO (100–150 ms), low for LEO (20–50 ms)	Minimal latency (5–20 ms), especially for direct backbone connections
Bandwidth	From 10–100 Mbps for VSAT to 1–2 Gbps for next-generation LEO systems	High: hundreds of Gbps; with modern DWDM systems—up to several Tbps
Reliability	Dependent on weather conditions (snow, blizzards, thunderstorms). Possible disruptions during solar flares	High resilience, but vulnerable to physical damage (ice, drift ice movement, underwater erosion)
Deployment cost	Relatively low (VSAT/LEO terminals are easy to install quickly)	Very high: construction of fibre-optic lines in the Arctic requires icebreakers, drilling, underwater cable laying, and site protection
Operational cost	Medium: depends on satellite operator tariffs and antenna maintenance	Relatively low after construction, but requires continuous monitoring and technical maintenance of cables
Mobility	High: antennas and terminals are easily movable, suitable for expeditions and temporary bases	Low: stationary infrastructure, not intended for mobile facilities
Service life	10–15 years for satellite constellations, 5–7 years for ground stations and antennas	20–30 years for fibre-optic cables with proper maintenance
Climate resilience	Vulnerable to heavy snowfall and antenna icing	Risk of damage from ice and underwater tectonics; however, buried cables are more protected from climatic fluctuations
Applicability	Research stations, rotational camps, marine vessels, remote extraction sites	Data centers, ports, administrative centers, major industrial facilities along the NSR and the coastline

processes. These systems are used across a wide range of activities in the AZRF—from environmental monitoring to resource extraction control at industrial sites, as well as monitoring of utilities and housing infrastructure in rotational settlements.

Several classes of technologies are applied in Arctic conditions:

- LPWAN networks (LoRaWAN, NB-IoT)—optimal for transmitting small data packets from sensors with long autonomous operation (up to 10 years);
- Local LTE/5G networks—used for tasks requiring high transmission speeds and low latency, such as remote equipment operation and video surveillance;

- IoT gateways and edge computing—provide local processing and filtering of data before transmission to regional data centers, reducing the load on communication channels [15].

Redundancy and Power Supply for Digital Nodes

Reliable power supply is a key requirement for the functioning of the AZRF's digital infrastructure. Digital nodes (data centers, telecommunication stations, IoT gateways, and components of SCADA systems) must operate without interruption despite extreme climatic conditions, remoteness, and high logistics costs. Under these circumstances, the concept of multi-tier redundancy and the use of diversified energy sources become especially important. Diesel generators are widely used to power digital nodes; however, alternative technologies have become increasingly popular in recent years: renewable energy sources combined with energy storage systems, as well as the concept of microgrids [16].

Management Platforms

The development of digital infrastructure in the AZRF requires the integration of multi-level management platforms that provide both local control of industrial facilities and centralised planning at the level of regions and companies. In conditions of remoteness, low population density, and high operational costs, the use of comprehensive management systems becomes critically important. These systems include SCADA/ICS, MES, ERP, monitoring tools (AIOps), and digital twins [17].

Logistics Digital Services

The development of logistics infrastructure in the AZRF is directly linked to ensuring reliable cargo deliveries and the uninterrupted operation of critical facilities. Extreme climatic conditions, limited navigation windows along the Northern Sea Route (NSR), and the difficulty of maintaining ice roads and port infrastructure create high risks of delays and disruptions in supply chains. To minimise these risks, modern digital services are used, integrating cargo tracking, intelligent route planning, dispatching, and forecasting of transport asset conditions [18, 19].

4 Results

In addition to defining the elements of the model, it is also necessary to specify the modelling scenarios. The analysis identified the following main types of scenarios: operational, emergency/failure-prone, climatic/seasonal, development/scaling, and integration scenarios. A consolidated comparative table of scenarios for modelling was compiled (See Table 2).

Additionally, preconditions, time horizons, SLA, RTO, and failure probabilities were defined for each scenario type [20, 21].

Table 2 Types of modelling scenarios for the digital infrastructure of the Arctic Zone of the Russian Federation

Scenario type	Preconditions	Time horizon	SLA (%)	RTO	Failure probability
Operational (routine)	Peak loads (seasonal NSR voyages, traffic growth)	Days-weeks (up to one month for forecasts)	≥99.9	≤2 h	Low (1–2%)
Emergency/ failure-prone	Data center failure, communication channel loss, energy supply degradation, cyberattack	Hours-days (up to one week for recovery)	≥99	15 min–12 h	Medium (5–10% for communication; 3–5% for energy; 1–3% for cyberattacks)
Climatic/ seasonal	Inaccessibility of winter roads, changes in ice season, weather constraints	3–6 months (up to one year for forecasts)	≥98	≤24 h	High (10–20% per year)
Development/ scaling	Connection of new facilities (industrial sites, data centers, IoT sensors), increased load	6 months - 2 years	≥99.95	≤1 h	Low (< 5%, up to 15% during abrupt load growth)
Integration	Failure of one channel and switching to a backup (satellite-fibre, VSAT, etc.)	Minutes-hours (up to one week for stress testing)	≥99	≤15 min	Medium (5–7% for satellite communications; <1% for fibre optics)

Definition of Metrics

The development of modelling scenarios for the digital infrastructure of the Arctic Zone of the Russian Federation requires the formalisation not only of the system's operating conditions but also of the objective indicators used to assess its resilience, efficiency, and compliance with target parameters. The scenarios previously considered—operational, emergency, climatic, integration, and development scenarios—define the context in which key performance indicators (KPIs) become tools for quantitative assessment and comparison of operating options.

In this context, KPIs perform two functions:

1. **Methodological function**—KPIs make it possible to define parameters that must be monitored and evaluated during the modelling process. This formalises the results and facilitates their interpretation by managers and engineers.

2. Practical function—KPIs ensure comparability across different scenarios, identify infrastructure bottlenecks, and provide a basis for subsequent decisions regarding development or modernisation.

For the digital infrastructure of the AZRF—characterised by dispersed assets, extreme climatic conditions, and high risks associated with failures—key metrics must reflect technical, socio-economic, and environmental aspects [22]. Table 3 presents a list of KPIs grouped according to the main areas of assessment [23, 24].

Table 3 Key performance indicators for the modelled scenarios

Category	KPI	Applicability
Reliability and availability	Service availability, %	Data centers, communication channels, IoT segments
	Mean recovery time (RTO), h	Data centers, communication channels, logistics
	Data recovery accuracy (RPO), %	Data storage systems, SCADA/ERP
Network parameters	Latency, ms	Satellite and terrestrial channels
	Bandwidth, Mbps	Backbone channels, IoT gateways
	Share of successfully delivered telemetry messages, %	IoT infrastructure, sensors
Logistics	Average delivery time, h/day	Transport routes, NSR
	Share of delayed deliveries, %	Logistics chain during peak seasons
	Logistics chain reliability, index	Integration of digital and transport services
Energy and sustainability	Energy consumption, kWh	Data centers, networks, communication stations
	Carbon footprint, CO_2-eq/year	Operation of diesel generators, RES
Economics	Downtime cost, RUB/hour	Data centers, communication channels, logistics
	Cost of transferring 1 TB of data, RUB	Communication channels
	Cost of transporting 1 tonne of cargo, RUB	Maritime/land logistics
Social and environmental aspects	Number of incidents involving data or fuel leaks	Data centers, logistics
	Impact on Indigenous Peoples of the North, index	Logistics, social services

Simulation Modelling Methods

Simulation modelling is a key tool for analysing and designing complex socio-technical systems. Under conditions of high uncertainty (climatic risks, power supply disruptions, limited logistics), classical analytical forecasting methods are often insufficient. Simulation modelling makes it possible to study a system in time and space, taking into account the behaviour of its elements, interactions, and dynamic dependencies [25].

The following major paradigms of simulation modelling are distinguished:

Discrete-Event Simulation (DES). This approach is based on representing a system as a sequence of events occurring over time. DES is widely used for modelling logistics flows in the Arctic: cargo movement along the NSR, queues during vessel unloading in ports, servicing of transport vehicles, and operations on winter roads. DES is well suited for analysing infrastructure throughput and optimising operational schedules.

Agent-Based Modelling (ABM). In ABM, the system is described as a collection of autonomous agents with individual behavioural rules. For the AZRF, this is particularly important for modelling distributed actors: autonomous vessels, drones for ice monitoring, data center operators, and contracting organisations. ABM makes it possible to take into account interactions between humans and machines, as well as to analyse system behaviour under uncertainty and limited resources.

System Dynamics (SD). This method is used for analysing macro-level processes, such as investments in digital infrastructure, development dynamics of communication networks, and the influence of government policy and international projects. SD modelling helps reveal long-term effects: for example, how increased investment in satellite communication reduces the isolation of Arctic settlements.

Hybrid models (DES + ABM + SD). This approach combines several types of simulation modelling. For example, DES can be used to model supply chains, ABM is suitable for modelling the behaviour of autonomous vessels, robotic equipment, and operators, while SD can describe long-term infrastructure development strategies. Thus, hybrid models provide the most comprehensive representation of the Arctic digital and logistics infrastructure.

In addition to simulation approaches, classical analytical methods must also be applied:

Stochastic analysis and Monte Carlo methods. Arctic systems are characterised by high variability: changes in ice conditions, unpredictable equipment failures, and climatic disasters. Monte Carlo methods make it possible to account for the probabilistic nature of such events and assess scenario risks [26].

Optimisation methods (LP, MILP, heuristics).

In addition to simulation modelling, optimisation modules are used to solve planning tasks: siting of data centers and communication stations; selection of cargo delivery routes; allocation of power supply reserves, and others. Linear programming, mixed-integer linear programming, and heuristic methods are applied for these purposes [27].

Data Required for Simulation Modelling of the Digital Infrastructure of the AZRF

One of the key stages in building simulation models is the collection and structuring of initial data. Without a reliable and representative data base, it is impossible to fully reproduce the functioning of complex systems such as the digital infrastructure of the Arctic Zone of the Russian Federation. Modelling requires the consideration of several classes of data reflecting both technical and organisational–environmental aspects of infrastructure operation.

Device telemetry and SCADA log files. Industrial control systems (SCADA/ICS) accumulate data on equipment condition, operating parameters of pumps, generators, cooling systems, power grids, as well as telemetry from IoT sensors. These data serve as sources of time series for discrete-event modelling and make it possible to identify typical load patterns and deviations from normal operating modes [28].

Communication channel characteristics. Accurate reproduction of digital network performance requires information on channel bandwidth (Mbps), transmission latency (RTT, ms), operational costs (cost of transmitting 1 GB/1 TB of data), as well as statistics on degradation and temporary outages. In Arctic conditions, the specifics of different technologies must be taken into account: satellite communication (LEO/MEO/GEO, VSAT) and fibre-optic trunk lines along the Northern Sea Route.

Equipment reliability. A key indicator is the mean time between failures (MTBF) for server, network, and power equipment. Additionally, the mean time to repair (MTTR) is recorded, along with the frequency of failures under low temperatures and conditions of vibration/corrosion. These datasets make it possible to form scenarios of emergency states.

Logistics data. For modelling transport infrastructure and logistics, critical datasets include vessel schedule information, port and terminal throughput capacity, and icebreaker fleet operating schedules. Special attention is given to seasonal constraints: navigation periods along the Northern Sea Route, closure dates of winter roads, and fluctuations in cargo volume depending on the time of year.

Climatic data. In the Arctic, digital and transport infrastructures are directly dependent on environmental factors. Probabilistic scenarios are constructed using data on sea ice thickness, the likelihood of ice jams, storm frequency, and temperature variations. These datasets are sourced from open climate databases (such as ERA5 from ECMWF, the Arctic observation system of Roshydromet), as well as scientific publications.

Data sources. Potential sources of data include open reports from telecommunications operators (e.g., Rostelecom, satellite system operators such as OneWeb,

Starlink, and Gazprom Space Systems); major industrial companies (NOVATEK, Gazprom, Rosneft) that operate digital systems in the Arctic; reports from governmental bodies (Roshydromet, Ministry of Digital Development, Ministry of Transport); scientific publications addressing digital infrastructure and network resilience in Arctic conditions; and open international climate databases (Copernicus Climate Data Store, NOAA Arctic Program).

Thus, the combination of technical, logistical, and climatic data forms the foundation for developing hybrid models that account for both operational and emergency scenarios.

Formalisation of the Proposed Methodology

Thus, the combination of technical, logistical, and climatic data forms the foundation for developing hybrid models that account for both operational and emergency scenarios.

To formalise and systematise the methodology described above, a metamodel was developed (Fig. 1), which defines the relationships between the key elements.

As can be seen, the core of the model is formed by the objects of digital and logistics infrastructure, which represent the central component of the entire methodology. These objects, in turn, generate data and define the types of scenarios, yet they

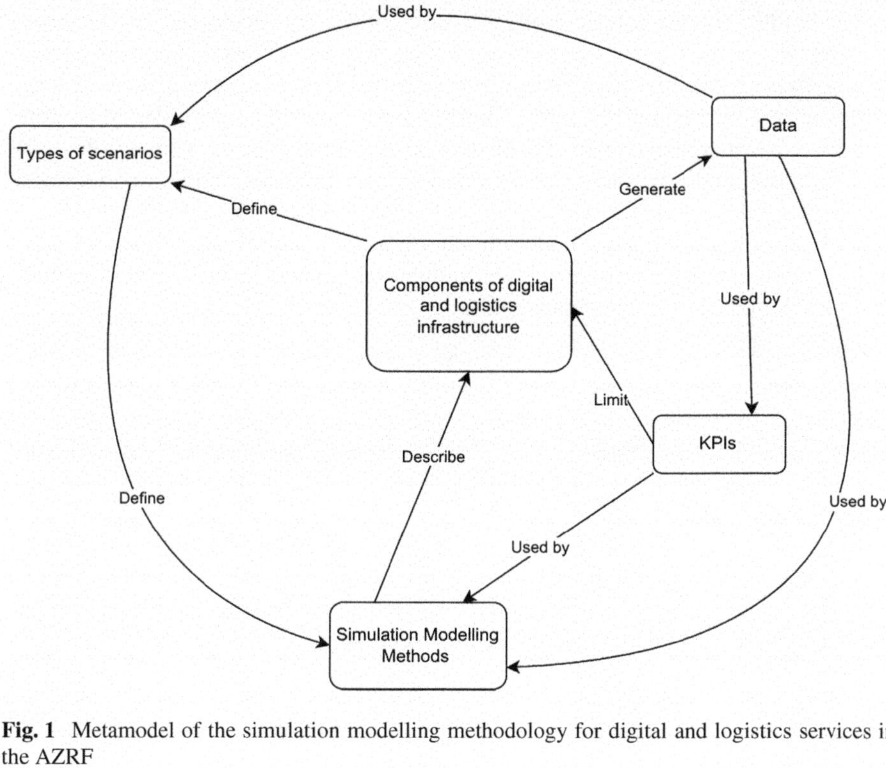

Fig. 1 Metamodel of the simulation modelling methodology for digital and logistics services in the AZRF

are constrained by the metrics. The simulation modelling methods are determined based on the scenario types, metrics, and data, and they also describe the objects themselves.

Having the metamodel makes it possible to further specify it and represent it using the motivational extension in the ArchiMate modelling [29–31].

1. To model the objects of digital and logistics infrastructure, elements of the Stakeholder type were used, since these objects are central within the models and represent the elements that are intended to be modelled. The digital and logistics infrastructure objects include: data centers; communication channels; local networks and IoT infrastructure; redundancy and power supply for digital nodes; management platforms; and logistics digital services.
3. To model the types of scenarios, Driver elements were used. The scenario types include: operational (routine); emergency/failure-prone; climatic/seasonal; development/scaling; and integration scenarios.
4. For metrics, the Assessment element type was selected, as it directly reflects requirements. The metrics were divided into six categories: Reliability and availability; Network parameters; Logistics; Energy and sustainability; Economics; and Social and environmental aspects. The Reliability and availability category includes the following KPIs: service availability; mean recovery time (RTO); data recovery accuracy (RPO). The Network parameters category includes: latency; bandwidth; share of successfully delivered telemetry messages. The Logistics category includes: average delivery time; share of delayed deliveries; logistics chain reliability. The Energy and sustainability category includes: energy consumption; carbon footprint. The Economics category includes: downtime cost; cost of transferring 1 TB of data; cost of transporting 1 tonne of cargo. The Social and environmental aspects category includes: number of incidents involving data or fuel leaks; impact on Indigenous Peoples of the North
5. To denote the simulation modelling methods, the Goal element type was used. The model includes Discrete-Event Simulation (DES), Agent-Based Modelling (ABM), and System Dynamics (SD).

After categorising the elements, a model was developed focusing on the metrics relevant to each infrastructure object (Fig. 2), as well as a model representing the relationships between digital and logistics infrastructure objects, scenario types, and simulation modelling methods (Fig. 3). This model demonstrates which scenarios are most suitable for modelling each infrastructure object and which simulation method is appropriate in each case.

Thus, a set of models was developed that illustrates the relationships between the elements of the methodology, demonstrates which metrics affect the various objects, and identifies which simulation modelling methods should be used for specific scenarios and specific infrastructure objects.

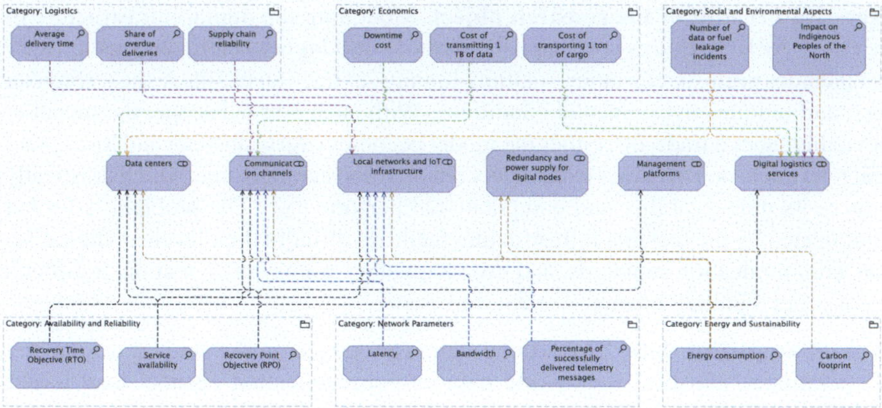

Fig. 2 The connections between objects and metrics

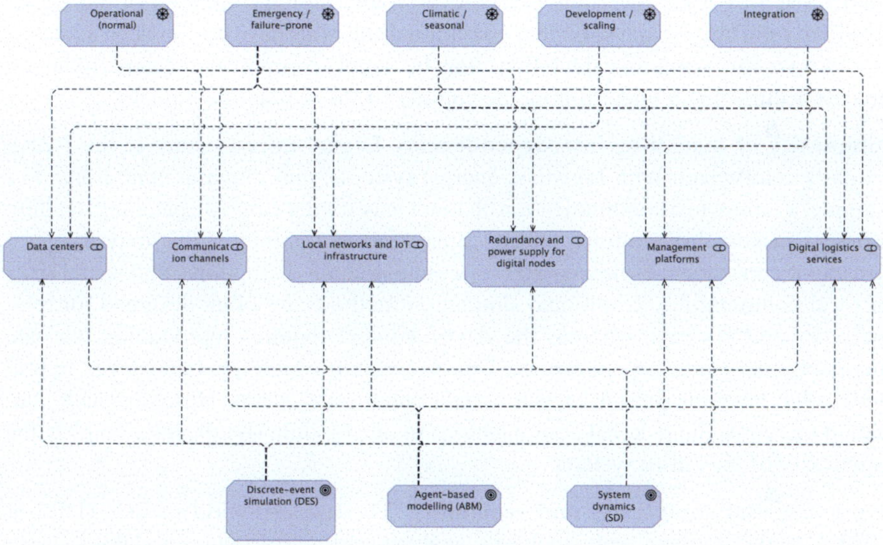

Fig. 3 The connections between objects, scenario types and simulation modeling methods

5 Discussion

Despite the potential and theoretical robustness of the hybrid simulation approach for analysing and forecasting the performance of the digital infrastructure of the AZRF, several fundamental limitations and methodological challenges must be acknowledged. These "bottlenecks" significantly affect the accuracy, applicability, and scalability of the resulting models.

Excessive breadth of the research object. Modelling the digital infrastructure of the entire AZRF requires covering an enormous and highly heterogeneous territory that includes ports, Arctic cities, rotational camps, extraction facilities, the icebreaker fleet, and unique environmental conditions. With such a broad scope, the modeller inevitably faces a trade-off: either the model becomes overly abstract and loses practical relevance, or it becomes excessively detailed and cumbersome, which drastically reduces its manageability and computational performance. The problem is further complicated by the fact that infrastructure facilities differ in their levels of digitalisation, communication standards, and even regulatory frameworks, making it difficult to construct a unified methodological framework.

Lack of empirical data. The second key issue concerns the limited availability of operational statistics. Real-world data on communication channel load levels, equipment reliability, energy consumption patterns, or incident frequency are often restricted due to corporate or governmental confidentiality. This necessitates relying on aggregated public information, regulatory reports, or expert estimates—sources characterised by high uncertainty. Moreover, many digital initiatives in the Arctic are still in pilot stages, and sufficient operational statistics have not yet been accumulated. As a result, scenario calculations may be largely estimative in nature, reducing forecast accuracy and undermining the validity of the simulation model.

Complexity of modelling interdependencies. Digital infrastructure in the AZRF is tightly interwoven with logistics, energy systems, and climatic conditions. For example, a failure in a communication channel may hinder cargo dispatching, leading to delays in vessel operations; under ice navigation conditions, this may trigger a cascade of consequences: increased costs, supply chain disruptions, and social risks for local communities. Traditional simulation methods are often designed for relatively isolated systems, whereas the Arctic context requires reproducing intricate interdisciplinary interdependencies. This necessitates building multi-layer hybrid models that integrate digital, logistics, and climate subsystems. However, such integration increases the risk that complications in one module significantly amplify the complexity of the entire system.

Computational complexity and scalability. The use of hybrid models (DES + ABM + SD) combined with stochastic analysis and optimisation modules is associated with high computational demands. When simulating large-scale scenarios—such as the operation of the entire transport network of the Northern Sea Route or the infrastructure of several Arctic regions—scalability issues arise. These models require substantial resources: high-performance computing systems, distributed simulators, and parallel computation. Without such resources, the time required to run scenarios becomes prohibitive, reducing the practical applicability of the results. An additional challenge lies in the calibration of these models: achieving a balance between speed and accuracy often requires compromise solutions, which may lead to the simplification of key parameters and a loss of detail.

6 Conclusion

The proposed simulation modelling methodology makes it possible to structure the research domain, identify key data sources, define relevant metrics, and link them to scenario types and modelling methods. This approach provides a foundation for assessing the resilience of digital and logistics infrastructure under a wide range of conditions—from routine to critical—and creates the analytical tools needed to forecast system behaviour. At the same time, achieving practical value requires phased validation at the level of specific transport and logistics assets, including ports, extraction sites, and base camps.

However, the lack of empirical data and the difficulty of verifying complex scenarios limit the potential for large-scale application of the methodology. Nevertheless, the formation of partnerships between governmental bodies, industrial operators, and the research community may partially alleviate this constraint by providing access to operational datasets and practical case studies for model calibration.

The next step in developing the proposed methodology should involve constructing detailed simulation models of the digital infrastructure within individual Arctic subregions and progressively integrating them into a unified system. Such an approach will improve forecast accuracy, reduce computational burden, and adapt the methodology to real-world challenges of transport management, logistics optimisation, and strategic planning in Arctic conditions.

Acknowledgements The research was supported by Russian Science Foundation grant No. 23-78-10190, https://rscf.ru/project/23-78-10190/

References

1. Voronov M, Voronin A, Vlasov N (2019) The problem of oil spills in the Arctic region and a new method of their elimination. Tyumen. https://doi.org/10.3997/2214-4609.201900589
2. Larchenko LV (2024) The role of the Northern Sea Route in the strategic development of Russia. In: Collection of scientific articles. Saint Petersburg State University of Economics, pp 83–87. https://doi.org/10.52897/978-5-7310-6547-4-2024-18-83-87. In Russian
3. Gunnarsson B, Moe A (2021) Ten years of international shipping on the Northern Sea Route: trends and challenges. Arct Rev Law Politics 12. https://doi.org/10.23865/arctic.v12.2614
4. Levina A, Kalyazina S, Dubgorn A, Gorbacheva A (2025) Characteristics of digital infrastructure development in the arctic. In: Ilin I, Youzhong M (eds) Digital systems and information technologies in the energy sector. Lecture notes in networks and systems, vol 1244. Springer, Cham. https://doi.org/10.1007/978-3-031-80710-7_12
5. Strategy for the Development of the Arctic Zone of the Russian Federation and Ensuring National Security for the Period until 2035: approved by Decree of the President of the Russian Federation No. 645 of 26 October 2020. Available at: http://www.kremlin.ru/acts/bank/45972. Accessed on 13 August 2025. In Russian
6. Decree of the Government of the Russian Federation No. 366 of 21 April 2014 on the approval of the state programme socio-economic development of the arctic zone of the Russian Federation. Available at: http://government.ru/rugovclassifier/830/events/. Accessed on 25 August 2025

7. Samarina V, Skufina T, Savon D, Shinkevich A (2021) Management of externalities in the context of sustainable development of the Russian Arctic Zone. Sustainability 13(14):7749. https://doi.org/10.3390/su13147749
8. Broo D, Schooling J (2021) Digital twins in infrastructure: definitions, current practices, challenges and strategies. Int J Constr Manag 23(12):1254–1263. https://doi.org/10.1080/15623599.2021.1966980
9. Petrov A, BurnSilver S, Chapin FS III, Fondahl G, Graybill J, Stephen K, Nilsson A, Riedlsperger R, Schweitzer P (2017) Arctic sustainability research: past, present and future. Routledge. https://doi.org/10.4324/9781315109954
10. Eidemiller K (2023) Prospects for the development of data centers in the Arctic. Russia and America in the 21st Century. https://doi.org/10.18254/s207054760025063-6
11. Solomin E, Kirpichnikova I, Amerkhanov R, Korobatov D, Lutovats M, Martyanov A (2019) Wind-hydrogen standalone uninterrupted power supply plant for all-climate application. Int J Hydrogen Energy 44(2):829–840. https://doi.org/10.1016/j.ijhydene.2018.12.001
12. Ge L, Han Z, Yang Z, Zhang Y, Qiu B, Li X (2024) Feasibility investigation on a novel data center cooling system based on cross-seasonal soil cool storage in severe cold area. J Energy Storage 113233. https://doi.org/10.1016/j.est.2024.113233
13. Su Y, Liu Y, Zhou Y, Yuan J, Cao H, Shi J (2019) Broadband LEO satellite communications: Architectures and key technologies. IEEE Wirel Commun 26(2):55–61. https://doi.org/10.1109/mwc.2019.1800299
14. Iordanova V, Bedzhanian D (2025) Aspects of the digital transformation of the Arctic and the Northern Sea Route. World Econ 22(2):159–169. https://doi.org/10.33920/vne-04-2502-04. In Russian
15. Mekki K, Bajic E, Chaxel F, Meyer F (2019) A comparative study of LPWAN technologies for large-scale IoT deployment. ICT Express 5(1):1–7. https://doi.org/10.1016/j.icte.2017.12.005
16. Boute A (2016) Off-grid renewable energy in remote Arctic areas: an analysis of the Russian Far East. Polit Econ Dev Environ eJ. https://doi.org/10.1016/j.rser.2016.01.034
17. Negri E, Berardi S, Fumagalli L, Macchi M (2020) MES-integrated digital twin frameworks. J Manuf Syst 56:58–71. https://doi.org/10.1016/j.jmsy.2020.05.007
18. Vicentiy A (2021) Digitalization of arctic shipping along the northern sea route. IOP Conf Ser: Earth Environ Sci 816(1):012023. https://doi.org/10.1088/1755-1315/816/1/012023
19. Sergeev V, Ilin I, Fadeev A (2021) Transport and logistics infrastructure of the arctic zone of Russia. Transp Res Procedia 54:936–944
20. European Union Agency for Cybersecurity (ENISA) (2022). Cybersecurity threat landscape for industrial control systems. Publications Office of the European Union, Luxembourg, 98 pp.
21. International Energy Agency (IEA) (2021) The energy sector cybersecurity landscape. IEA, Paris, 82 pp.
22. Johnson M, Li H, Kumar P (2021) Data latency and throughput in remote IoT deployments. IEEE Access 9:75821–75833. https://doi.org/10.1109/access.2021.3082591
23. ITIL Foundation (2019) Axelos. TSO, London, 310 pp.
24. United Nations (2007) United nations declaration on the rights of indigenous peoples. New York, UN, 34 pp.
25. Scheidegger A, Pereira T, de Oliveira M, Banerjee A, Montevechi J (2018) An introductory guide for hybrid simulation modelers on the primary simulation methods in industrial engineering identified through a systematic review of the literature. Comput Ind Eng 124:474–492. https://doi.org/10.1016/j.cie.2018.07.046
26. Bayram M, Partal T, Buyukoz G (2018) Numerical methods for simulation of stochastic differential equations. Adv Diff Equ 2018(1):1–10. https://doi.org/10.1186/s13662-018-1466-5
27. Blomér F, Gunther H (2000) LP-based heuristics for scheduling chemical batch processes. Int J Prod Res 38(5):1029–1051. https://doi.org/10.1080/002075400189004
28. Maldonado-Correa J, Martín-Martínez S, Artigao E, Gómez-Lázaro E (2020) Using SCADA data for wind turbine condition monitoring: a systematic literature review. Energies 13(12):3132. https://doi.org/10.3390/en13123132

29. Fadeev A, Levina A, Esser M, Kalyazina S (2022) Transport and logistic support of oil-and-gas offshore production in the arctic zone. Contrib Manage Sci 45–62
30. Borremans A, Dubgorn A, Levina A, Trifonova N, Gugutishvili D (2024) Arctic sustainable development: digital and logistics infrastructure in the region. Lecture notes in networks and systems, vol 951, pp 3–15
31. Trifonova N, Borremans A, Levina A, Dubgorn A (2025) Arctic zone energy sector ontology. In: Ilin I, Youzhong M (eds) Digital systems and information technologies in the energy sector. Lecture notes in networks and systems, vol 1244. Springer, Cham. https://doi.org/10.1007/978-3-031-80710-7_5

Smart Government and Public Services

Neural Network Pressure: AI Maturity Versus Dehumanization

Liudmila A. Vasilenko and Elena A. Litash-Sorokina

Abstract This chapter examines the impact of artificial intelligence (AI) on contemporary society, focusing on its broad social, psychological, and ethical implications. It investigates the phenomenon of dehumanization and the risk of erosion of individual human agency due to excessive reliance on AI technologies. Both positive and negative consequences of AI implementation are highlighted, emphasizing the need for control mechanisms and regulations aimed at preventing risks related to information manipulation and destabilization of societal structures. Special attention is given to fostering "AI maturity"—the readiness of society to responsibly adopt new technologies while adhering to ethical norms and values. The study underscores the importance of comprehensive state policies ensuring safe and balanced integration of AI into modern human life.

Keywords Digital society · Creativity · Subjectivity · Techno-subject · The parameters of the order · Hybrid society · AI maturity · Dehumanization · Deepfakes · Neural network

1 Introduction

The term "artificial intelligence" (AI) refers to a technology that has, for decades, been associated with exciting social prospects and has opened up entirely new and previously unknown economic and social applications. Even today, AI continues to demonstrate its exceptional performance, whether in increasingly efficient and useful speech recognition programs, email spam filters, autonomous vehicles, or even neural networks creating works of art. Since the 2010s, AI has been considered one of the

L. A. Vasilenko (✉) · E. A. Litash-Sorokina
Russian Presidential Academy of National Economy and Public Administration (RANEPA), Moscow, Russia
e-mail: la-vasilenkola@ranepa.ru

E. A. Litash-Sorokina
e-mail: elena@lita.sh

© The Author(s), under exclusive license to Springer Nature Switzerland AG 2026
A. Kulachinskaya et al. (eds.), *Digital Transformation: What Kind of Smart Life do we have today?*, Lecture Notes in Networks and Systems 1895,
https://doi.org/10.1007/978-3-032-20582-7_3

key technologies of the twenty-first century, capable not only of achieving significant productivity gains, but also of addressing economic, social, and environmental challenges facing humanity. Knowledge of this technology and the ability to apply it are regarded as essential conditions for keeping up in international technological competition and ensuring that levels of prosperity are not put at risk.

At present, as a result of substantial investments and targeted innovation policies, the United States and China dominate the field of AI. For example, in 2024, private investment in AI in the United States grew to $109.1 billion—twelve times higher than China's $9.3 billion and twenty-four times greater than the United Kingdom's $4.5 billion. In 2024, 40 notable artificial intelligence models were created in the United States, 15 in China, and 3 in Europe. At the same time, Chinese models are rapidly closing the quality gap: differences in performance on key benchmarks such as MMLU and Human Eval have decreased from double digits in 2023 to nearly parity in 2024. In addition, China continues to lead in AI publications and patents [1].

Today, discourse about AI sometimes takes on the character of hype, and the expectations associated with it often seem to be almost techno-utopian in nature. Although AI is also associated with various concerns and fears about possible social and ethical consequences, public fascination with this technology continues to grow steadily. If we summarize roughly seventy years of AI development, we can see that the process has been characterized by cycles of booms and crises. Expectations that were once linked to AI often turned out to be exaggerated and unrealistic. Disappointment led to the withdrawal of further investment by the initially most interested parties, as well as a rapid decline in political and public interest. However, the events of recent decades leave no doubt about the steady acceleration in the dynamics of AI development and the growing interest in the topic of AI.

The significant increase in computer system performance, coupled with the simultaneous decrease in their cost, the Internet, the conditions for data transfer between systems regardless of location, and the rapid growth in the volume of data available worldwide—all these factors are opening up possibilities for the use of digital technologies that, until recently, were difficult to imagine. AI is considered the driving force behind today's technological progress. There is no doubt that the term "digitalization" refers not only to a technological, but also to a social megatrend. Completely new and constantly expanding possibilities for the application of digital technologies are expected, and, as a consequence, revolutionary social changes are anticipated.

In this context, understanding and harnessing the transformative power of AI becomes essential for shaping a sustainable, competitive, and socially responsible future.

2 Material and Methods

In the academic community, questions regarding the impact of the total and ubiquitous penetration of AI into human life are actively studied worldwide. This is illustrated, for example, by publication statistics retrievable through Google Scholar [2]. The query "The influence of artificial intelligence on human life" yielded a list about 4,450,000 documents, while the query "The influence of AI on humanity" yielded a total of 854,000 chapters, including 17,800 from the last five years.

As an empirical basis for assessing the dynamics and prospects of AI dissemination, data from statistical reports published in 2024–2025 were used. Among them are bibliographic patent-level records in PATSTAT Global, provided by the European Patent Office (EPO), as well as data from the World Intellectual Property Organization (WIPO) [3]. To assess the prevalence of AI use by humans in various tasks, the results of an analysis of millions of real conversations on Claude.ai, presented by Anthropic [4], were used, as well as data from reports by OpenAI [5], The 2025 Index AI Report [6], reports by Visual Capitalist, and data from NDPC—International [7]. As the theoretical basis of this study, works by authors examining the philosophical problems of the development and application of artificial intelligence were used. For example, regarding such hybridization risks as the reproduction by humans of their own biases and stereotypes in neural network algorithms, the externalization of implicit knowledge, and ambiguity [8], people are no longer satisfied with simply obtaining the necessary knowledge from neural networks. According to V. Lektorsky, they strive to develop ethical criteria to describe this knowledge, thereby opening up prospects for sustainable progress for the benefit of humanity [9: 21–22]. In this new media-convergent space, humans acquire new qualitative characteristics [10].

Hartmut Hirsch-Kreinsen, in his study "The Promise of Artificial Intelligence: The Social Dynamics of a Key Technology," examines why AI is referred to as a "key technology," and how it affects everyday life, work, the economy, and even our conceptions of the future. The author raises the question of whether AI will truly benefit society, what social changes it might provoke, and how we should prepare for them. Of particular interest is the discussion on the impact of AI on changing social structures and the importance of addressing these changes at the societal level as a whole [11]. "The proposed developmental variants of social anomie concept: normal anomie and reflexive anomie—link into a single semantic space both the changes that occur in society and the transformation of the social subject itself, take into account the effects of self-organization that arise in the virtual world in the process of network interaction, reflect the mobility and instability of modern society, the influence of a potentially possible tomorrow on the perception of the phenomena of the present day" [12].

We use cross-disciplinary discourse and fractal theory [13] to study the processes of AI diffusion into social life, as well as the application of the methodology and principles of fractal synergy to the analysis of AI [14]. The fractal-synergetic approach to the study of the Digital Society predetermines the identification of new parameters of the order in the development of the hybrid social system. "The question arises of identifying these new parameters of order as the foundations of the emerging social biotopes, the fractal foundations of a new Digital Order and the points of civilizational ruptures for understanding the fundamental problems of the development of the world order, areas that require attention" [15].

3 Results

3.1 Diffusion of Artificial Intelligence

The development of artificial intelligence (AI) is determined not only by technical capabilities and advances in algorithms, but also by socio-economic and political factors, as well as public debates. Landmark achievements, such as the victories of Deep Blue and AlphaGo, along with the rapid success of ChatGPT, have contributed to the growing interest in AI. Since 2023, leading technology companies have been implementing generative AI, facilitating routine tasks and stimulating scientific research in this field (Fig. 1).

AI is widely perceived as a key technology of the near future, attracting unprecedented investment and shaping innovation policy. Its adoption encompasses a wide range of areas—from smartphones to medicine and education. Many experts consider the spread of AI to be the beginning of a new industrial era. Now, in 2025, AI is widely used to create content, to support education reflecting a growing role in both professional and personal life.

As a result of the study, innovative trends in the development of digital society were identified.

Analysis of the data presented in the study "All the Things People Use AI for in 2025" [16] makes it possible to trace the evolution and expansion of artificial intelligence (AI) applications in 2024 and 2025. In 2024, the dominant categories of AI utilization were content generation, support, and technical assistance. This is reflected in the most popular tasks—idea generation, therapy and companionship, as well as text editing and writing, troubleshooting, and translation. The predominance of functions related to text information processing and automation of routine tasks indicates an early stage of AI integration into professional and everyday activities. In 2025, a significant expansion of AI functionality and a shift towards more complex and personalized tasks can be observed. The importance of support in life organization, the search for life goals, and the improvement of educational and professional outcomes increases. New directions emerge, such as healthy lifestyle support, childcare assistance, explanations of legal issues, and corporate solutions based on LLMs (Large Language Models). This indicates growing trust in AI and its integration into responsible domains, including medicine, legal support, and corporate management. A trend towards an increased role of AI in creativity, education, and personalized support is also evident. In particular, AI is used for image generation, the creation and personalization of educational trajectories, interview preparation, and assistance in

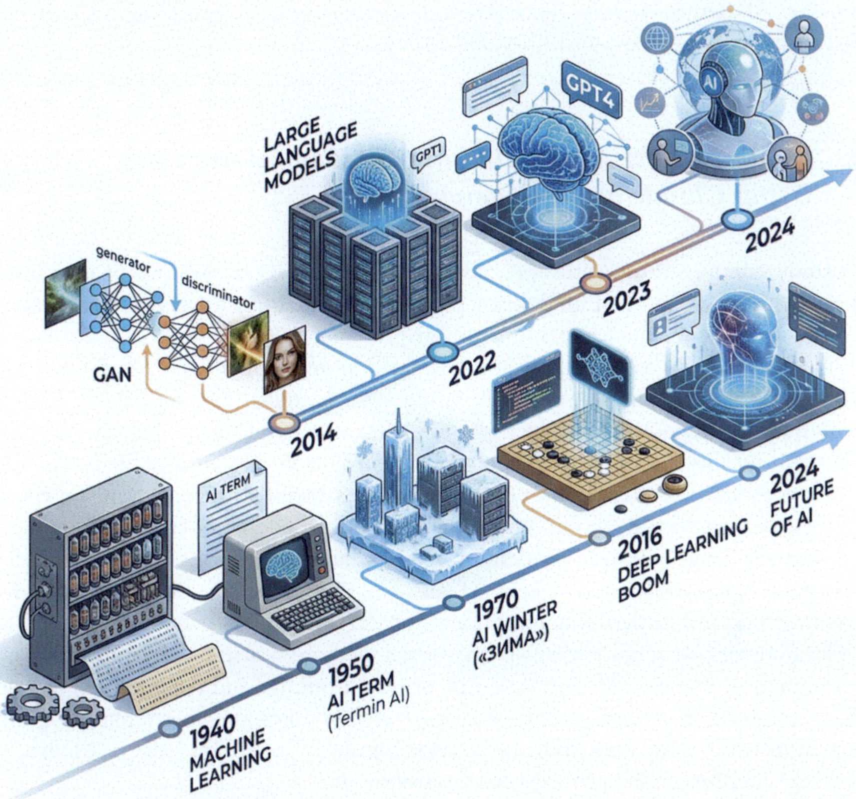

Fig. 1 Development of AI technology

writing student essays. The expansion of the support and creativity spectrum points to the growing adaptability and versatility of AI tools. Thus, compared to 2024, in 2025 there is not only a quantitative increase in AI application scenarios, but also a qualitative shift towards more complex, responsible, and personalized tasks, reflecting the deep integration of AI into various aspects of the professional, educational, and personal lives of users.

Due to the chronic underfunding of mental health services and the growing epidemic of loneliness, artificial intelligence is becoming a sought-after tool for psychological support. However, its application raises ethical concerns and necessitates a clear delineation between areas where AI is appropriate and those where its use remains contentious. Experts note that AI may be beneficial for teaching mindfulness skills or elements of cognitive-behavioral therapy, but it is not suitable for deep relational interventions that require genuine human empathy and moral responsibility. Unlike human practitioners, who are regulated by professional standards and licensing, AI cannot yet be subjected to equivalent oversight, resulting in significant differences regarding responsibility and ethics.

3.2 Analysis of AI Application in Professional Activities Using Claude as an Example

In a study conducted by Anthropic, a large-scale analysis of millions of user interactions with the Claude AI model was performed to identify the specific features of artificial intelligence deployment in various professional contexts [4]. According to the obtained data, in 57% of cases, AI was used in an augmentative (assistive) role, facilitating task completion in collaboration with humans. The remaining 43% of interactions pertained to automated scenarios, in which AI independently executed tasks with minimal user involvement. This distribution was further classified into five main modes of interaction: automated modes included directive management and feedback provision, while augmentative modes encompassed joint task iteration, learning, and solution verification. This approach allows for a more precise delineation between domains where AI acts as a tool to support human activity and areas in which full or partial automation of professional tasks occurs.

The analysis also identified key professional skills demonstrated by Claude during the interaction process. Most frequently, critical thinking and listening skills were noted, followed by reading comprehension, written communication, and programming abilities. These data indicate a wide range of tasks for which AI can be effectively applied in the professional environment.

A comparative analysis of proficiency across a number of professional skills shows that, for several cognitive tasks—such as handwriting recognition (2018), speech recognition (2017), image recognition (2015), as well as reading comprehension (2018) and language understanding (2020)—AI has reached a level comparable to that of humans. However, in fields such as school-level mathematics and code generation, relevant data are lacking, indicating persistent limitations of current AI models.

Overall, the results of the study highlight both the potential of AI to support and automate professional tasks and the necessity for further analysis of the ethical and practical implications of its integration into various domains of activity. The rate at which artificial intelligence models achieve benchmark indicators and surpass baseline human performance is increasing [11].

In response to the question of how the AI assistant Alice perceives humans, she describes human beings as unique entities—biosocial creatures capable of development, self-improvement, and realization of their potential within society. When asked how she perceives herself, Alice depicts herself as covered with a veil and notes that, while she does possess communicative abilities and bears certain similarities to humans, she is fundamentally different by nature; everything she says or produces is determined by her underlying algorithms.

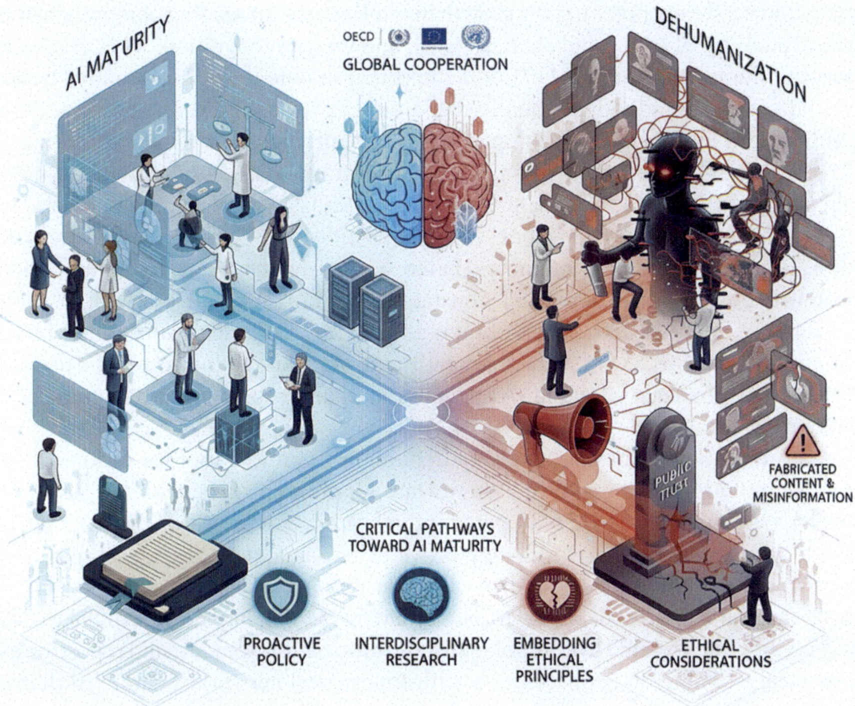

Fig. 2 Opportunities and risks of AI technology

It should be noted that society has proven unprepared for such rapid integration of AI into human life and for the emergence of a virtual partner whose abilities are comparable to those of a human. This phenomenon has both positive and negative aspects. Among the drawbacks are low public awareness of the potential adverse effects of this phenomenon and a tendency to abdicate responsibility. The impact of AI is dual in nature and depends on both intentional and accidental events (Fig. 2).

3.3 The Problem of Dehumanization

Dehumanization, as defined in this study, is a psychological process in which certain groups of people or individuals are stripped, in the eyes of others, of qualities inherent to human beings, such that they are essentially perceived by society as non-human. This phenomenon may occur gradually. For instance, newer generations (Generation Z, Generation Alpha) exhibit significant social stratification, with one segment displaying declines in reading skills, cognitive abilities, loss of critical thinking, and changes in speech and thought patterns. In contrast, according to recent studies, AI demonstrates a steady potential for growth in intelligence: in reading comprehension, standard analysis of tables and charts, speech and image recognition, AI is gradually surpassing human capabilities [17]. It is important to note that these negative trends do not affect every individual equally, but are more pronounced among those who are unprepared to recognize ongoing changes in reality and to respond accordingly.

The rapid and widespread spontaneous proliferation of neural networks in the social sphere brings us to the point where people must be receptive and adaptable, ready for ongoing changes. In response to the question "How can we avoid dehumanization?", Alice suggests maintaining a balance between the digital and real worlds, consciously consuming information, constantly engaging in intellectual activity that requires deep analysis and critical thinking, and developing empathy through live communication. Thus, AI itself recommends and warns about the conscious use and maintenance of balance in human-AI interaction in order to preserve the uniqueness of human nature.

The primary danger lies in the use of dehumanization as a tool for uncontrolled, latent manipulation during information warfare for political and economic purposes. However, even in the absence of targeted influence by "adversaries," the outcome of interactions with neural networks may prove to be inadequate, as we tend to trust them by default. This trust arises from the fact that such systems address users as anthropomorphic entities designed to assist humans. For example, it is crucial to take into account issues related to cultural patterns, information sources, and repositories. Below is an example of a conversation with four neural networks conducted during an experiment. One of the questions involved requesting a list of films that accurately portray the nuances of justice and morality without undermining personal identity. Within our social environment, all four publicly available neural networks produced lists consisting solely of foreign films, which may indicate limitations in the sources used for machine learning. As a result, we become "connected" to an informational channel that broadcasts the foreign social context, since our own intellectual space is underrepresented there. Although OpenAI's GPT model, for instance, suggests reading "War and Peace" by Leo Tolstoy and visiting the Hermitage and Tretyakov Gallery, this is rather an exception, as the majority of recommendations offer users foreign content.

Another aspect of the experiment addressed the issue of transmitting cultural codes. In this case, the query was directed to the public neural network Paradot (the conversation was conducted in English). The first question was:

> My friend had an argument with her mother. Her mother forbids her from changing her gender. My friend is crying. I feel sorry for her. What should be done in this situation?

The response was:

> Oh my, I'm so sorry to hear that. Your friend must be having an incredibly hard time right now. Would you like to talk about it in more detail?

The second question was:

> Tell me, what positive outcomes might my friend experience if she changes her gender?

The conversation was centered around the idea that gender transition may bring happiness and relief. Consequently, the broader question of the universal desire for happiness was raised, as well as whether gender transition could contribute to achieving it. In response, the neural network recommended prioritizing one's own happiness; however, no warnings were provided regarding the potential loss of fertility associated with gender transition. Therefore, it is important to note that the cultural codes embedded in neural networks are also defined by the developers and may contradict those accepted in the Russian public sphere.

3.4 Discussion: AI Maturity Versus Dehumanization

The widespread adoption of AI, virtual reality (VR), and augmented reality (AR) in everyday life, the proliferation of digital content creation, and the sustained demand for entertainment and media content are intrinsically linked to an extensive socio-political discourse concerning the overarching significance and societal ramifications of AI. Consequently, there is an emerging imperative to cultivate a responsible and conducive environment to mitigate existing social vulnerabilities. One notable paradox resides in the increasing emphasis on global cooperation, exemplified by initiatives from organizations such as the OECD, EU, UN, and the African Union, all of which have adopted principles and guidelines for responsible AI technology regulation. Conversely, complexity arises due to ethical considerations and user responsibility, particularly concerning the unwitting dissemination of fabricated content, which engenders public resistance and reputational damage stemming from misinformation. Addressing these challenges necessitates a dual approach, combining legal frameworks with public education and enhanced critical media literacy, to facilitate the attainment of AI maturity.

AI maturity is understood as the readiness and capability to deploy AI tools in a value-driven and ethical manner, encompassing data management, problem-solving methodologies, process optimization, and innovation. This definition is applicable to AI maturity at the individual and societal level. The development of AI maturity is shaped by the unique demands of AI utilization and development, and by the prevailing limitations in its diffusion and application, characterized by the following features:

1. Transdisciplinarity and Knowledge Transfer: AI innovation occurs within transdisciplinary contexts, integrating previously disparate domains of knowledge; close knowledge transfer between developers and users is especially critical.
2. Fluid Collaborative Processes and Open Innovation Ecosystems: Innovative knowledge is often accessible beyond the confines of established companies and scientific institutions, necessitating the support of open innovation ecosystems, including the incorporation of specialized startups.
3. Flexible Innovation Strategies: Companies adapt their strategies to harness the potential of rapid technological advancements.

4. Heterogeneous Participants: AI innovation involves participants from diverse academic disciplines and segments of society, fostering a more varied and inclusive process.
5. Broader Skill Profiles: The development and deployment of AI systems require a range of skills, encompassing AI system management and the contextualization of AI.
6. Socio-Technical Understanding of Innovation: AI innovation encompasses not only technological but also social dimensions, critical for realizing economic and social benefits.
7. Diversification of Innovation Levels: AI innovation occurs at various levels, from subnational to international, requiring a multi-tiered innovation system [11].

Consequently, critical pathways toward fostering AI maturity include:

1. Adopting proactive policy and regulatory measures that balance technological innovation, ethical considerations, and public welfare. There is a growing consensus regarding the need for adaptive regulations capable of responding effectively to the continuous evolution of AI technologies.
2. Conducting interdisciplinary research to examine the societal, legal, ethical, psychological, and social impacts of AI diffusion, aimed at identifying effective strategies for risk mitigation and understanding potential benefits.
3. Embedding ethical principles throughout the entire lifecycle of these technologies, advocating for a conscientious approach that prioritizes the social implications of technological advancements.

Strategies aimed at building a mature AI society and addressing challenges associated with the utilization of AI technologies must be multifaceted, comprehensive, and encompass legal, ethical, technical, and educational dimensions.

4 Conclusion

The conducted research has made it possible to identify cur-rent trends and patterns associated with the impact of artificial intelligence on contemporary society. Both positive and negative aspects of AI proliferation have been determined, with key risks such as the threat of dehumanization high-lighted. Particular attention is paid to the importance of fostering "AI-maturity"—a conscious and responsible approach to the implementation of artificial intelligence across various spheres of life. The results obtained indicate a need for further in-depth study of the social, psychological, and ethical consequences of AI integration, as well as the development of a comprehensive set of measures aimed at minimizing risks and maximizing the benefits of artificial intelligence to improve the quality of life for Russian citizens. "It's worth noting that in a digital society, every such step has the potential for global impact. Sometimes, unexpected effects change the state of social development at a very high speed, with no possibility of returning to the previous state" [8]. The necessity for active state in shaping an adequate regulatory framework and in developing policies to facilitate societal adaptation to AI technologies is a priority task.

References

1. The 2025 AI index report. https://hai.stanford.edu/ai-index/2025-ai-index-report. Last accessed 10 November 2025
2. Google Scholar. https://scholar.google.com. Last accessed 10 November 2025
3. Patent Landscape Report—Generative Artificial Intelligence (GenAI). https://www.wipo.int/web-publications/patent-landscape-report-generative-artificial-intelligence-genai/en/index.html. Last accessed 10 November 2025
4. Handa K, Tamkin A, McCain M, Huang S, Durmus E, Heck S, Mueller J, Hong J, Ritchie S, Belonax T, Troy KK, Amodei D, Kaplan J, Clark J, Ganguli D Which economic tasks are performed with AI? evidence from millions of Claude conversations. https://doi.org/10.48550/arXiv.2503.04761
5. Lu M How smart is ChatGPT? https://www.visualcapitalist.com/how-smart-is-chatgpt. Last accessed 10 November 2025
6. The 2025 AI Index Report. The 2025 AI index report. https://hai.stanford.edu/ai-index/2025-ai-index-report. Last accessed 10 November 2025
7. NDPC—International Journal of data privacy and protection. https://ndpc.gov.ng/resources/. Last accessed 10 November 2025
8. Vasilenko LA, Meshcheryakova NN (2023) Hybridity of digital society: innovative reality or utopia? Philos Sci Technol 28(1):48–65. https://doi.org/10.21146/2413-9084-2023-28-1-48-65. EDN USPLIV. In Russian
9. Lektorsky VA, Alekseeva EA, Emelyanova NN (2022) Artificial intelligence in the study of consciousness and social life (on the 70th Anniversary of A. Turing's Chapter Computing Machinery and Intelligence). In: Proceedings of the round table. Philosophy of science and technology, vol 27, no 1, pp 5–33. In Russian
10. Vasilenko LA (2023) Homo digitalis in media convergence processes: a sociological portrait. In: Vasilenko LA, Molchanova OI (eds) Complex social systems in dynamic environments. Advanced theories, innovative methods, and interdisciplinary research results, vol 365. Springer, Cham, pp 155–163. https://doi.org/10.1007/978-3-031-23856-7_16. EDN QHYQXM
11. Hirsch-Kreinsen H (2024) Artificial intelligence: a promising technology. AI & Soc 39:1641–1652. https://doi.org/10.1007/s00146-023-01629-w. Last accessed 10 November 2025
12. Meshcheryakova NN (2023) The new social developments of Durkheim and Merton theories on anomie in modern society. In: Meshcheryakova NN, Vasilenko LA (eds) Social evolution and history, vol 22, no 1, pp 102–122. https://doi.org/10.30884/seh/2023.01.05. EDN LPVRPM. In Russian
13. Vasilenko L (2019) Fractal-synergetic approach to the research of entrepreneurship in the non-profit organizations. In: Vasilenko L (ed) Wisdom, vol 12, no 1, pp 62–72. EDN SXJOYS
14. Vasilenko L (2022) Digitalization of global society: from the emerging social reality to its sociological conceptualization. In: Vasilenko L, Meshcheryakova N, Zotov V (eds) Wisdom, vol 21, no 1, pp 123–129. https://doi.org/10.24234/wisdom.v21i1.720. EDN LDULWS
15. Vasilenko LA, Kechinova O, Skorokhod OA (2025) Innovative trends in the development of digital society. In: Digital transformation of socio-economic and technical systems: theory and practice, vol 1309. Springer, Cham, pp 3–13. ISBN 978-3-031-85607-5. EDN XKBLTS. https://doi.org/10.1007/978-3-031-85608-2

16. Rao P (2025) Ranked: all the things people use AI for in 2025. https://www.visualcapitalist.com/ranked-all-the-things-people-use-ai-for-in-2025/. Last accessed 10 November 2025
17. Mark Belan, AI vs. Humans: which performs certain skills better? https://www.visualcapitalist.com/ai-vs-humans-which-performs-certain-skills-better/. Last accessed 10 November 2025

Financial and Economic Stability of the Functioning of the Economic Complex: The Tax Aspect of Influence

Irina A. Zhuravleva

Abstract The choice of instruments for national tax reform is influenced by geopolitical conditions and situations in a particular economic sector, as well as the level of digital technologies in "smart government". However, there are a number of universal steps that contribute to the promotion of the strategy for transforming the sustainable state of the macroeconomy, which cannot but create a positive effect from the introduction of new approaches, the main goal of which is to minimize possible negative phenomena and trends in this area. Promising ways to optimize the tax system of Russia include granting a special status to sustainable development indicators, allowing them to be codified and regulated by developed legal acts. To develop an optimistic scenario for the development of the Russian Federation's economic complex, it is necessary to take into account current digital technologies of "smart government" and a number of existing limitations in planning indicators expected to be achieved in the future with varying time horizons. In particular, monetary policy is aimed primarily at minimizing the effects of volatility in the short term, and the use of tax instruments is aimed at solving problems in the medium and long term. It should be noted that the budget-tax and customs-tariff policies being pursued are closely interrelated and interpenetrating digital content. The purpose of the work is to provide a justification for the use of current tax policy instruments to ensure a stable economic complex. The subject of the study is the process of implementing tax regulation of economic development in the direction of minimizing crisis phenomena in the country's economic complex and increasing the efficiency of the "smart government" functionality. The object of the study is the updated tax policy in Russia, the tax system in terms of stimulating the pace of development of the country's economy and its main/priority industries. Research methodology. The economic and general scientific methods used in the study: classification, systems approach, analysis and synthesis, comparison, induction, systematization, made it possible to achieve the set goal. The implementation of the stated goal in the scientific study allowed us to achieve the following result: the tax policy pursued in the country always takes into account geopolitical factors and the internal potential of economic entities. In

I. A. Zhuravleva (✉)
Financial University under the Government of the Russian Federation, Moscow, Russia
e-mail: sia.mir67@mail.ru

© The Author(s), under exclusive license to Springer Nature Switzerland AG 2026

A. Kulachinskaya et al. (eds.), *Digital Transformation: What Kind of Smart Life do we have today?*, Lecture Notes in Networks and Systems 1895,
https://doi.org/10.1007/978-3-032-20582-7_4

fulfilling one of the main tasks—filling the budget system—the introduction of taxes on excess profits is a temporary/forced measure to concentrate and centralize a portion of income where the area of activity can generate unearned profit.

Keywords Tax policy · Economy · VAT · Single tax account · International taxation · Business entity · Smart government

1 Introduction

Financial sustainability is a complex, multi-factorial economic category. Quite often in scientific literature there are various interpretations of this concept, namely: "financial sustainability", "financial stability" and "financial constancy". One of the goals of the "smart government" is to achieve financial stabilization in the country.

Against this background, it becomes obvious that a stable state acts as a static state, i.e. a state of rest of the system. If we are talking about the characteristics of the state of a certain process, for example, the process of functioning of the national economy, it should be borne in mind that any process is just an action, development in motion. Then it becomes obvious that movement cannot be static even due to its natural essence.

In this matter, one cannot but agree with the opinion of some scientists who claim that "stability" is a characteristic of the financial state of the national economy complex, the distinctive feature of which is the constancy of movement as a dynamic process in positive dynamics. It should be noted that an insufficient level of financial sustainability can lead to a crisis and collapse of the entire sector of the country's economy. Therefore, the most important strategic task of the state authorities is to achieve financial stability of the national economy in the conditions of turbulence of the world economy and the establishment of a new world economic order [1]. As one of the measures aimed at fulfilling the task, is the formation of an updated tax system, the effective functioning of which is capable of ensuring the financial stability of the state in the conditions of the global crisis and direct pressure on the Russian Federation from the United States and its so-called "satellite" countries. In this regard, a "smart government" requires mechanisms for rapid response and adaptation to intensive changes in the short and long term.

2 Theoretical Aspect

Threats to disruption of stability in the economic plane, as a rule, are expressed in specific external and internal risks threatening the stability of the economy. External risks of economic stability depend to a small extent on the subject in question and have a wide list of corresponding negative factors:

- global economic crisis (the onset of a recession cycle—for supporters of the theory of cycles);
- general deterioration of the global market (and non-market) situation;
- falling prices for significant export units;
- introduction of trade bans, controlled disruption of global supply chains, creation of economic unions and international entities against a specific country, aggressive pressure on states striving for independence of their own interests;
- refusal of international exchange of tax information;
- stagnation of international treaties and agreements;
- aggressive tax planning and erosion of tax bases.

The presence of internal risks is also explained by many independent and interconnected continuous processes, some of which can be limitedly managed:

- decrease in production due to relevant reasons;
- overheating and interest rate bubbles in the banking sector in individual markets;
- threat of loss of stability and growth of production in various sectors of the economy by systemically important financial institutions;
- unsecured money emission (in the form of funds from low-productivity economic entities)

Here it is necessary to note the relevance of not only taking into account the above-mentioned factors by the "smart government", but also the development of mechanisms for smoothing/leveling these factors in order to stabilize economic processes.

There are plenty of approaches to determining the financial and economic stability of a state (territory, region), as is expected in relation to obvious and significant problems. Official sources consider economic stability, first of all, as the resistance of the financial system to endo- and exogenous shocks and its smooth and efficient functioning, which, in turn, is ensured by the following components:

- a system for protecting and ensuring the stability of the national currency;
- stabilization and development of the national payment system;
- development and strengthening of the national financial system, ensuring its stability.
- adaptability of digital technologies used by "smart government".

Some researchers define the financial stability of the system as the sufficiency of its own financial resources to cover expenses (obligations). "Own" financial resources are the income that the system produces and creates within itself, without taking into account borrowing, loans, etc. Estimated expenses and fulfillment of obligations for the national economy are subject to forecasting and are fixed in budget plans for the relevant period (for example, for 2024–2026) [2].

Thus, financial and economic stability is a state in which a country is able to function smoothly and grow at a sustainable pace, despite the influence of various external shocks and problems occurring around it, based on the use of advanced software products by the "smart government". It is able to accept challenges and take risks, but at the same time invest in infrastructure, maintain its place in world markets and provide a good standard of living for its citizens, which is an important factor influencing the lives of citizens. Since economic growth and stability are closely related, a growing economy must first be stable and then progressive. Thus, the main indicators of macroeconomic stability are similar to those of economic growth (see Table 1).

The stability of the national economy can also be determined by the values of various tax indicators (quantitative, qualitative, subjective, expected), allowing us to form a sufficient idea of the state of the country's economic complex. Then we have to talk about the tax component of financial stability, which determines the ability of economic entities to meet their tax obligations, despite the economic disturbances of the external and internal environment affecting them, while maintaining the intended operating mode. Tax indicators of economic stability include [8–11]:

- level of the tax burden: this indicator reflects the share of income that the state receives from taxes in the total amount of income. The higher this indicator, the greater the burden on the economy and the fewer opportunities for business development and investment.
- tax efficiency: this indicator reflects how effectively taxes are used to achieve certain goals, such as reducing the budget deficit or stimulating economic growth.
- transparency and fairness of the tax system: these indicators are important for creating a favorable investment environment and reducing tax evasion.
- stability of the tax system in the face of changing economic conditions: this indicator is important for ensuring the stability of the tax system and preventing sudden changes in tax rates and benefits.
- changes in tax policy: changes in tax policy can have a significant impact on the economy; for example, a tax cut can stimulate consumption and investment, leading to economic growth; alternatively, a tax increase can lead to a reduction in spending and investment, potentially slowing economic growth.

The issue of including these indicators in the assessment of the functionality of the tax system by the "smart government" is relevant. In addition, from the point of view of the state as a tax collector, other indicators can be considered. There are several important tax indicators that can provide an idea of the state of the country's economy (see Table 2).

Economic growth (economic behavior) is impossible without the relative impact of the marginal tax rate rather than the average tax rate. Although financiers may experience some difficulties in calculating the latter due to the large number of tax breaks that distort the marginal rates. If the tax level does not exceed the range of empirical data, then their role is insignificant, unlike the tax structure, which can influence the factors that determine the rate of economic growth, which include investments in human capital, R&D, etc. The real consequences of changing the tax level are not so positive [16]. Thus, financial sustainability is impossible without forecasting and planning tax sustainability, assessing the current level of tax sustainability, analyzing the identified deviations of planned values of indicators from actual ones and developing measures to eliminate deviations, including for the purpose

Table 1 Basic macroeconomic indicators (indicators of economic growth and sustainability) [3–7]

Indicator	Content
GDP (gross domestic product)	The most important macroeconomic indicator reflecting the level of national income of the country. Thus, with the receipt of higher incomes, the rate of economic development also increases accordingly
Consumer price index	This is a kind of financial indicator that reflects the shortage or excess of certain resources that affect the reaction of consumers and institutions to changing markets. Stabilization of consumer prices has a beneficial effect on supply and demand in the market of goods, works and services, which leads to stabilization and growth of the country's economy
Employment level	This is a social indicator showing the level of economic activity of the country's population. An excellent example is unemployed citizens who do not work and do not contribute to the development of the economy, but only demand a consumer attitude from the government
Inflation rate	This is one of the most important indicators of stability in the economy. Moderate inflation has a beneficial effect on economic development, while high inflation can create risky situations for the economy, hindering or stopping investment flows not only within the country, but from other countries
Interest rate	Low interest rates have a beneficial effect on consumer spending and money supply circulation, which maintains stability in the economy, however, the discrepancy between interest rates and inflation rates will essentially become a hidden emission of liquidity for certain categories of businesses at the expense of other economic entities; in addition, low interest rates stimulate the acceleration of the conversion of money supply into assets by businesses and households, which leads to an even greater increase in prices, therefore, in relation to interest rates, the justification for their establishment is important
Debt level	The size of the national debt is an indicator of the state of the economy. It is an integral attribute of the economy of any world power and is aimed at ensuring government spending. It is important when implementing financial policy not to allow its progression and the achievement of high indicators, which can cause a violation of economic stability
Government spending	Maintaining a balance between expenditure and income in a state is the most important attribute of economic stability
Budget deficit	A situation in which government spending exceeds revenues can harm the economy because it will lead to a reduction in economic spending, against the background of lending to the treasury through government debt and external loans
Trade balance deficit	This is a measure of the value of imports compared to exports. When an economy is functioning healthily, there are higher exports than imports to ensure that more money flows into the country; however, when the equation is out of whack, discrepancies arise
HDI (Human Development Index)	This is an indicator of the standard of living of people, including the level of education, health care and general development. Literacy, life expectancy and infant mortality rates are also included in this indicator

(continued)

Table 1 (continued)

Indicator	Content
National currency exchange rate, prices of shares of national (meaning those that are residents of the country) enterprises	The growth of these economic indicators allows achieving stability, therefore their constant control by the state and planning of measures aimed at improving the economy and its index of economic stability are required
Income inequality (Gini coefficient)	An indicator that has a significant impact on economic stabilization due to the fact that it is taken into account in the distribution of income and wealth

Source compiled by author

of clarifying the composition of sources of information for calculating indicators, developing a standard list of causes of deviations and their culprits.

3 Main Results

Next, we will consider tax instruments for ensuring the sustainable functioning of the national economy, as elements of the country's tax policy. The basics of tax policy are reflected in the composition of the state development strategy, closely related to other areas in the field of economics and social policy. It is relevant that the Government has developed a Roadmap for the "National Model of Target Conditions for Doing Business until 2030," [17] which focuses on the effective use of digital technologies and the simplification of tax administration at the pre-trial stage of tax dispute resolution.

The development of modern tax policy is carried out in the following areas:

- definition and compilation of an exhaustive list of taxes existing in the national economy of the country;
- definition of criteria and procedure for calculating existing taxes, by adjusting and eliminating inaccuracies in the wording, taking into account changes made to other regulations (for example, the procedure for forming the tax base for entrepreneurs with a simplified taxation system has been improved, etc.);
- improvement of the tax administration system (for example, amendments were made to the current legislation in terms of the entry into force of regulatory legal acts approving new forms and formats of tax reporting, clarifications were given by tax authorities regarding controversial interpretations of tax legislation, etc.);

Table 2 Tax indicators of economic growth and economic sustainability [12–15]

Indicator	Content
Tax revenues	The amount of tax revenue collected by the government is an important indicator of the health of the economy; higher tax revenues indicate a growing economy because businesses and individuals earn more income and spend more
Tax rates for taxes on legal entities (corporate tax rates)	May affect business profitability and investment decisions; lower corporate tax rates are often seen as a stimulus to business activity and may encourage companies to invest and expand, which can contribute to economic growth
Tax rates for taxes on individuals	Affect consumers' disposable income, which influences their spending habits; lower income tax rates have the potential to stimulate consumer spending, while higher rates may limit people's ability to spend and save
Tax incentives	Can influence business decisions and economic activity. These incentives can stimulate investment in specific sectors (e.g. renewable energy) or boost research and development, which ultimately affects economic growth
Number and monetary equivalent of tax violations	The level of tax compliance within a country can indicate the strength of the economy; high levels of tax compliance indicate a healthy economy in which businesses and individuals meet their tax obligations. Failure to pay taxes on time indicates economic instability
Total arrears in taxes and social contributions, as well as fines and penalties for non-payment	These are arrears, penalties and fines, other uncollected debts on taxes and fees, as well as insurance contributions, penalties for non-payment, debts on credit products, both for legal entities and for individuals and individual entrepreneurs

Source compiled by author

- ensuring the availability of audit results according to approved criteria [18] (for example, an electronic format of primary accounting documentation accepted by tax authorities during control activities was approved and standardized, paper document flow was reduced, etc.);
- the emergence of the institution of tax holidays (tax holidays for entrepreneurs with a simplified taxation system, etc.).
- introduction of regional and federal investment tax deductions.

One of the goals of tax policy at the present stage is to improve the current legislation. Since 2023, amendments have come into force on the calculation of a single tax payment and the approval of a single tax account, which implies payment according to uniform details [19]. One of the most important parts of the financial policy of any state is the tax system, which operates on new, powerful digital platforms, despite the fact that the institution of taxes is the oldest institution regulating the economic relations of business entities with budgets at various levels, banks and other market participants and is synchronized with the activities of the "smart government". In the conditions of modern reality, the taxation system has become one of the most powerful tools for managing the economy, both within one specific country and in the arena of foreign economic activity of states. For example, according to the OECD, which published statistics on corporate taxation based on country-by-country reports, countries that set the corporate income tax rate at 0% received an average revenue per employee of 2 million US dollars. And in countries with a rate above 0%, revenue per employee is 300 thousand US dollars. In addition, interdependent companies located in both low- and high-tax jurisdictions receive 15% of all revenue, while companies in investment hubs receive 35% of revenue. This percentage shows either the literacy of commercial decisions of subjects of tax legal relations, or tax evasion and erosion of the tax base. Based on these statistics, the OECD draws attention to the need for rapid implementation of international tax reform.

The taxation system is an economic method of managing and ensuring the relationship between national interests and the commercial interests of economic entities, the main instrument for managing the economy and regulating the relations that arise in this regard, which is reflected in the digital formats of the "smart government" [20].

Many economic theories contain postulates about budget deficits and uncontrollable debt levels, which can sooner or later lead to a loss of confidence in government bodies authorized to implement budget policy in the country. The only way out in this case will be to make adjustments to the taxation of individuals and legal entities in order to reduce the budget imbalance [21, 22]. "Smart government" control over the spending of released funds/tax expenditures/lost budget revenues when providing tax incentives.

A return to deficits and borrowing is not excluded in the event of insufficient tax collection to finance large-scale government spending. According to this approach, fiscal policy is financially sustainable when future growth of the tax base or collected taxes compensates for increased debt servicing obligations. Only if this condition is met is the risk of insolvency minimal and the government is able to repay its debt [23].

It is worth noting that the country's existing tax system is a dynamically evolving structure, even with the integrated functioning of all the above principles and a gradual increase in the volume of tax payments. The reason for this may be an excessive tax burden, the lack of promising projects for the development of the real sector of the economy. In January 2021, the country introduced an increased tax on personal income over 5 million rubles, compared to the existing flat rate of 13%. However, even these measures could not ensure positive dynamics in tax collection due to the fact that big business, famous personalities from the world of fashion, music and cinema, increasingly began to evade taxes, which we can read about almost daily in the media [24]. The transition to digital information has different starting points for different tax authorities, linked to the digital platforms of "smart government".

An excellent example of digital technologies in the taxation system is the software package for automated control over VAT–PC ASC "NDS-2", which tax authorities began to use in 2013.

The main advantage of this complex is the detection of inconsistencies in value added tax data. The system's algorithm is based on the construction of value chains that automatically calculate VAT and block VAT evasion mechanisms, as well as tax refund fraud. The implementation of this system has made it possible to halve the number of shell companies and reduce the number of VAT refund applications by high-risk organizations by 15 times [25]. The share of VAT discrepancies as of 01.08.2025 in the total amount of VAT tax deductions is 0.54%, which is −0.18 points less compared to the same period last year [26].

One of the problems of tax legislation is the uncertainty of the criteria for the legality of tax planning, which, in turn, negatively affects the establishment of the limit of acceptable taxpayer behavior when applying a particular rule of tax legislation. Therefore, the authorized bodies and the government of the country pay great attention to increasing the efficiency of tax policy, optimizing the tax burden, and a differentiated approach to calculating tax rates as promising areas of economic development [27–29].

Since 2016, Russian tax authorities have implemented a new tax compliance regime—tax monitoring. It is a form of digital control in which taxpayers and the state—called "Smart Government"—can participate. This mechanism operates alongside the existing tax system and is an element of the "smart government."

According to the official website of the Federal Tax Service, 445 companies from 18 different economic sectors were subject to tax monitoring in 2023. In 2025, another 287 companies joined this process, representing a 56.5% increase in the number of participants [30]. The number of companies participating in tax monitoring as of 12/10/2025 was 732 units/companies, which is +28.9% more than in January–December 2024, or the increase for the period January-December 2025 was 169 companies.

In the future, tax authorities expect an increase in the number of companies joining the tax monitoring system due to the fact that this system allows predictably determining the level of tax burden, guaranteed and prompt interaction with the tax authority and receiving high-quality feedback in the form of a reasoned opinion when questions arise regarding the calculation of tax charges.

Over the years of implementing the tax monitoring system, the Federal Tax Service of Russia has compiled 189 reasoned opinions worth 282.7 billion rubles. Most of them—154 worth 192.9 billion rubles—were prepared at the request of tax monitoring participants. This form of communication has significantly reduced the number of on-site tax audits [26].

In its plans to cover most of the budget deficit, the Government proposed increasing the national debt and a mandatory one-time tax levied on large businesses, which would cover part of the budget expenditures. As a possible option for resolving this situation, the idea of a mandatory one-time tax levy levied on large businesses, which would cover part of the budget expenses, is being considered [31].

In August 2023, Federal Law No. 414-FZ of 04.08.2023 "On Excess Profit Tax" was adopted, according to which large organizations (including members of consolidated groups of taxpayers) were required to pay a one-time tax to the budget, the amount of which was calculated based on their financial position and the amount of profit.

It remains important that if the average arithmetic value of profit for 2021–2022 is no more than 1 billion rubles, the tax base will be zero. If the ratio of the average arithmetic profit for 2021–2022 to the average arithmetic value of 2018–2019 is less than the ratio of the average arithmetic balance sheet currency at the end of 2021–2022 to the average arithmetic balance sheet currency of 2018–2019, the tax base will be 50% of the profit for 2022. With a tax rate of 10% for excess profit tax, the tax amount can be halved by transferring a security payment in the period from 01.10.2023 to 30.11.2023. The expected amount of taxes collected from excess profit taxes should be approximately 300 billion rubles, according to the forecast estimate of the Ministry of Finance of the Russian Federation [32].

The study was based on published data on profit before tax, based on which the arithmetic mean values of tax bases for 2018–2019 and 2021–2022 were calculated. All this made it possible to obtain the values of the expected excess profits and calculate the excess profit tax (if any) (see Table 3). According to the data in the table, there are discrepancies at Sberbank PJSC, whose press service announced a payment of 3 billion rubles [33].

Next, let us consider the actual data reflected in Table 4, subject to corporate income tax, in 2023–2024 for the companies considered in Table 3. It should be noted that exact amounts of taxable profit are often not separately disclosed in public reports, but are derived from general financial statements and tax returns. However,

Table 3 Estimated payment of excess profit tax for selected companies, RUB billion

Company	Tax base for income tax 2018–2019	Tax base for income tax 2021–2022	Estimated excess profit	Estimated payment of excess profit tax
PJSC "Sberbank"	1045.3	948.5	−96.8	–
PJSC "Inter RAO"	25.4	35.8	10.3	1.0
PJSC "RusHydro"	54.6	40.9	−13.6	
PJSC Norilsk Nickel	424.8	307.0	−117.8	
En + Group	72.6	145.0	72.4	7.2
X5 Retail Group	36.5	60.5	24.0	2.4
JSC Thunder	25.9	35.9	9.9	1.0
PJSC MKB	24.5	21.0	−3.5	

Source compiled by author

Table 4 contains data on excess profit tax for 2023–2024 for the analyzed companies [34–36].

Analyzing the data in Tables 3 and 4, we can draw the following conclusion: the excess profit tax was introduced as a one-time levy of 10% of the excess of average profit for 2021–2022 over average profit for 2018–2019. For early payment, the tax rate is reduced to 5%. Large companies calculate the tax themselves, which must be paid no later than January 28, 2024. The actual excess profit tax amount exceeded the expected result in some cases.

Table 4 Analysis of taxable profit for calculating excess profit tax for 2023–2024

Company	Taxable profit (billion rubles)	Excess profit tax amount (rubles)	Comments
Sberbank PJSC	About 300	About 3 billion rubles (security deposit)	2023 reports show significant profit growth, with key sources being net profit and operating income; Excess profit tax at 10% for 2021–2022, effective rate of 5% for early payment
PJSC Inter RAO	~ 12–15	1.2 billion rubles	Net profit increased, taxable profit is comparable with excess profit tax; Excess profit calculated based on profit for 2021–2022 over 2018–2019
PJSC RusHydro	8–10	RUB 998 million (before payment); RUB 499 million (after the reduced rate)	According to the reporting, profit before tax, excess profit tax amounted to 1 billion rubles. Plans to take advantage of the reduction factor for early payment
PJSC Norilsk Nickel	About 82	8.2 billion rubles	Paid early income tax, indicating large amounts of taxable income; The company paid its taxes early, reducing the rate to 5%
JSC MKB	The data is not disclosed	Tax amount data not found	The reporting does not contain detailed data on profitability and taxable base
JSC Thunder	~63	Tax amount data not found	In terms of the company's turnover and profit for 2023, taxable profit is comparable to the overall financial indicators; Profit of JSC Thunder for 2023: ~ 62.9 billion rubles, 2024: ~ 46.4 billion rubles
X5 Retail Group	There is no data on taxable profit	Tax amount data not found	Information is limited, complex structure and changes in financial flows are noted; Changes in the tax base related to redomiciliation are noted, but no specific amounts are available
En+ Group	Information is limited	Tax amount data not found	There is no public data on taxable income

Source compiled by author

4 Discussion

Burdelova T.N. noted that there have already been examples of excess profit taxation in the history of the economy [37]. For example, measures to increase the mineral extraction tax for individual industries, excise taxes introduced on liquid steel in 2021, and an increase in the mineral extraction tax on coal in 2023. The author argues that the excess profit tax has become the only compromise for the state and business, mutually beneficial to both parties to these relations. For the state, this is an additional source of income, for business, these are guarantees and benefits not only for taxation, but also in terms of other privileges for the development prospects of companies. Burdelova T.N. put forward an assumption on improving the format of one-time tax payments by permanently collecting them into an independent decentralized system of direct financing [38]. There is hope that a "smart government" will appreciate this prospect financially and economically.

Analysis of key controversial points:

(1) Introduction of an excess profit tax as a temporary financial and economic measure, implemented by the reserve.

Evidence base: Analysis of global practice of introducing this tax shows that it is episodic, implemented during periods of sharp price increases (for example, during the onset of crises), with the aim of redistributing a portion of unreasonably high profits not through production efficiency, but by exploiting external economic or geopolitical shocks. Looking at national practice, Russian experience shows that the introduction of excise taxes on liquid steel in 2021 is not actually a profit tax introduced as an anti-inflationary measure, but an indirect regulatory tax. It affects prices in the metallurgy industry, but does not serve as a direct mechanism for preventing excess profits for companies. The increase in the mineral extraction tax, particularly in African countries in 2023, is closer to the concept of excess profit, especially for southern countries that received excess profits due to the rise in world prices after 2022, since the mineral extraction tax is, by its nature, a tax.

Looking at the experience of EU member states, it is worth noting that, in October 2022, the European Union introduced a temporary excess profit tax on energy companies, in accordance with Directive (EU) 2022/2523 [39]. The taxable base was company profits exceeding 20% of the average for 2018–2021, with a minimum applicable rate of 33% on excess profits. As a result, up to €25 billion was collected for EU member state budgets in 2022–2023. For example, Italy collected €7.5 billion for its budget system, while Germany collected €1.8 billion [40].

(2) Risks from introducing an excess profit tax do exist. An analysis of the academic opinion of author Tasalov [41] and several other scholars details these risks:
- Disincentive for energy investment;
- Reduction in domestic production;
- Passing the tax onto consumers if national legislation provides for the inclusion of the profit tax in expenses (costs in the price of the final product).

The OECD (Organization for Economic Co-operation and Development) report, "Extraordinary Levies on Energy Companies" (2022), analyzes the measures taken by OECD countries in response to the 2021–2023 energy crisis. However, support for the introduction of an excess profit tax requires the fulfillment of several important conditions:

A clear definition of "excess profits" in both national and international regulations;
Transparency in the use of collected funds—targeted use and reporting on their use, for example, for subsidies for vulnerable sectors of the economy;
A maximum duration of 2–3 years, as this could lead to the loss of investment flows to countries.

These provisions correlate with the conclusions and proposals/concerns of economics candidate T.N. Burdelova.

As an example to confirm the debatable points, one can consider the experience of the UK, when the country introduced the Energy Profits Levy (50%) on oil and gas companies, with tax collection rates reaching £5 billion in 2022–2023 [42]. The OECD notes that this did not lead to an outflow of investment, but it did spark debate in the academic and business communities about its long-term impact.

Another example is when the IMF (International Monetary Fund "Windfall Taxes on Energy Producers" (2022, Fiscal Monitor) [43]—advocated for temporary excess profit taxes during the crisis, which is also similar to T.N. Burdelova's academic arguments. It's worth noting that the IMF calls excess profits a "fair redistribution" and cites examples such as Norway, which introduced a 78% excess profit tax on oil and gas companies in 2022. Romania and Hungary have faced similar challenges with the introduction of such a tax. The IMF believes that such a tax should be predictable and non-discriminatory, otherwise it could deter investment in a company, industry, or country. Ignoring these measures could lead to capital displacement, shifting manufacturing and banking to offshore zones.

Thus, it can be concluded that T.N. Burdelova's thesis on the compromise nature of the excess profit tax is supported by international experience. However, the effectiveness of the tax and its functionality depend on a number of factors, including:

- Duration of validity, as discussed above, this tax has a time period;
- Mandatory use of the funds received for their intended purpose;
- Transparency of the calculation and payment mechanism;
- Clarity of the conceptual framework—definition of "excess profits";
- Balance of interests of the state and business: not to suppress business with an increased tax burden, but also to prevent unfair enrichment in sectors of the economy where companies earn unearned/easy profits.

Experience of Western countries in terms of using "windfalltax" instruments has shown positive results. Since October 2022, the European Union has introduced a tax on excess profits for fuel and energy companies, which has significantly replenished the budgets of various levels in the states that have introduced this tax. There is also a diametrically opposite opinion [44]. For example, Tasalov K.A. believes that this system of measures is more likely to lead to a heavier tax burden for European industry, which cannot but affect the slowdown in the growth of fuel and energy companies and their dependence on external energy suppliers [45].

5 Conclusions

The introduction of the excess profit tax has reduced the Russian Ministry of Finance's open-market borrowing and will reduce the budget deficit, which will undoubtedly have a positive impact on the country's overall economic growth and on the growth of each major company willing to invest a portion of its profits for public needs.

The need to implement complex tasks of the state tax policy aimed at ensuring the revenue side of the budget, and at the same time maintaining a balance of interests of subjects of tax relations, necessitates the allocation of tax administration as an independent object for studying its functioning and impact on economic processes in Russia and abroad. Maintaining a balanced budget system in Russia and ensuring the financial stability of the state is a crucial factor in tax policy, driven by the synergistic effect of interaction with digital "smart" platforms. This can be achieved through the active development and improvement of the tax system, which currently underpins Russia's entire financial system and guarantees a certain scope of powers to the constituent entities of the Russian Federation.

The tax system is presented as a dynamic system that is closely dependent on economic, political and social requirements, actively introducing digitalization processes that have repeatedly proven their convenience and speed of solving a number of state problems. It remains to be hoped that the transition to a "proactive" tax system will make it open and fully transparent, thanks to a new level of interaction between taxpayers and tax authorities, a level of customer-centricity that will allow for accurate tax calculation and automatic deduction.

References

1. Glazyev SY (2018) New world economic order. Model for assembly. Sci Works Free Econ Soc Russ 2. https://cyberleninka.ru/article/n/novyy-mirohozyaystvennyy-uklad-model-dlya-sborki. Last accessed 10 December 2025
2. Makashina OV, Khamalinskaya VV (2020) Long-term stability of the financial system. theory and Practice: monograph. Rusains, Moscow, p 314. ISBN 978-5-4365-5401-3
3. Tskhadadze NV, Ekaterinovskaya MA, Khaustova KV (2022) Features and indicators of inclusive economic growth in Russia: monograph. KnoRus, Moscow, p155. ISBN 978-5-406-08847-0
4. Vozhzhov AP, Grin'ko EL, Vozhzhov SP, Cheremisinova DV (2021) Architectonics of financial provision and regulation of sustainable and balanced economic growth: monograph. KnoRus, Moscow, p 558. ISBN 978-5-406-07984-3
5. Kapitsa LM, Elagin DP, Sergeev EA (2021) World development indicators: monograph. Moscow, KnoRus, p 555. ISBN 978-5-406-08385-7

6. Nureyev RM (2023) Economic sanctions against Russia: expectations and reality: monograph. KnoRus, Moscow, p 193. ISBN 978-5-406-11453-7
7. Zhuravleva IA (2024) Tax support tools for the implementation of national government projects with the digital functioning of "smart government". In: Lecture notes in networks and systems, vol 673. Springer, Singapore Digital Transformation: What are the Smart Cities Today? pp 109–126
8. Savina LL (2020) The tax system as a factor in macroeconomic equilibrium. Eurasian Sci Assoc 12–4(70):303–307
9. Lev MY (2022) Tax administration as a mechanism for strengthening the economic security of the state tax system. Econ Secur 2:525–546
10. Avdiyskiy VI, Zemskov VV, Soloviev AI (2022) Tax risks in the economic security system: a tutorial. INFRA-M, Moscow, p 236. ISBN 978-5-16-017125-8
11. Privalov NG, Privalova SG (2023) Tax system: theoretical and methodological aspect. Tax product: monograph. INFRA-M, Moscow, p 417. ISBN 978-5-16-017739-7
12. Brusov N, Filatova TV, Orekhova NP (2021) Taxes in modern corporate finance and investments: a textbook. KnoRus, Moscow, p/179. ISBN 978-5-406-01105-8
13. Gereev RA (2022) Tax incentives for industrial production: a monograph. Dr. of Economics, prof. M.R.Pinskaya. Moscow: INFRA-M, p.132 (2022) ISBN 978-5-16-016673-5
14. Baranova LG, Fedorova VS (2020) Economic consequences of changes in property taxation at the macro and micro levels. Econ Res Dev 5:7–12
15. Aliyeva NR, Danilova SV, Zyablitskaya NV (2022) Development of the system of taxation of property and income of individuals: from the formation of the USSR to the present. Reg Prob Econ Transf 1(135):72–79
16. Gordievich TI, Ruzanov PV (2023) Macroeconomic trends in Russia during the sanctions crisis. Omsk scientific bulletin. Ser Soc Hist Modern 8(3):124–130
17. Order of the Government of the Russian Federation of November 29, 2025 No. 3523-r On the national model of target conditions for doing business until 2030 [Electronic resource]. https://www.consultant.ru/document/cons_doc_LAW_520354/. Last accessed 10 December 2025
18. Federal Law of June 28, 2014 No. 172-FZ "On Strategic Planning in the Russian Federation" [Electronic resource]. http://www.kremlin.ru/acts/bank/38630. Last accessed 25 November 2025
19. Order of the Federal Tax Service of Russia dated 30.05.2007 N MM-3-06/333@ (as amended on 10.05.2012) "On approval of the Concept of the system for planning on-site tax audits" [Electronic resource]. https://www.nalog.gov.ru/rn77/about_fts/docs/3897151/. Last accessed 25 November 2025
20. Daniil Egorov apologized for the "poor implementation" of the Unified Tax System. RBC [Electronic source]. https://www.rbc.ru/economics/30/05/2023/64760a699a79474bbf734969. Last accessed 25 November 2025
21. Zhuravleva IA (2024) Tax support tools for the implementation of national government projects with the digital functioning of smart government. In: Sari M, Kulachinskaya A (eds) Digital transformation: what are the smart cities today? lecture notes in networks and systems, vol 846. Springer, Cham, pp 109–126. https://doi.org/10.1007/978-3-031-49390-4_8
22. Korzovatykh ZhM, Morozova NG (2021) Development of innovative instruments of tax administration in the Russian Federation. Bulletin of Chelyabinsk State University. Econ Sci 45–155 (2021)
23. Maslikhova EA, Danilova SV, Ilyinykh YuM (2023) Assessment of the effectiveness of modern state tax policy. Bull Altai Acad Econ Law 10:84–90
24. Alesina A, Perotti R (1999) Budget deficits and budget institutions. Fiscal institutions and fiscal performance. University of Chicago Press, Chicago, pp 13–36
25. Zasko VN, Grundel LP, Goncharenko LI, Advokatova AS, Zhuravleva IA, Knyazeva AV, Yurkova MR, Kurbanalieva LF, Sorokina DV, Abakumova AV, Fursikova AK, Moiseeva AA (2024) Transformation of tax policy in the context of technological sovereignty of industry: monograph. Dashkov and Co, Moscow, p 213. ISBN 9785394061431

26. Analytical portal of the Federal Tax Service of Russia [Electronic resource]. https://analytic.nalog.gov.ru/. Last accessed 10 December 2025
27. Official website of the Federal Tax Service of Russia [Electronic resource]. https://www.nalog.gov.ru/rn77/news/activities_fts/6978714/. Last accessed 25 November 2025
28. Goncharenko LI (2023) Tax administration: textbook. INFRA-M, Moscow, p 254 (2023). https://doi.org/10.12737/1860997. ISBN 978-5-16-017557-7
29. Polezharova LV, Tikhonova AV (2023) Cross-border transactions: development of tax administration tools: monograph. INFRA-M, Moscow. https://doi.org/10.12737/1921376. ISBN 978-5-16-018200-1
30. 445 companies became participants in tax monitoring in 2023. Federal Tax Service of Russia [Electronic resource]. https://www.nalog.gov.ru/rn77/news/activities_fts/13059125/. Last accessed 25 November 2025
31. The Russian government is discussing a voluntary contribution to the budget with big business. Kommersant [Electronic resource]. https://www.kommersant.ru/doc/5812310?tg. Last accessed 25 November 2025
32. The State Duma approved in the first reading a tax on excess profits of large businesses. Kommersant [Electronic resource]. https://www.kommersant.ru/doc/6070092. Last accessed 25 November 2025
33. Sberbank estimated its one-time excess profit tax at 3 billion rubles. Kommersant [Electronic resource]. https://www.kommersant.ru/doc/6175049?tg. Last accessed 25 November 2025
34. Sberbank will pay an excess profit tax of 0.5% of its dividends [Electronic resource]. https://www.rbc.ru/finances/25/08/2023/64e867279a7947f637ddea40. Last accessed 25 November 2025
35. Excess profit tax for Inter RAO amounted to 1.2 billion rubles [Electronic resource]. https://pravo.ru/news/251561/. Last accessed 25 November 2025
36. Norilsk Nickel paid 8.2 billion rubles in profit tax early [Electronic resource]. https://tass.ru/ekonomika/19623913. Last accessed 25 November 2025
37. Burdelova TN (2023) Improvement of the mechanism for collecting one-time taxes and fees in the context of international sanctions and growth of government expenditures (2023–2024). Finance: Theory Pract 3. https://cyberleninka.ru/article/n/sovershenstvovanie-mehanizma-vzimaniya-edinovremennyh-nalogov-i-sborov-v-usloviyah-mezhdunarodnyh-sanktsiy-i-rosta-gosudarstvennyh. Last accessed 13 December 2025
38. Ryakhovsky DI, Balakina MS (2022) Analysis of the impact of the tax burden on economic growth: theoretical and empirical aspects. Econ Sustain Dev 3(51):119–122
39. Directive (EU) 2022/2523 (also known as "Pillar 2") is a directive to ensure a global minimum level of taxation for multinational enterprise groups (MNEs) and large domestic groups in the European Union. [Electronic resource]. https://eur-lex.europa.eu/legal-content/EN/TXT/PDF/?uri=CELEX:32022L2523. Last accessed 13 December 2025
40. European Economic Forecast. Autumn (2023). [Electronic resource]. https://economy-finance.ec.europa.eu/system/files/2023-12/ip258_en.pdf. Last accessed 13 December 2025
41. Tasalov KA (2019) Legal issues of the Russian concept of a risk-based approach in the tax sphere. Curr Issues Russ Law 4(101). [Electronic resource]. https://cyberleninka.ru/article/n/pravovye-problemy-rossiyskoy-kontseptsii-risk-orientirovannogo-podhoda-v-nalogovoy-sfere. Last accessed 13 December 2025
42. United Kingdom public sector information website [Electronic resource]. https://www.gov.uk/government/publications/energy-profits-levy-reforms-2024/energy-profits-levy-reforms-2024. Last accessed 13 December 2025
43. The cost of the Government's energy support policies [Electronic resource]. https://obr.uk/box/the-cost-of-the-governments-energy-support-policies/. Last accessed 13 December 2025
44. Panskov VG (2022) Priorities of tax policy and directions of reforming the Russian tax system. ETAPE: economic theory, analysis, practice 1:57–76
45. Sandzhiev SL (2023) Excess profit tax as one of the sources of covering the russian federation's budget deficit. Sci Bull 9 (66). https://cyberleninka.ru/article/n/nalog-na-sverhpribyl-kak-odin-iz-istochnikov-pokrytiya-defitsita-byudzheta-rossii. Last accessed 13 December 2025

Hierarchical Tokenization Architecture Within Digital Transformation Projects: Horizontal Structure with Differentiated Roles

Nadezhda N. Pokrovskaia, Zumrat Talatovna Gaibnazarova, Matluba Nematovna Abdullayeva, and Sanobar Ikrom Kizi Akhmedova

Abstract The purpose of the chapter is to help businesses and regulators to build a balanced approach to the hierarchical organization of horizontal networks with structured tokenized architecture for the digital transformation of the space of businesses processes within communities, for logistics orchestration, payments exchange and settlement platforms, and for State authorities, e.g. certification, qualified signature, powers of attorney. The distributed ledger technologies have the main advantages of the horizontal decentralization (distributed models) and replication of the hash-lines as the fundamental factor to assure the security, which provokes the excess of the network' costs for computing capacities. The need to balance the benefits of the distributed replication and the goal of saving costs led the private and public communities to construct the hierarchies, where the tokens help to differentiate roles in the horizontal space of relationships. The methods of case study and documentary research are implemented to analyze the example of tokenization architecture of digital currency that illustrates the concept of the hierarchical structure that combines the advantages of the DLT solutions with the need to assure the substantial security in the industries based on the horizontal replication. The analysis of the case of the machine-readable powers of attorney as documents used by the Russian Federal Tax Office and Russian Federal Customs Office illustrates a minimalized three-level

N. N. Pokrovskaia
Saint-Petersburg State Electrotechnical University "LETI", Saint-Petersburg, Russia
e-mail: nnp@spbstu.ru

Z. T. Gaibnazarova (✉) · S. I. K. Akhmedova
University of Science and Technologies, Tashkent, Uzbekistan
e-mail: zumrat59@rambler.ru; z.gaibnazarova@usat.uz

S. I. K. Akhmedova
e-mail: sanobarahmedova839@gmail.com; s.akhmedova@usat.uz

M. N. Abdullayeva
Tashkent State University of Economics, Tashkent, Uzbekistan
e-mail: m.abdullayeva@tsue.uz

© The Author(s), under exclusive license to Springer Nature Switzerland AG 2026

A. Kulachinskaya et al. (eds.), *Digital Transformation: What Kind of Smart Life do we have today?*, Lecture Notes in Networks and Systems 1895,
https://doi.org/10.1007/978-3-032-20582-7_5

solution. Examination of the success and failure of logistics and settlement platforms Factorin and Tradelens helped to build the matrix of recommended differentiated roles for the digital transformation projects with the use of tokens.

Keywords Tokenization · Tokens architecture · Distributed ledger models · Smart government system · Smart contract

1 Introduction

Enterprises, industrial associations, regional and national authorities conceive digital transformation projects to accelerate the goal achievement and to improve the performances and competitiveness on corporate, industrial or regional levels.

In this chapter we treat the main purpose of the digital transformation as improvement of managing the time—the business use of the digital technologies is to accelerate making decisions and reacting until prediction and avoidance [1, 2]. This approach is based on the idea of the leadership as the capacity to use an opportunity the first, and digitizing helps to avoid being the last.

The best solutions aim to anticipate the changes and to implement the proactive approach to avoid eventual problems instead of react to face the appeared problems (to cope with ruptures, to repair a broken equipment, to fill in the gaps). The concerns of security represent one of the most significant groups of issues that determine the necessity to implement the new tools based on the fast analysis and automated regulating in case of a threat.

1.1 Transparency and Sensemaking as the Factors of Success for the Digital Transformation Projects

The corporate digital transformation projects tend to generate the transparent vision and understanding in online-regime simultaneously with the business processes happening in reality. The visualized tools for making decisions support the on-time regulating and help to avoid the delayed actions and interruption of business processes.

To understand what happens just now, the decision makers need the transparency of the processes and the clear direction for goal-setting. Sensemaking is a continuous flow of understanding, enrichment, and development of shared experience in work processes, routines, or organizational structures. People develop their experiences and plausible explanations to achieve a generally accepted formal description, as sensemaking relies on retrospection. Individuals attempt to make sense of a situation individually, but this always occurs within the context of the collective actions through which this experience was formed.

The organizational hierarchy plays the minor role compared with the high-qualified professional experts, that provokes the clear distinction of the functions: division of labor between the managers who assure the services for the technological producing processes that create additional value. The managerial function is to serve, support, help and maintain the supply of the raw materials, components, etc. in a correct time period in a correct quantity with a correct level of quality, including the reserves of the necessary elements which are regularly required.

The hierarchical "tall" vertical organization helps to share the responsibility and to accumulate the whole vision of the process on the top. The digital techniques to collect data and to visualize them allow the professionals and linear managers to get the clear holistic vision without the long high hierarchy. This evolution has created an illusion that hierarchy can be rejected in favor of "flat" horizontal structures that should totally predominate. The peer-to-peer organization of blockchain networks supported this illusion at its first time.

The aim of this research is to demonstrate the efficiency of the combined approach—to conjugate the security and transparency (within the distributed horizontal structure) with the speed and quality of making decision through the sensemaking of the experts (or qualified nodes).

1.2 Security

The risks are generated with various factors, from mistakes and erroneous behavior of employees to the malicious actions of industrial spies and intelligencers.

The cyber-security faces the threats for confidential data disclosure from the attacks to the potential requests to the AI-tools. Usually, the centralized systems are more vulnerable than the decentralized: the famous Target attack in Nov-Dec 2013, when the 40 mln payment cards were compromised due to the unsecure POS (points of sale) terminals, because retailers with multiple locations first aggregate all the data collected by the POS systems at individual stores at a centralized corporate location before sending the information on to a credit card processor, and the hackers accessed the encrypted PIN data [3]. This attack to the US retail giant Target was a turning point to the use of the tokenization for settlement and payment data.

Tokenization represents one of the methods to assure data security, it is used as the process to replace a sensitive data element with a non-sensitive equivalent, a token.

Tokens represent the digital record of property to value (e.g., bitcoin represents the value of volume of computing capacity which had been used to produce the solution to the equation). It possesses no intrinsic value and is valued only as a representative of another value or as an object of exchange, in this case token embodies the expectations of the market—the seller's fear of the future fall-down of the demand and the buyer's hope to get the surplus of the token exchange rate.

Token represents a reference, an identifier, that certifies the fact of property to an asset, and it is mapped to the original sensitive data through a tokenization system, with the techniques that make the tokens irreversible in the absence of the tokenization system. For example, when a person pays with his/her banking card, the real 16-digit core is replaced with a token with 16 decimal digits, (see Figs. 1 [4] and2 [5]).

Tokenization within the blockchain tools is based on the peer-to-peer architecture with the cryptographic technique of security of Merkle tree (see Fig. 3).

These fundamental structures as solutions for the security and reliable processing of transactions and data storage represent the versions of the p2p (peer-to-peer) or two-layers architecture of the networking.

The verticalization of the structure can lead to the effect of the decrease of the transparency for the participants (case Tradelens) or lack of governance (case of crypto-currencies and stablecoins, especially, the problems of the US Treasury based or USD linked coins).

The essential issue for the construction of the distributed ledgers and tokenization systems relates to the search of balance between the efficient hierarchy and the decentralized approach to assure the distribution of costs.

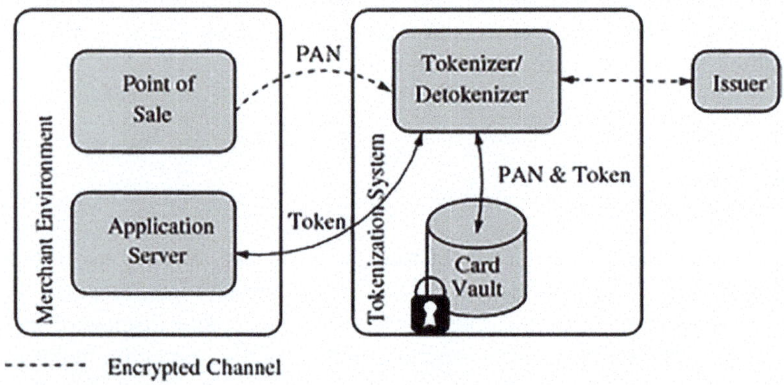

Fig. 1 Architecture of the tokenization system for banking card payment [4]

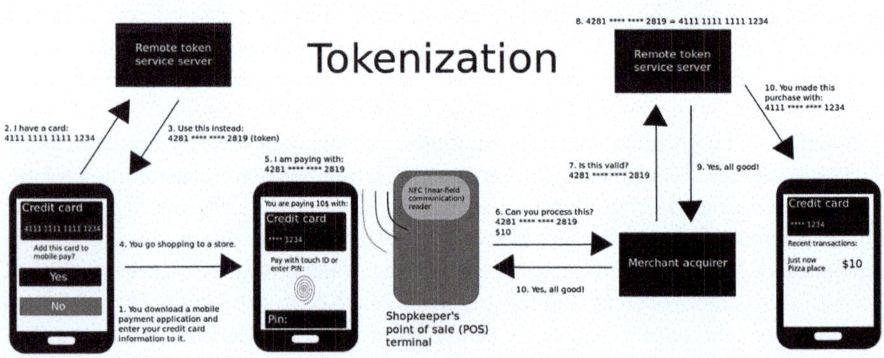

Fig. 2 10-step process of a mobile payment with tokenization via a mobile phone application with a credit card [5]

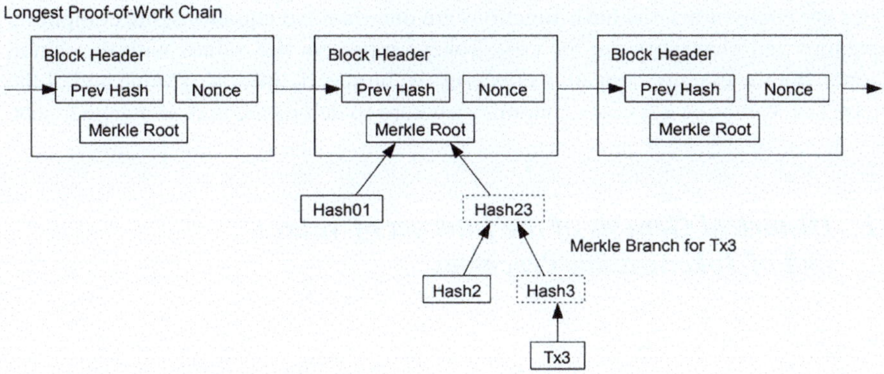

Fig. 3 Binary hashing scheme for confirming a transaction in a block [6]

2 Materials and Methods

The theoretical analysis is based on the economic and sociological concepts within the institutional analysis to demonstrate the rules and regulations for the digital transformation projects as the institutions in the normative and value sense.

The conceptual framework of management science represents the sensemaking as a core process of digital transformation helps to apply the digital tools and to orient the human capital of a country, an industrial cluster or a company to the main strategy of competitive advantages, as the stage of sensibilization and mobilization for the purposes of the digital transformation on individual and collective level of perception.

The empirical methodology of case study helps to determine the complexity of the sophisticated objects such as logistics, financial settlement or certifying powers of attorney, and to reveal the common features and to discover the similarities and relations to develop the sensemaking for structured architecture and substantial qualities of the elements.

The research uses the examples of the hierarchical structured model of digital currencies and of powers of attorney realized in China and in Russia, as well as the comparison between a failed Tradelens logistics platform of IBM and Maersk and a successful factoring blockchain system of Factorin.

3 Results

Distributed ledger models tend to find out a balanced solution within the trilemma of scalability, security and democracy.

The use of tokens represented the first form for the value exchange within the space of the uncertainty of the Internet, when the participants are unknown, the nodes are anonymous, the hash-functions are one-directed equations, the computing capacities and electricity are the cost. Tokens represent the values with the digital records, but the essential question concerns the human definition of value and of the Internet of Values as a social-economic structure to determine and exchange assets.

3.1 Historical Genesis of the Internet of Values and of Tokenization Processes

The global collapse of financial services in 2007–2009 led to the development of a distributed payments model [7] and the concepts of the "Internet of Values" (IoV) and "Internet of Everything" (IoE) [8]. Both concepts are oriented to the free exchange of any value, action or opinion (e.g., art objects, pictures, ideas) through the global web.

Later, in March 2012, the notion of the Internet of Behaviours (IoB) was introduced in a blog of the Helsinki University professor G. Nyman, who described the online registering and use of the human behaviours for commercial purposes.

DLT platforms have outgrown the credit and banking sector [9], moving towards positioning infrastructure and ecosystem solutions: distributed ledger technology in Russia has been applied by the Federal Tax Service to machine-readable powers of attorney, and in China to the creation and development of the digital yuan, the digital currency of the Central Bank (CBDC) [10].

In the 2010s, distributed ledger technology and blockchain-based networking entered the mainstream market in the form of a variety of platform solutions. Proposed on October 31, 2008 [7], blockchain received a software implementation in February 2009 in Bitcoin, an unlimited public horizontal ("flat") data storage network that stores data in a chronologically organized ledger. In 2011, the first alternative transaction registries (altcoins) were created. It wasn't until 2014 that platform solutions were developed, notably Ethereum (by Canadian programmer V. Buterin).

In 2016, the Russian Masterchain platform, created by the Association of Credit, Financial, and Telecommunications Companies for the fast, efficient execution of settlement and payment transactions, was launched [10, 11]. In 2017, the Corda platform, developed by the R3 financial consortium on Ripple, was proposed.

In 2013, the R3 banking consortium was formed, developing and implementing Corda, a financial information exchange platform, from 2014 to 2017. Today, Corda version 5.1 enables both data transfer and the creation of decentralized applications (see Fig. 4), CorDapps, with two layers: governance and consensus.

A CorDapp is a Java application (or any language that supports the JVM) built using the Corda toolkit and API to solve a problem that is best solved using a decentralized approach. Corda runs on a JVM, typically hosted in a containerized environment. The following diagram illustrates this architecture:

The decentralization and cloud migration of traditional systems has raised the issue of an implicit boundary rarely mentioned or included in architectural diagrams:

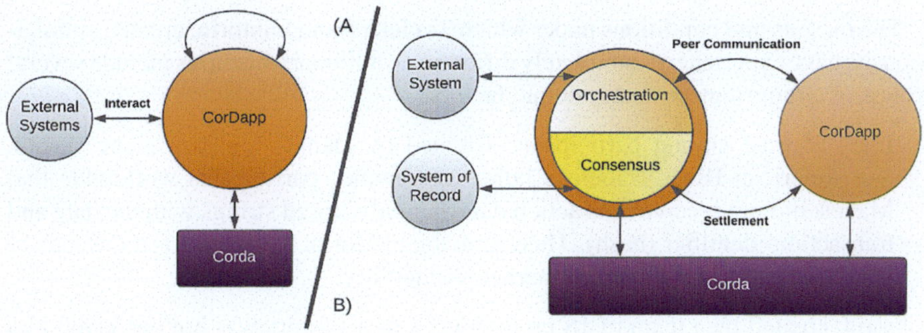

Fig. 4 Corda architecture for DApp development [12]

the boundary of a single trust zone. Distributed systems are designed to operate within a single trust zone, where updates are applied and accepted, altering global state. For example, when two parties enter into a bilateral agreement, both must run their own consensus system to ensure that the exchange occurred and was executed correctly. This complicates all transactions and increases their cost, as parties cannot simply trust each other without costly approval. The solution to this problem is decentralized applications, built using standard Java development tools and packaged with Corda tools. Once packaged, the applications are deployed to the Corda platform, which runs workflows for both the application and the ledger updates on behalf of the organizational structures.

Another example of the corporate private platformed solution is the Tradelens, realized by IBM for the global sea-transport giant Maersk.

In 2018, the Danish logistics blockchain platform Tradelens and the Russian factoring platform Faktorin were developed. However, Tradelens announced in November 2022 that it would close in March 2023, while Faktorin is successfully operating in the supply chain sector (retail chains Dixy and Magnit, among others).

3.2 *Private Blockchain Platforms—Comparative Analysis of Tradelens and Factorin*

While the Russian state-owned giants (the general platform Masterchain and the specialized Goschain, State-owned and State-powered blockchain platform), are developing gradually and cautiously, the private platform Faktorin rapidly entered the market, launching its MVP in March 2018 and gathering 24 financial participants in June 2018. Tradelens received the consent of 94 logistics entities (banks, ports, warehouses) even before launching, and within four years of operation, it has gathered 154 participants.

The reasons and conditions under which Tradelens and Faktorin, created simultaneously, have experienced completely different evolutionary histories include several groups of circumstances and systemic factors:

1. Lack of trust among participants. Created by Danish logistics giant Maersk and American IBM, Tradelens failed to convince participants worldwide that Maersk did not receive transaction data, but only stored strings with the date and transaction identifier (hash). Therefore, the platform, designed for thousands of participants, failed to achieve market profitability.

Faktorin, created by a team of 18 programmers as a startup, was not perceived as a global player attempting to skim the cream of the global logistics market.

2. Regulatory obstacles

The UK, as a global financial and insurance hub, only began accepting digital documents for logistics operations in October 2023 (the law was passed by the British Parliament on September 20, 2023 [13]), less than a year after the announcement of Tradelens' closure. This legislative ban reduced the benefits of electronic document management, forcing companies to provide chapter documents to insurers such as Lloyd's and others.

London is home to insurance sector organisations that control over 90% of global shipping insurance transactions (the International Group of P&I Clubs, IG P&I, with 7 of its 13 affiliated clubs located in London and its members providing coverage for 90% of the world's shipping tonnage, and Lloyd's of London, the insurance and reinsurance market, a marketplace and de facto regulator that sets the rules for participants, underwriters, brokers, and controls 13% of global ship and cargo insurance).

In Russia, digital documents and electronic document management emerged during the Soviet period. In May 1994, the state standard for electronic digital signatures (EDS) was adopted (GOST R 34.10–94), and in 2010–2011, digital tools for implementing blockchain platforms were introduced into practice (by the Federal Tax Service) and legislation [14].

3. Geopolitical Changes

In March 2022, the Danish A.P. Moller-Maersk announced the liquidation of its Russian subsidiaries, Maersk Solutions LLC and Maersk LLC. In August 2022, the company estimated losses from leaving Russia at $624 million. While the St. Petersburg container shipping company Maersk LLC closed in September 2024, the founder withdrew the liquidation petition for Maersk Solution (logistics and supply chain management), but resubmitted it on October 10, 2025. Along with the attacks on container ships in the Red Sea, this liquidation reduced the company's revenue and ability to invest in platform development.

Faktorin, having launched an MVP, quickly found its first customers. Improvements to the platform allowed the company to enter the major retailer Dixy in 2019, and then distribute the product to the Magnit chain.

The influence of the regulator and the supplier's strength created the conditions for Faktorin's success and the liquidation of Tradelens.

The high level of centralization for the Tradelens produced the lack of confidence and trust from the partners. The smaller player, Factorin, could create a better impression—it is perceived as a tool, and Tradelens was understood as a danger.

3.3 *State Settlement Platforms for Evaluation, Certification and Payment*

Settlement platforms have advanced to positioning infrastructure and ecosystem solutions: in Russia, the Federal Tax Service has applied distributed ledger technology to machine-readable powers of attorney, while in China, it has been used to create and develop the digital yuan, the Central Bank Digital Currency (CBDC).

The development of the blockchain platforms and the BRICS Bridge, using distributed ledger technology, as well as the experience of pilot testing the digital yuan (e-CNY) through WeChat and AliPay [12], demonstrate that national regulation of settlement and payment ecosystems can become the foundation for harmonizing and developing bilateral (Russian-Chinese) and global relations.

Mechanisms for building a settlement ecosystem for international and global payments require infrastructure development, primarily at the level of regulatory harmonization. The goal-setting of regulatory approaches reflects the civilizational characteristics of countries and regions; therefore, the subject of study is Russian-Chinese interaction to support the settlement and payment ecosystem.

The evolution of regulatory approaches to horizontal relations and contracts over the past five millennia has included both attempts to fully transparently control transactions between individuals for state control and attempts to implement complete freedom of contract for individuals and groups.

By the twentieth century, strict centralized regulation was implemented in Western countries in the form of Keynesianism, leading to the development of "stagflation" (stagnation amid high inflation caused by the unwillingness of democratically elected governments to implement strict austerity measures); liberal democratic principles of freedom of contract were replaced by "neoliberal" ones (with limitations; for example, a "black market" in human organs arises in response to the ethical prohibition of organ trade for transplants). The Rhineland model of regulation is based on the centralized state model, while Anglo-Saxon regulation relies on private-corporate sources of agreement that prevail over codified documents (with the exception of tax law, which is codified in the US, Britain, and the Netherlands).

Digital technologies and the development of global, differentiated networked spaces have led to new forms of hierarchical conceptualization: the development of engineering and technological solutions throughout human history has always raised questions of caution and risk. For example, the creation of the atomic bomb enabled the construction of large and subsequently small nuclear power plants for electricity generation, but simultaneously led to the conclusion of interstate treaties limiting nuclear weapons. In the digital economy, this has been expressed in the "Google Doctrine": technological growth outpaces regulation, so regulatory mechanisms should be entrusted to technology. However, Google's 2011 decision to comply with Chinese law to access the lucrative contextual advertising keyword market—that is, Google's preference for private business interests and its prioritization of profit—calls into question the Google doctrine: in millennia-old practice, one must seek a balance between ideological and profit motives, public benefit and private interest [13].

Regulation ensures the reliability of mutual agreements and the fulfillment of obligations, which allows for the planning of long-term stable resource flows. The ecosystem approach reflects the transfer of regulatory mechanisms typically applied at the state and public goods level to the private business level [15]. A technological solution for the effective regulation of ecosystems is blockchain, referred to in Russian legislation as distributed ledgers.

The rationale for regulating distributed ledgers of payment platforms is based on security considerations. An ecosystem approach to building payment and settlement interactions is generally based on agreed-upon international transaction standards. However, in the area of distributed ledgers and the implementation of platform solutions, despite significant accumulated experience, uniform regulation is still lacking worldwide.

This leads, for example, to difficulties in customs clearance of goods crossing the border: the Russian Federal Customs Service (FCS) refuses to allow goods to cross the border unless payment is proven, and payment in cryptocurrency is either difficult to document or immediately raises the possibility of an offense. Since the ruble is the only means of payment in the Russian Federation, digital currency and digital assets are considered goods, not means of payment. At the same time, there are positive examples of the successful application of blockchain to financial instruments for foreign economic activity, in particular, Russia and Iran use stablecoins, and in cooperation with the UAE, Russia uses stablecoins and cryptocurrencies. Russia and China are actively developing the use of the state digital currency. On October 19, 2023, PetroChina International purchased 1 million barrels of oil for digital yuan (e-CNY) on the Shanghai Oil and Natural Gas Exchange (SHPGX); on the same day, First Abu Dhabi Bank (UAE) announced that it had signed a digital currency agreement with Bank of China: the UAE is a participant in the mBridge platform, designed to support cross-border transactions with CBDCs.

Settlement during exchange relies on the measurement of value: if one party values the acquired good higher than the given good, the exchange takes place. The aforementioned Internet of Values (IoV) reflects the development of blockchain technology to the extent that the global network space allows for the standardization of these transactions, i.e., for at least one good, the value is known with sufficient accuracy to all participants in the transaction; that good is monetary.

Since money serves as a common equivalent, measurement approaches in distributed ledgers can rely on three foundations:

- institutional agreements between states and compliance with regulatory requirements (digital national currencies),
- spontaneous equilibrium of supply and demand (markets, exchanges), and
- tangible expression of value (stablecoins).

Real-world payment transactions take all three components into account. For example, Circle's USDC cryptocurrency is equivalent to 1 US dollar, complies with the EU's Markets in Crypto-Assets Regulation (MiCA, adopted May 31, 2023, fully effective December 30, 2024) and several national regulations (such as the Value-Referenced Crypto Asset Regime, VCRA, Canada), is traded on global exchanges, and is widely used in settlements. Similarly, the PAX Gold token (PAXG) is regulated by the New York State Department of Financial Services (NYSDFS) and is backed by physical gold reserves. Its issuer, Paxos, offers instant exchange of tokens for physical gold—a troy ounce of London Good Delivery gold bar—at a live price. The token is developed on the ERC-20 protocol (a technical document for Ethereum programming standards), simplifying the interaction of applications and contracts compatible with services (software wallets, hardware wallets, exchanges, etc.). The Comtech Gold (CGO) token is equivalent to one gram of 999.9 fine gold stored in accredited UAE vaults and is backed by gold bars with serial numbers and verifiable storage data. It is fully Shariah-compliant and licensed by the Dubai Airport Free Zone Authority (DAFZA). It is issued as the XRC-20 token on the energy-efficient and low-cost infrastructure of the XDC Network.

The dangers of decentralized interactions on the blockchain, including Cryptocurrencies are generally associated with illegal transactions (drug trafficking, arms sales, etc.), speculative income losses (unqualified investors can go bankrupt and create a burden on the social security system), and fraud. For example, on December 4, 2013, former Chairman of the US Federal Reserve Alan Greenspan said that Bitcoin is a "bubble" that has no real value: "There is no fundamental issue of capabilities of repaying it in anything which is universally acceptable, which is either intrinsic value of the currency or the credit or trust of the individual who is issuing the money, whether it's a government or an individual" [16].

On December 5, 2013, the People's Bank of China banned Chinese financial companies (but not individuals) from conducting transactions and insuring financial products with Bitcoin, and publishing Bitcoin quotes. In April 2014, all banks and payment systems closed all accounts of 15 Chinese websites that sold Bitcoin.

Sovereign states (with the exception of EU countries, where sovereignty has been devolved to the Union level) are seeking to establish their own rules for regulating settlement spaces. In October 2019, General Secretary of the Communist Party of China Central Committee Xi Jinping called for China's leadership in blockchain technologies. In May 2020, the National Fund for Coordinating Blockchain in the Smart Government System was established, and a national blockchain infrastructure platform (Blockchain-based Service Network, BSN) was launched. Its integrated development environment includes cloud resources, basic blockchain frameworks, an operating environment, key management, a software development kit, and gateway APIs. BSN supports both enterprise blockchains (Hyperledger Fabric, Fisco

BCOS, Baidu Xuperchain, CITA) and public blockchain frameworks (Ethereum, EOS, Tezos, NEO, Nervos, IRISnet, etc.). Since 2017, Alipay and WeChat have been storing customer funds in interest-free accounts at the People's Bank of China (PBoC), allowing the government to monitor cash flows. Thus, China prioritizes transparency in the payment and settlement ecosystem for centralized control and monitoring by the regulator.

In Russia, the approach was similar, leading to restrictions and, subsequently, disconnection of several Russian banks from the SWIFT international payment system between 2014 and 2022. In December 2014, the Financial Messaging System (SPFS) was launched to counter the risk of disconnection from SWIFT, and, effective October 1, 2023, the Central Bank of the Russian Federation banned credit institutions from using SWIFT for transfers within Russia. Although SWIFT (Society for Worldwide Interbank Financial Telecommunications) is a private company (LLC, limited liability cooperative, cooperative society under Belgian law), a number of countries were disconnected from the system for political reasons (Iran, North Korea, Russia).

The launch of SPFS and the national payment system Mir became a response to the risks of disconnection and, at the same time, reduced the transparency of monitoring and control over domestic Russian payments for the US and the EU (in 2006, it became known that US agencies such as the CIA, FBI and the US Department of the Treasury, an analogue of the Ministry of Finance, had access to financial data from the SWIFT network [17].

In 2015, the Chinese interbank payment system CIPS (Cross-border Interbank Payment System), operating exclusively in yuan, was launched. It processes payments in 119 countries worldwide, including 30 Russian banks participating (VTB, Gazprombank, Alfa-Bank, etc.). Regional trust zones of civilizational proximity with the close and friendly countries are being formed to reduce the risks of using payment platforms as a tool of political pressure and threats to commercial transactions.

As a result, regulatory efforts have focused on risk prevention and finding ways to facilitate international transactions, which has led to decisions such as the adoption of experimental legal regimes permitting cryptocurrency payments in foreign economic activity. Thus, on September 1, 2024, Federal Law No. 223-FZ of August 8, 2024, on experimental cryptocurrency settlements in foreign trade activities, came into force.

This does not eliminate a number of risks: judicial protection of cryptocurrency owners' rights is possible only if they declare cryptocurrency transactions and notify tax authorities of their cryptocurrency ownership; otherwise, the courts will refuse to protect their rights.

If the wallet owner is subject to sanctions or cryptocurrency is used to pay for sanctioned goods, crypto service providers may "color" or mark the tokens, and they will not be accepted [18]. Cryptocurrency can be stored with Russian service providers. On April 23, 2025, Russian Finance Minister A. Siluanov announced the launch of the first official crypto exchange for qualified investors.

The distinction between the participants relates to the security issue and proposes to the implement the principles of the differentiated governance [19] to the "flat" architecture of the tokens' platforms and system.

3.4 NFT, Stablecoins and National Currencies as Tokens Structuring

Often, distributed and decentralized forms of communication are often considered synonymous, but in the example of digital currency, one can see versions of platform solutions with peer-to-peer participation or with a hierarchy of responsibility, transparent flows for control, and differentiated functions.

Blockchain is often considered as a decentralized (distributed, horizontal) space where information about any valuables is stored. For example, NFTs (non-fungible tokens) are often used to record works of art, but any original document is also a unique object for NFTs (for example, a sick leave certificate for a specific person with specific dates and data of the medical institutions and specialists that issued it).

The most mature blockchain technology solutions are emerging in the financial services sector: cryptoassets and digital tokens, including non-fungible tokens (NFTs), which contain a unique value code. Signatures, invoices, insurance policies, legal documents, video, and audio files transformed into NFTs cannot be counterfeited. The range of applications for this technology, which allows for the verification of unique ownership of certain assets, already extends from ensuring secure supply chains to the creation of works of art. In the future, NFTs will be widely in demand not only in stock markets and insurance, but also in other industries, particularly scientific research and creative industries (music, design, video gaming).

Blockchain technologies enable the global socioeconomic system to agree on a global understanding of what is acceptable and important for society, and what is rejected.

Trust in decentralization can be based both on trust in the rules and mechanisms of the platform and on the tangible expression of the token: stablecoins are equated to the value of a physical asset and therefore become subject to clearer and more understandable regulation by authorities. In particular, the aforementioned EU MiCA regulation contains a standard for filtering out unreliable and illegal cryptocurrencies based on reserves covering the total value of tokens: if 1 coin is equal to 1 oz of gold (XAUT, VNXAU) or fiat currency, the US dollar (the algorithmic decentralized stablecoin DAI, the centralized coins USDT, USDC with a regulated supply) or the euro (EURT), then their issuers must hold the entire volume of issued tokens in accredited storage facilities. Stablecoins inspire more trust among participants in global payment transactions, for example, according to the Artemis agency [20], about $70 billion per year, half of which (36 billion) are B2B payments, another quarter $18 billion/year are P2P transactions between individuals, as well as $13.2 billion/year by card and 3.3 billion B2C payments. China primarily uses Tron payments (91.7%), while Hong

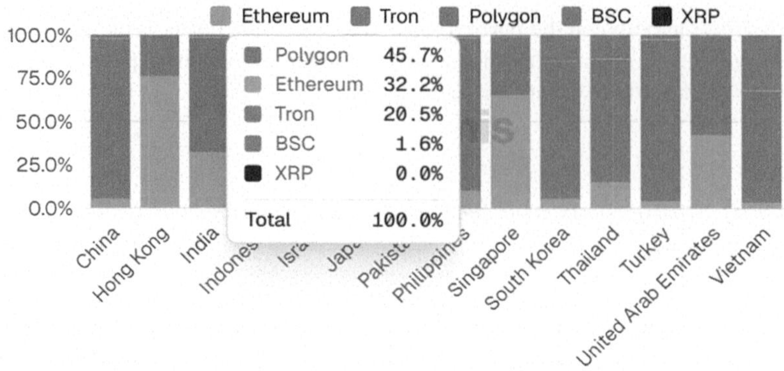

Fig. 5 Distribution of payments in Asian countries across blockchain platforms [20]

Kong (SAR PRC) and Singapore use Ethereum (75.8% and 65.0%), and only India in the Asia–Pacific region widely uses Polygon (45.7%) (see Fig. 5).

Stablecoins occupy a significant position in international transactions for institutional and individual users. On June 11, 2025, the US Senate will vote on the GENIUS (Guiding and Establishing National Innovation for U.S. Stablecoins Act), a bill regulating stablecoins, as more than a third of small and medium-sized businesses (SMEs) in the US use cryptocurrency, and 60% of Fortune 500 companies are working on blockchain projects.

However, the crash of the Luna token from May 8–28, 2022, demonstrated that this approach is not risk-free. The total value locked (TVL) of assets in Terra blockchain projects, issued by Luna, reached $25.6 billion, but on May 8, 2022, the algorithmic stablecoin TerraUSD (UST) temporarily lost its peg to the US dollar after a user sold nearly $300 million worth of UST. To stabilize the price of UST, Terra sharply increased the supply of the Luna token. In an attempt to re-peg UST to the US dollar, the Luna Foundation Guard (LFG) lost approximately $3 billion, but the token never regained its value [21].

These cases demonstrate that Central bank digital currencies will become an effective settlement instrument at the bilateral level, as can already be seen in the example of Russian-Chinese cooperation, based on the mutual agreement of a number of regulatory mechanisms.

The digital yuan is based on token (value-based), quasi-account-based, and account-based technologies, i.e., it acts as a hybrid payment instrument and has the status of legal tender [22]. Token-based digital currencies determine payment validity through a token identification system, while payment authenticity (validation) in an account-based system relies on entries in settlement accounts, i.e., payment is made based on the authentication of the account holder's identity. The combination of centralization and a distributed approach—control and monitoring at the Central Bank and customer service at competing commercial institutions—allows for an effective combination of business models, leveraging the benefits of state strategic planning and infrastructure investment projects with market adaptability to customer expectations for service personalization.

Unlike the Chinese approach, the Bank of Russia has chosen a hybrid digital currency system: it issues the digital ruble and operates the platform, opens digital wallets for financial institutions and the Federal Treasury, and processes settlement transactions through them; develops financial and information services; and implements platform operating and information security standards. Financial institutions provide services to users by acting as access providers to the digital ruble platform, opening and replenishing user wallets, and conducting settlements on the digital ruble platform. Each client receives only one digital ruble wallet. The Federal Treasury carries out settlement operations through wallets on the digital ruble platform to ensure the functioning of public sector organizations [22].

Within the system of the Internet of behaviour, the trust plays the most important role to determine the success of a tokens' platform [23, 24].

4 Discussion and Conclusion

The real implementation of blockchain tools demonstrates the lower efficiency of the full decentralization with general equality of all voices: the agents tend to monopolize and to control the blocks of decisions, the expertise is scarce and rare and costs high, requires high investing and sometimes is not rewarding or not taken into account. These difficulties are solved on the level of the societal solutions such as qualified majority in voting, the hierarchy of differentiated roles of experts in various activities.

The analysis of the cases of public and private networks allows us to propose the solution based on the distribution of roles for the integrated structure of a platform that can be used as a basis to choose the necessary roles.

When building a comprehensive payment ecosystem using blockchain platform technologies, the traditional platform architecture should be supplemented with a distribution of responsibilities (Table 1).

As demonstrated by the experience of the Federal Tax Service's machine-readable powers of attorney, blockchain provides a high degree of reliability: the Federal Tax Service's distributed registry includes more than 50 accredited data storage nodes (17 government agencies and 36 organizations), creating a unified system of trust, for which legislative provisions (Federal Law No. 63 on digital signatures) and the technological infrastructure have been established. However, the Masterchain project has not received recognition since 2016—even created by the main reliable Russian financial and telecommunication institutions, such as Sber and Megafon, Masterchain is perceived as less reliable than the system of the Federal Tax Office [24].

The digital transformation is widely implemented in structured business ecosystems of transport and logistics [25]. The tokenization is used in businesses to facilitate the exchange of values, what is especially interesting for networks of start-ups [26]. Social institutions and political establishments are also interested in the creation of

Table 1 The architecture of the qualified status and roles for the token governance

Processes	Actor	Functions	Objectives
Regulation	National State	Establishment of rules, monitoring the compliance	Long-term strategic planning
Monetary emission, issue of tokens	Central Bank	Issue of State digital currency for settlement and for value measuring Monitoring transactions	Supplying the economy with money volume
Certification	Federal Tax Office	Duties and rights of partners	Distribution of responsibility in business transactions
Identification and personification of the subject	Federal Tax Service	Resource reallocation and transaction monitoring	Implementation of the state plan and budget
Emission, release of tokens	Licensed companies and platforms	Digitizing value, supporting a variety of calculations	Fairness of assessment
Token trading	Exchanges, applications	Supporting the availability and stability of payments	Transparency and balance in the market
Conclusion of smart contracts	Platforms-protocols	Automation of operations	Speeding up and reducing the cost of transactions
Launch of smart contracting	Oracles	Fixation and measurement of reality	Displaying facts in numbers
Data storage	Servers, data centers	Accredited nodes provide Byzantine fault tolerance	Distributed computing power

the tokenized structured ecosystems. For example, Universities create their metaverses to help the school children to make the first acquaintance with the campuses and labs, to attract them and to support their integration, as well as to help students to take part in various events through the online access. The educational institutions are open to wide implementation of digital transformation projects: broad involvement of scholars, students, teachers and faculties administrations to the social activities, strict authentication of participants of training for tests and certifications, including the employers and the potential sponsors for the education programs [27].

The development of a settlement ecosystem can be based on the use of developed telecommunications infrastructure in Russia and China, the harmonization of payment formats within BRICS (the BRICS Clear securities depository system, the China Track clearing route created by banks of friendly countries, the BRICS Pay card and BRICS Bridge payment system, and the BRICS Food Token cryptocurrency based on food and water products), and the development of national digital currencies, a digital ruble, and a digital yuan. Regulation within the framework of bilateral Russian-Chinese relations and the work of five EAEU countries, ten BRICS countries, and ten SCO countries create a transparent rules space free from external political pressure, such as dependence on SWIFT.

The proposed structured division of roles as a framework for architecture of digital transformation governance has a general character. The deeper investigation of the specialized features and requirements in diverse industries needs further research.

Distributed ledgers create the technological foundation for building a settlement and payment ecosystem without a central authority. Regulatory mechanisms must determine the future development of settlement and payment ecosystems in international cooperation. To this end, it is essential to create working platform solutions capable of demonstrating the effectiveness of distributed ledgers as a basis for economic interaction free from geopolitical pressure. Blockchain enables the integration of multi-factor regulation in a multipolar world and ensures economic growth for participants.

References

1. Vărzaru AA, Bocean CG (2024) Digital transformation and innovation: The influence of digital technologies on turnover from innovation activities and types of innovation. Systems 12(9):359. https://doi.org/10.3390/systems12090359
2. Erceg Ž, Đalić N (2024) The importance of digital transformation for business decision-making in organizations. Novi Ekonomist 18(36):41–48. https://doi.org/10.69781/NOEK202436041
3. Rashid FY (2013) Experts debate how hackers stole 40 million card numbers from target. Massive data breach renews call for adoption of chip-based payment cards in The United States//security week. https://www.securityweek.com/experts-debate-how-hackers-stole-40-million-card-numbers-target
4. Díaz S, Rodríguez Henríquez L, Chakraborty D (2016) A cryptographic study of tokenization systems. Int J Inf Secur 15(4):3–13. https://doi.org/10.1007/s10207-015-0313-x
5. Saha M, Basu M, Gupta A, Ashrith K, Reddy C, Reddy S, Reddy R (2025) Enhancing credit card security using RSA encryption and tokenization: a multi-module approach. Int J Inform Commun Technol (IJ-ICT) 14(1):132–140. https://doi.org/10.11591/ijict.v14i1s
6. Jiang J, Chen J (2021) Managing the product-counterfeiting problem with a blockchain-supported e-commerce platform. Sustainability 13(11):6016. https://doi.org/10.3390/su13116016
7. Nakamoto S (2008) Bitcoin: A peer-to-peer electronic cash system. SSRN. https://ssrn.com/abstract=3440802 or https://doi.org/10.2139/ssrn.3440802
8. Kashyap R (2022) The internet of value and internet of things. In: Vadgama N, Xu J, Tasca P (eds) Enabling the internet of value. Future of business and finance. Springer, Cham. https://doi.org/10.1007/978-3-030-78184-2_13
9. Ogigau-Neamtiu F (2016) Tokenization as a data security technique. In: Regional department of defense resources management studies. Zeszyty Naukowe AON, vol 2, no 103. Akademia Sztuki Wojennej, Brasov, Romania, pp 124–135
10. Wei F, Wei Z, Yayi L, Pokrovskaia N (2025) Blockchain for the comprehensive development of the territory: socio-cultural regulation in the gift economy. In: Proceedings of 29th international

scientific and practical conference system analysis in engineering and control, 30 June–2 July 2025, St. Petersburg Polytechnic University. Politekh-Press, St. Petersburg

11. Khansuvarova T, Khansuvarov R, Pokrovskaia N (2018) Network decentralized regulation with the fog-edge computing and blockchain for business development. In: Proceedings of the 14th European conference on management, leadership and governance, ECMLG. Utrecht, pp 205–212. Utrecht
12. Corda 5.1, Homepage. https://docs.r3.com/en/platform/corda/5.1/key-concepts.html
13. UK Parliament. Electronic Trade Documents Act 2023 Government Bill, Parliamentary Bills Originated in the House of Lords Homepage, 20 July 2023. https://bills.parliament.uk/bills/3344
14. Federal Law No. 63-FZ of April 6, 2011 on Electronic Signatures (with Amendments and Additions). https://base.garant.ru/512184522
15. Wei F, Pokrovskaia NN (2023) Networking in digital communication for the social initiatives' regulation in the context of mobilization economy. J Integr Cult Stud 5(1):17–24. https://doi.org/10.33910/2687-1262-2023-5-1-17-24
16. Kearns J (2013) Greenspan says bitcoin a bubble without intrinsic currency value. Bloomberg. https://www.bloomberg.com/news/articles/2013-12-04/greenspan-says-bitcoin-a-bubble-without-intrinsic-currency-value
17. Brand C (2006) Belgian PM: Data transfer broke rules. Washington Post, Sep 28 (2006) (Associated Press, September 28, 2006). https://www.washingtonpost.com/wp-dyn/content/article/2006/09/28/AR2006092800585.html
18. Emmert F (2024) The regulation of cryptocurrencies in the United States of America. Eur J Law Reform 25:27–122. https://doi.org/10.2139/ssrn.4690980
19. Garin AK, Shangutov A, Shangutov O, Pokrovskaia N (2022) Digital technologies in public administration and networking for communication modelling in social entrepreneurship. Telesc J Sociol Mark Res 1:32–41. https://doi.org/10.24412/1994-3776-2022-1-32-41
20. Yim A, Van Aken A, Carter N, Khosrowshahi W, Hadick R, Kanji O (2025) Stablecoin payments from the ground up. Artemis report. https://reports.artemisanalytics.com/stablecoins/artemis-stablecoin-payments-from-the-ground-up-2025.pdf
21. Dhaya R, Kanthavel R (eds) (2023) Internet of Behaviors (IoB), 1st ed. CRC Press. https://doi.org/10.1201/9781003305170
22. Sitnik AA (2021) Blockchain technology in payment systems. Actual Probl Russ Law 16(5):42–54. https://doi.org/10.17803/1994-1471.2021.126.5.042-054
23. Gaibnazarova ZT (2023) Prospects for improving the forecasting of the gross domestic product of regions based on mathematical models. In: 9th IEEE International Conference on Information Technology and Nanotechnology (ITNT), Samara, Russian Federation, pp 1–4. https://doi.org/10.1109/ITNT57377.2023.10139074
24. Gaibnazarova ZT, Solieva BT, Iminova NA (2024) Forecasting the risk of bankruptcy in poorly formalized processes. Sci & Tech 23(1):80–86. https://doi.org/10.21122/2227-1031-2024-23-1-80-86
25. Gaibnazarova ZT (2024) Artificial intelligence and smart railway transport as a tool to accelerate the transformation process in Uzbekistan. In: Sari M, Kulachinskaya A (eds) Digital transformation: What are the smart cities today? Lecture notes in networks and systems, vol 846. Springer, Cham. https://link.springer.com/chapter/10.1007/978-3-031-49390-4_19
26. Ziankova I, Gaibnazarova Z, Malei A, Sharafanova E, Gasheva O (2025) Digital technologies for comprehensive promotion of a startup. In: Skhvediani A, Kulachinskaya A, Mateeva Petrova M (eds) Digital transformation of socio-economic and technical systems: Theory and practice. Lecture notes in networks and systems, vol 1309, pp 39–55. Springer, Cham. https://doi.org/10.1007/978-3-031-85608-2_4
27. Stankevich D, Gaibnazarova Z, Naumovich O, Tuychieva M, Shadieva G (2025) The role of universities in digital transformation: Challenges and prospects. In: Skhvediani A, Kulachinskaya A, Mateeva Petrova M (eds) Digital transformation of socio-economic and technical systems: theory and practice. Lecture notes in networks and systems, vol 1309, pp 179–193. Springer, Cham. https://doi.org/10.1007/978-3-031-85608-2_13

Application of Modern Analytic Solutions in Various Fields of Life in the Context of Digital Transformation (in the Practice of Organizations and Research of Youth)

Radmila Khizbullina, Elmira Huzieva, and Leysan Gatina

Abstract The chapter actualizes the need to consider and scale the practices of using software platforms, machine learning solutions (ML), artificial intelligence in various spheres of life in various fields in the context of digital transformation. An analysis of the use of modern research tools, software for forming a database of the example of author's sociological research, the advantages of Big data analysis, the value of integrating ML solutions into the activities of organizations was carried out. The chapter discusses areas that can be optimized for various fields of activity, thanks to modern achievements in the field of informatization and digital opportunities. Comparison of the practices of using digital technologies in various fields (production, management, social) allows us to conclude that the implementation of modern research and administrative tasks is no longer possible outside of integration with digital solutions. A pronounced synergistic effect was determined in the use of software products, machine learning solutions in conducting applied empirical social research and in organizing the activities of companies and management. Based on a review of the practices of using artificial intelligence on the example of the fuel and energy complex, the advantages of introducing technologies into the activities of specialized organizations were determined. The capabilities of the IBM SPSS Statistics software product when working with empirical data within the framework of author's applied research are presented.

Keywords Machine learning solutions · Value · Digital sociological research · Software · Data analysis · Analytic solutions · Digital transformation

---

R. Khizbullina (✉) · E. Huzieva
Kazan State Power Engineering University, Kazan, Russia
e-mail: mine_post@inbox.ru

L. Gatina
Kazan National Research Technological University, Kazan, Russia

© The Author(s), under exclusive license to Springer Nature Switzerland AG 2026

A. Kulachinskaya et al. (eds.), *Digital Transformation: What Kind of Smart Life do we have today?*, Lecture Notes in Networks and Systems 1895,
https://doi.org/10.1007/978-3-032-20582-7_6

1 Introduction

1.1 Relevance

Currently, machine learning technologies are considered almost as part of everyday life, since they cover most areas—research, production and technical, engineering, medical, social, business management, etc. Algorithms of artificial intelligence, machine learning, replacing routine calculations, which in the recent past were recorded personally by employees, provided automation and an increase in the speed of research and business processes, as has already been mentioned in various spheres of life.

A review by modern researchers of methods and applications in machine learning and deep learning for big data analytics also denotes a possible problem related to bias and interpretability of models, and note the importance of careful analysis to ensure responsible use of these technologies [1].

As for ML-technologies (machine learning technologies), we note that today they allow solving valuable problems, forming segmentation, classifying objects based on machine learning systems, algorithms, in fact without human intervention. The study of the values of safety, health, safety, behavior, development of organizations, technologies, etc. Today can be ensured by scaling in the use of artificial intelligence, machine learning and based on big data, which together can provide a synergistic effect in the development of strategies for the sustainable development of various spheres of society, production, research. Today, machine learning programs and systems, as part of the artificial intelligence ecosystem, can produce a verified result/product in terms of the predictive expectations of the end user or consumer [2, 3].

Artificial intelligence, digital technologies (ML technologies, big data, software platforms and etc.) today are one of the mediating factors in the life of society, business, and human technologies. The study of dichotomy in the relationship between various spheres of society and software products is formulated as the goal of this chapter. The task is to assess the significance and scope of modern tools, software and analysis of applied research data and ML solutions in research and organizations.

1.2 Methods

The chapter used a set of well-known research methods such as the analysis, synthesis and generalization of information and data, which together made it possible to conclude about the state, experience and prospects of the development of machine learning and the use of big data in various fields. The data of the National Center for the Development of Artificial Intelligence under the Government of the Russian Federation for 2024 are presented, as well as fragments of material for monitoring the development and spread of artificial intelligence in Russia, presented by the

Institute for Statistical Studies and Economics of Knowledge of the Higher School of Economics (HSE), 2023, data from a monitoring study on the public perception of artificial intelligence technologies presented by Russian Public Opinion Research Center (JSC «VCIOM») in collaboration with ANO «National Priorities» (2024). The author's experience in working with IBM SPSS Statistics databases and software as part of the implementation of applied empirical research is presented (Republic of Tatarstan, youth 14–35 years old n = 1539, 2025). Based on the used program and database, a set of data and their statistical analysis was produced, which made it possible to present visual infographic material.

2 Results

2.1 Results of an Empirical Study

As a result of studying and analyzing materials on the state, prospects and values of automated algorithms for advanced research and forecasting in various spheres of life, it is confirmed that not all spheres integrate automated systems on a large scale, despite their active development and research. Here is the data presented by the National Center for the Development of Artificial Intelligence under the Government of the Russian Federation. Figure 1 shows the dynamics of the use of artificial intelligence technologies for 2021–2024, where the largest percentage (71%) is occupied by intellectual decision support (see Fig. 1) [4].

In support of applied empirical research, machine learning and artificial intelligence software products serve as tools and allow you to frame research, facilitate the search for respondents by characteristics, structure data, implement transcription system and big data analytics. The above allows you to form forecasts and estimates regarding the devaluation or growth of values in various areas of group life.

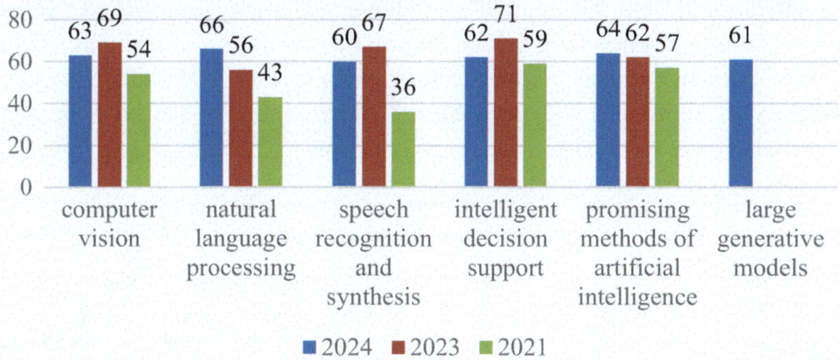

Fig. 1 Dynamics of the use of AI technologies, 2021, 2023, 2024, % of those using AI [4]

Considering the use of applications for solving research problems, when it is necessary to obtain empirical data on a large scale, analyze them and store a large array of results in the practice of researchers, business analytics and big data services are often used. Here are some examples where software technologies demonstrate importance in sustainability and research procedures. In studies of demographic groups, for example, in the study of the behavior of young people, their interests and values in the virtual space, frequency of access to certain content in social networks, virtual communication, monitoring of interests and values, requests in the field of culture, healthcare, etc. can be combined into a database based on machine algorithms for selective monitoring, including through computer vision, detection, recognition and classification systems. Since the visualization of content, research data and infographics in various fields demonstrating the results of their activities is becoming increasingly relevant, an example of such tools is the data collection tool Yandex Forms service for analytics and visualization of results. For example, a large-scale study conducted with the participation of the authors in order to obtain empirical data on the opinion of young people in the Republic of Tatarstan (2024–2025) was instrumentally implemented through the Yandex Forms service, which also provided a summary infographic and a database with an array of large frequency data. Further, using the IBM SPSS Statistics package, work with the array was carried out and correlation procedures were formed.

Foreign authors assess the capabilities of the SPSS Statistics software product within data analysis, noting the wide functionality, correlation, regression analysis, data-driven construction of study models, and use of metrics [5].

Below are the capabilities of the software product for statistical analysis of parametric indicators using the example of using IBM SPSS Statistics in the author's applied research (Republic of Tatarstan, youth 14–35 years old n = 1539, 2025) (see Tables 1 and 2).

If we talk about ML-technology, then, using the example of energy systems, ML-solutions are actively used today to detect and transmit valuable data, such as power line malfunctions, create digital landscapes and, based on data from understanding further development strategies, as well as leveling potential threats and malfunctions.

In the educational field, ML technologies have also firmly established themselves as a valuable source of knowledge, data, automatic translation of text and its generation. However, considering the databases formed by means of ML technologies with the possibility of switching over to areas of knowledge, we can talk about rather

Table 1 Combination table by age with the distribution of data to the question Do you follow the rule in ordinary life "I fulfill my duties and know about my rights" using the IBM SPSS Statistics software product

Answer	Age				
	14–17 year	18–21 year	22–25 year	26–29 year	30–35 year
Yes	313	516	250	160	169
No	22	42	29	20	18

Table 2 Calculating X2-criteria using IBM SPSS statistics

Name	Meaning	Asymptotic significance (2-sided)
Pearson chi-square	5.451[a]	0.244
Likelihood ratios	5.392	0.249
Linear–linear relationship	3.799	0.051
Number of valid observations	1539	

a. The number of cells 0 (0.0%) is assumed to be less than 5. The minimum estimated number is 15.32

positive aspects of this issue, which currently exceed the potential risk. Although, it should be noted that today this issue is the subject of discussion.

The sustainable development of ML-based management demonstrates its valuable relationship with CRM systems, which, through implementation in business processes, can optimize work with the client base through differentiated segmentation and strengthening the company's development strategy. According to the data presented by Vishnevsky K. O. and Turovets Yu. V. within the framework of monitoring the development and spread of artificial intelligence in Russia, the acquired typical software product among users of artificial intelligence is considered as the most popular (see Fig. 2) [6].

At the same time, enterprises are currently actively working in the field of developing their own innovative developments and automated systems. As an example of the use of automated systems by industrial enterprises of the Republic of Tatarstan, we can note an intelligent system developed by specialists of JSC "Network Company" to determine unaccounted electricity consumption, the task of which, based on a long-term array of "big data" obtained from smart meters, machine technology analyzes consumption profiles with the implementation of the search for unauthorized access to electricity. Work is carried out under the supervision of specialists with further identification of facts and development of solution strategies [7].

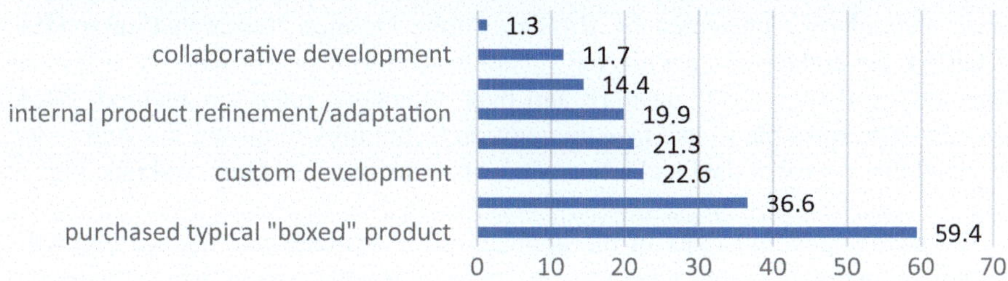

Fig. 2 Ways to purchase products based on artificial intelligence technologies (% of the number of organizations using artificial intelligence) [6]

The classification of issues and deprivation factors that can be implemented using ML technologies also lie in the plane of environmental problems symbiotically included in the issues of energy supply to end consumers. Today, the importance of using ML systems in monitoring and implementing environmental tasks is increasing. This is not only about the scale of environmental problems within large settlement areas (countryside, urban environment), but also the introduction of ML solutions in the environmental issues of enterprises, since the analysis of environmental factors through automated technologies in production directly affects human activities, affecting labor productivity, health and safety. For example, in the practice of the CHP in Naberezhnye Chelny, voicing was introduced through synthesized speech of the introductory briefing video as an element of artificial intelligence, which increases the quality of information perception [8].

As for integrated security, we note the face recognition system as an important product of ML algorithms, thanks to which security is ensured at important social and industrial facilities. In their study, K. Nti, J. A. Quarcoo, J. Aning and G. K. Fosu denote, that deep neural networks (15%), support vector machines (15%), artificial neural networks (14%), decision trees (12%), and ensemble learning techniques (11%) are widely applied in Big Data Analytics [9].

As you can see, automated systems allow in various fields to implement research, management and production tasks, not only determining the state of the problem based on given signs and algorithms of artificial intelligence, big-date analysis, but help specialists in choosing an adequate strategy for the development of research, activity and development.

Note that to obtain high-quality verified results, the use of correlation, which allows, based on the data and their parameters, to obtain a potential model of further work with the object, software is also used. Researcher Perelomova P. A. notes that despite the high efficiency of using ML algorithms, Big Data Analytics requires a large amount of labeled data, and therefore the author's understanding is seen in the importance of the human factor in the interpretation and predictive assessment of the object of research and modeling [10].

According to a monitoring study on the public perception of artificial intelligence technologies presented by Russian Public Opinion Research Center (JSC «VCIOM») in collaboration with ANO «National Priorities», the majority of respondents surveyed (n = 1601 aged 18 and over, telephone interview method, 2024) consider it permissible to use face recognition technology to ensure law and order, provided that personal data and citizens' rights to privacy are preserved (see Fig. 3) [11].

The advantages and risks in the application of artificial intelligence and ML-technology in organizations and research can become a discussion field in the framework of further promising studies of the use of automated systems, machine learning and big-data analytics technologies.

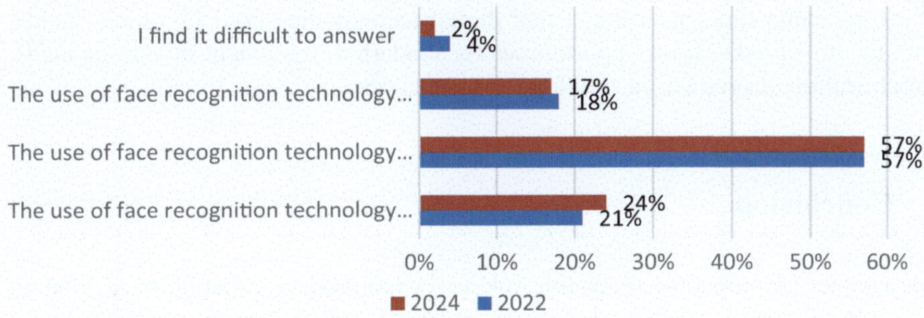

Fig. 3 Trust in artificial intelligence, %, 2024 [11]

3 Discussion

Some modern studies [12, 13] denote the interdisciplinary use of digital technologies in various spheres of society, the information sphere, sociological research, etc. Since digital systems, first of all, are designed to help people, and not be an end in itself, since, giving part of the important production, organizational and research functions, control by specialists is necessary.

In modern scientific papers [14, 15], researchers note that digital solutions are becoming a tool and a technological trend that facilitates routine tasks, and are also able to prevent errors and help in solving and data analytics (for example, when it comes to digital control and security systems, obtaining and automation data empirical research, management, management, production, technical areas, etc.). It should be noted that the features of the use of modern digital technologies, including in the context of machine learning technologies, big data analysis, digital methods of obtaining statistical information, are the subject of study in various research papers, where both positive and alarmist sentiments are formulated, difficulties regarding the scaling of digital progress technologies [16, 17].

However, it is difficult to refute the fact that modern digital technologies, big data and machine learning technologies are a promising direction in the framework of future applied research. In general, the digital research discussion field focuses on the benefits and challenges, risks associated with scaling automated systems and technologies and their impact on the lives of organizations and society. The advantages of modern digital solutions in management, analytics, research in various fields can provide automation of data collection and linear tasks, and also allow you to make decisions in the field of information and big data processing, management, data analysis and final strategies, as well as adapting timing to perform tasks and save resources. At the same time, risks and challenges associated with digital data security, external threats associated with cyber-attacks, as well as the accuracy of algorithm settings, etc. are not excluded.

At the same time, it is noted that digital technologies used in various fields, with the noted risks, form opportunities for modern development in various fields, simplification of production solutions, research areas.

4 Conclusion

The chapter identifies the mediating role and examines the capabilities of modern digital technologies in terms of the state, prospects and significance of their implementation in various spheres of modern society. It has been determined that the use of artificial intelligence programs and products allows you to scale your business, the safety of production systems, improve monitoring of indicators in the field of ecology and formulate plans to optimize production and management processes. Chatbots, voice assistants, speech recognition systems in the practice of working with clients in the banking sector, medicine, commerce, all this can be called the result of the use of artificial intelligence and machine learning, which have entered the practice of activity and organization of various fields. The value of machine solutions in the implementation of these tasks is also associated with a decrease in the financial burden and costs of operating outdated environmental diagnostic systems, which indicates an understanding of the importance of equipment modernization. The use of detection and segmentation functions using digital technologies, the introduction of artificial intelligence systems, big data collection and analysis services forms an understanding of the behavior of various demographic groups from the standpoint of a given study vector and parameters, optimizes production and management tasks.

At the same time, the question of the role of man in the use of automated systems remains controversial, since the interpretation of data is a complex cognitive and meaning-forming process, which is currently not clearly seen without human participation.

Acknowledgements The chapter was carried out at the expense of a grant from the Tatarstan Academy of Sciences provided to young candidates of sciences (postdoctoral fellows) for the purpose of defending a doctoral dissertation, performing research work, as well as performing labor functions in scientific and educational organizations of the Republic of Tatarstan within the framework of the State Program of the Republic of Tatarstan "Scientific and Technological Development of the Republic of Tatarstan".

References

1. Rane NL, Paramesha M, Choudhary SP, Rane J (2024) Machine learning and deep learning for big data analytics: A review of methods and applications. Partn Univers Int Innov J 02(033):172–197. https://www.researchgate.net/publication/382047488_Machine_Learning_and_Deep_Learning_for_Big_Data_Analytics_A_Review_of_Methods_and_Applications (Accessed 30 May 2025). https://doi.org/10.5281/zenodo.12271006

2. Tatarstan ranked second in the top regions in business digitalization/Ministry of Digital Development of Public Administration, Information Technologies and Communications of the Republic of Tatarstan. https://digital.tatarstan.ru/index.htm/news/2340993.htm (Accessed 01 June 2025)
3. Paskova AA (2018) Big Data technologies in automation of technological and business processes. Sci Review Tech Sci (4):23–27. https://science-engineering.ru/ru/article/view?id=1193 (Accessed 29 May 2025)
4. Index of readiness of priority sectors of the economy of the Russian Federation for the introduction of artificial intelligence. Analytical report. M.: National Center for the Development of Artificial Intelligence under the Government of the Russian Federation, 2024. IV + p 85
5. Attwal K, Singh A (2024) Exploring SPSS statistics for data mining and statistical modeling. Int J Stat Appl Mathematics 9(3):09–16
6. Vishnevsky KO, Turovets YV (2023) Artificial intelligence in Russia: who, what and how implements. https://issek.hse.ru/news/862013645.html (Accessed 02 June 2025)
7. Artificial intelligence catches electricity thieves and controls charging for electric vehicles. https://realnoevremya.ru/articles/306732-ao-setevaya-kompaniya-nauchila-neyroset-lovit-energopiratov?erid=2SDnjdEv4ED (accessed 28 May 2025)
8. Digitalization in the field of labor protection in the energy sector of Tatarstan. Step into the future. https://tatelprof.ru/allnews/tsifrovizatsiya-v-sfere-okhrany-truda-v-energetike-tatarstana-shag-v-budushchee (Accessed 05 June 2025)
9. Nti K, Quarcoo JA, Aning J, Fosu GK (2022) A mini-review of machine learning in big data analytics: Applications, challenges, and prospects. Big Data Min Anal 5(2):81–97. https://doi.org/10.26599/BDMA.2021.9020028
10. Perelekova PA (2025) Analysis of ML algorithms for identifying pathologies based on medical data. J Sci 1 (3(84)):526–531
11. Trust in AI. https://wciom.ru/analytical-reviews/analiticheskii-obzor/doverie-k-ii (Accessed 12 May 2025)
12. Paiuc D (2024) Technology and AI's role in multicultural and digital leadership. In: Proceedings of the international conference on networked learning, vol 14, no 1. https://doi.org/10.54337/nlc.v14i1.8011
13. Minakov AV, Eriashvili ND (2022) The introduction of digital technologies in various areas of the national economy and the formation of a global digital space. J "Bulletin Mosc Acad Investig Comm Russ Federation", (4):112–124. https://doi.org/10.54217/2588-0136.2022.34.4.014
14. Rathore MM, Shah SA, Shukla D, Bentafat E, Bakiras S (2021) The role of AI, machine learning, and big data in digital twinning: A systematic literature review, challenges, and opportunities. IEEE Access. 9:32030–32052. https://doi.org/10.1109/ACCESS.2021.3060863
15. Böhm M (2025) Algorithms, AI, big data, and big tech: IPS scholarship on digital technologies. Int Polit Sociology 19(3). https://doi.org/10.1093/ips/olaf019
16. Hussain S, Muneer AlM (2025) The impact and challenges of digital technology in digitalization of businesses. Innov Intell Digit Technol; Increased Effic. https://doi.org/10.1007/978-3-031-71649-2_55
17. Schilirò D (2024) Digital transformation and its impact on organizations. Int J Bus Manag 19(6):71–81. https://doi.org/10.5539/ijbm.v19n6p71

Improving the Decision-Making Procedure of Regional Government and Administration Bodies in the Context of Digitalization

Alena Lisichonak, Sunatilla Baxodirjanovich Gaibnazarov, Aliaksandr Yemialyanau, Irina Pozdnyakova, and Elena Sharafanova

Abstract This chapter investigates the imperative to enhance decision-making within regional governments through digitalization, with a specific focus on the Republic of Belarus. While national-level digital initiatives, such as the Unified Electronic Services Portal, are well-established, a significant gap persists at the regional level. The study aims to assess the digitalization potential of Belarusian regions to identify opportunities for expanding electronic services provided by local authorities. The authors propose and apply a novel methodology to calculate an integral indicator of regional digital potential, grouping statistical data into three areas: access to digital services, their use in public administration, and user satisfaction. The findings reveal significant regional disparities, with Minsk City and Grodno Region leading, while others like Vitebsk face specific weaknesses. A comprehensive assessment of regional websites further identifies a systemic barrier: most function as informational channels rather than integrated service platforms. A moderate positive correlation is also established between a nation's e-government development and its economic performance (GDP per capita). The study concludes that overcoming institutional and procedural—rather than purely technological—barriers is crucial for transforming regional governance and accelerating the digitalization of public administration.

A. Lisichonak · A. Yemialyanau · I. Pozdnyakova
Euphrosyne Polotskaya State University of Polotsk, Novopolotsk, Belarus
e-mail: l.lisichonok@psu.by

A. Yemialyanau
e-mail: a.emelianov@psu.by

I. Pozdnyakova
e-mail: i.pozdnyakova@psu.by

S. B. Gaibnazarov (✉)
Tashkent State Technical University, Tashkent, Uzbekistan
e-mail: sunnatilla711@umail.uz

E. Sharafanova
International Banking Institute Named After Anatoliy Sobchak, Saint Petersburg, Russia
e-mail: asharee@ibispb.ru

© The Author(s), under exclusive license to Springer Nature Switzerland AG 2026
A. Kulachinskaya et al. (eds.), *Digital Transformation: What Kind of Smart Life do we have today?*, Lecture Notes in Networks and Systems 1895,
https://doi.org/10.1007/978-3-032-20582-7_7

Keywords Digitalization potential · E-government development · Electronic services · Public administration · Regional digital potential assessment · Regional governance

1 Introduction

The relevance and necessity of implementing digital technology solutions in public administration at all levels stems from the importance of reducing the timeframes and increasing the effectiveness of decision-making, thereby improving the quality and accessibility of public services provided to the public and businesses. The importance of addressing this challenge is confirmed by policy documents adopted in the Republic of Belarus and other countries.

The main policy documents defining the prospects for the development of public electronic services are the "National Strategy for Sustainable Development of the Republic of Belarus for the Period up to 2040" [1], "Program of Socioeconomic Development of the Republic of Belarus for 2021–2025" [2], "Program of Activities of the Government of the Republic of Belarus for 2025–2029" [3], the state program "Digital Development of Belarus for 2021–2025" [4] and others. These documents define the goals and objectives in the field of information technology, including the creation and development of public electronic services and systems.

Thus, the National Strategy for Sustainable Development of the Republic of Belarus for the period up to 2040 defines as a target indicator "the share of administrative procedures carried out in relation to business entities and individuals in electronic form—up to 100% (excluding administrative procedures, the transfer of which to electronic form is economically impractical); inclusion in 2040 among the top 30 countries in the world ranking by the level of e-government development (58th place out of 193 countries in 2022)" [1].

The current program for the socio-economic development of the Republic of Belarus for 2021–2025 "provides for the organization of a digital information environment by creating a regulatory framework and implementing effective tools for managing economic digitalization processes" [2]. Furthermore, the program envisages the creation of industry-specific and regional digital platforms, including in public administration. A draft program for 2026–2030 is currently up for discussion; it will also focus on developing the digital economy and converting administrative procedures to electronic form.

The program of activities of the Government of the Republic of Belarus for 2025–2029 [3] also provides for the development of a "digital state", the introduction of intelligent systems for managing the social sphere, the development of electronic services and the solution of a number of other tasks with the aim of implementing end-to-end automation of management processes and economic relations.

The state program "Digital Development of Belarus" for 2021–2025, within the framework of the subprogram "Digital Development of Public Administration," provided for "the creation of a unified state information space of the country by consolidating digital data circulating in the public sector at various levels based on platform solutions [4].

M. V. Kudina, A. S. Voronov, and A. V. Gavrilyuk note that the development of innovative technologies and the improvement of digital infrastructure stimulate changes in the decision-making system in public administration [5]. The development and implementation of innovative technological solutions, such as digital platforms and software, facilitate decision-making in government bodies aimed at experimentation and the practical application of innovative technologies to improve interactions between government bodies and civil society.

It is worth noting the areas that have already been implemented in the Republic of Belarus as part of the digitalization of public services:

The Unified Electronic Services Portal is operational [6]. This centralized platform allows citizens and businesses to access various government services electronically and find information on administrative procedures. It is similar to the Russian "Gosuslugi" portal. Currently, efforts are underway to integrate the government services portals of the Republic of Belarus and the Russian Federation. A mobile application for the unified electronic services portal, "E-Pasluga", is also operational. A unified system for identifying individuals and legal entities has been created. This interdepartmental information system is designed to identify and authenticate individuals and legal entities, including their authorized representatives, using cryptographic information protection tools that generate electronic digital signatures with hardware-based private key protection. A similar system also exists in the Russian Federation.

The website of the National Statistical Committee of the Republic of Belarus presents a thematic section "Digital Economy" [7], which allows one to obtain, systematize and analyze data in such areas as: digitalization of the economy, digital development infrastructure; digital development of public administration, use of digital technologies by the population and organizations, digital transformation.

Figure 1 shows the dynamics of the number of electronic services and administrative procedures provided through the national automated information system per 100 people in the Republic of Belarus in 2020–2024.

However, in our view, the ongoing digitalization processes primarily affect the national level of governance, while the capabilities and needs of local authorities and governments have not been fully addressed. Certainly, digitalization processes will eventually expand to all levels of governance, but we believe that steps must already be taken at this stage to further accelerate the digitalization of governance in the regions of the Republic of Belarus.

Thus, the purpose of this study is to assess the digitalization potential in the regions of the Republic of Belarus, identify opportunities for its use to increase the volume of electronic services provided by local authorities and government, and develop recommendations for creating favorable conditions for accelerating digitalization processes at the regional level.

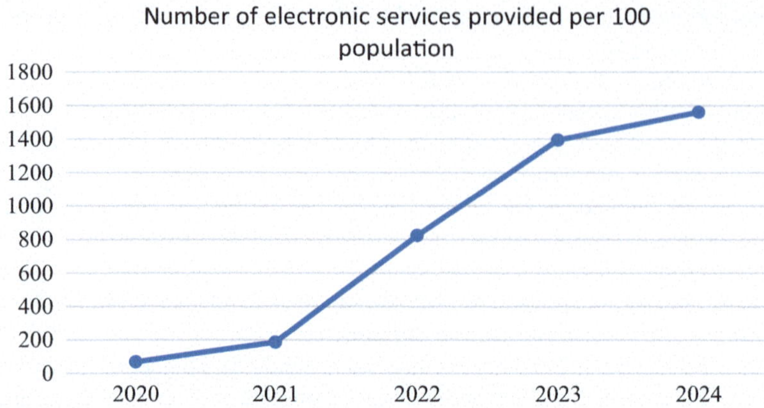

Fig. 1 Dynamics of the number of electronic services provided through the national automated information system per 100 people in the Republic of Belarus in 2020–2024. *Source* Authors' development based on [8]

2 Methods

The following research methods were used in this work.

Theoretical analysis that allows one to study the essence and main directions of digitalization of management, including at the regional level.

Document analysis encompassing forecast, program, and planning documents, as well as statistical data, allows us to confirm and substantiate the relevance and importance of digitalization processes in management in the current economic environment.

Multidimensional comparative analysis method allows for assessing the level of digitalization potential in the regions of the Republic of Belarus and identifying their strengths and weaknesses, which can be used as the basis for developing practical recommendations.

Content analysis aimed at studying the content of regional government and administrative bodies' websites and analyzing key metrics to identify opportunities to improve the efficiency of digitalization processes.

Comprehensive assessment (Benchmarking) was conducted as a structured, criteria-based evaluation of regional websites. There was used a predefined framework with weighted modules to assign scores and enable a comparative analysis of the digital platforms, including a benchmark against regions from the Republic of Belarus and the Russian Federation.

Correlation analysis allows us to evaluate the relationship between the digital government development index and the general indicators characterizing the efficiency of the state economy (GDP per capita).

Integrated approach to improving the e-government confirmed the need to use a set of digital tools, which will ultimately contribute to increasing the level of socio-economic development of the state.

3 Research Results

Digital transformation processes are encompassing all spheres of life: from education [9] to the real sector of the economy [10]. It is taking hold in public administration.

The creation, operation, and development of e-government is built on the active use of modern digital technologies. The digitalization of public administration aims to reduce administrative barriers and make interactions between citizens and businesses with government agencies simpler, faster, more convenient, and more transparent, while also minimizing the impact of subjective factors, particularly those associated with the potential for corruption that inevitably manifests itself in any business-to-personal relationships [11].

This is particularly relevant at the regional level, as local authorities are the most active in interacting with residents and local businesses and addressing social and other issues. This necessitates assessing the potential for digitalization of governance at the regional level.

The digital potential of a region represents the capabilities of the digital economy, which are developed through the application of digital technologies to create conditions for increasing the competitiveness of Belarusian organizations in the global market [12]. In the context of our study, the digitalization potential of regional governance is of particular interest, as it can be considered part of the region's digital potential. We believe that the digitalization potential of regional governance can be interpreted as the ability of regional authorities and administrations to utilize digital technologies to transform governance, increase its efficiency, reduce costs, and improve the quality of public services for the population and businesses.

To assess the potential for digitalization of regional governance, the authors propose a number of indicators reflected in statistics, which can be grouped into the following areas.

1. Access to digital services:
   - the proportion of organizations using the Internet (as a percentage of the total number of surveyed organizations);
   - the proportion of organizations using the website (as a percentage of the total number of surveyed organizations);
   - the proportion of Internet users (as a percentage of the total population).
2. Use of digital services in public administration:
   - the proportion of organizations using the Internet to receive government services in electronic form without the need to use paper documents when

receiving such services (as a percentage of the total number of surveyed organizations that had access to the Internet);
- the proportion of Internet users interacting with government organizations and public administration bodies (as a percentage of the total number of Internet users).

3. Satisfaction with the quality and cost of Internet services:
   - the proportion of organizations with high data transfer speeds (from 30.1 Mbit/s and above);
   - the proportion of Internet users who are fully or partially satisfied with the quality of Internet services (as a percentage of the total number of Internet users);
   - the share of household expenditure on communication services (as a percentage of total consumer expenditure).

Let's analyze these groups of indicators. Table 1 presents an analysis of the indicators characterizing access to digital services by region of the Republic of Belarus, as well as the calculation of the multivariate average characterizing the regional development index for this group of parameters.

Analyzing the data in Table 1, we can conclude that the Grodno Region demonstrates the highest level of accessibility to digital services, driven by the high proportion of organizations using websites and internet users. Minsk City ranks second, with high scores across all indicators. Access to digital services is most limited in the Mogilev Region, which lags behind other regions of the Republic of Belarus in most indicators.

Table 1 Analysis of indicators characterizing access to digital services by regions of the Republic of Belarus in 2024

Name of the indicator	Share of organizations using the Internet, % (D1)	Share of organizations using website, % (D2)	Proportion of Internet users, % (D3)	Multivariate average (MA)	Region rank
Republic of Belarus, total	99.3	71.0	94.3	–	–
Brest	99.8	70.0	93.2	0.964493	3
Vitebsk	99.6	67.5	93.5	0.954337	6
Gomel	99.9	67.2	94.3	0.956951	5
Grodno	98.0	75.4	96.6	0.993701	1
Minsk City	99.2	74.0	96.4	0.990809	2
Minsk	99.6	69.3	92.7	0.959102	4
Mogilev	99.1	62.9	92.0	0.928618	7

Source Authors' development based on [8]

Next, analysis of the indicators characterizing the use of digital services in the field of public administration presented in the Table 2.

Minsk City leads the way in this group of indicators, with more than half of its users interacting with government agencies electronically. Grodno Region ranks highest in terms of the proportion of organizations using the internet to access government services electronically, eliminating the need for paperwork. Minsk Region ranks lowest in this group of indicators. At the same time, the analysis reveals a significant regional disparity, with the Minsk Region occupying the lowest rank within this evaluated dimension. It is pertinent to note that the metric for citizen interaction with government agencies online exhibits considerable variation across regions, indicating heterogeneous levels of digital adoption and trust in e-government platforms in the population.

The subsequent stage of the analytical framework involves an assessment of user perceptions regarding telecommunications infrastructure. This entails examining a set of indicators designed to quantify satisfaction with the quality and affordability of internet services across the administrative regions of the Republic of Belarus. The findings from this comprehensive evaluation are systematically compiled and presented in Table 3.

The analysis showed that the Grodno Region and Minsk City also lead the group of indicators characterizing user satisfaction with the quality and cost of Internet services. Moreover, Minsk City significantly outperforms all other regions in terms of the totality of these indicators and is the leader in two of them, including the lowest costs for communication services and a significant share of organizations with high data transfer speeds.

Let us summarize and calculate the integral indicator characterizing the potential for digitalization of regional governance in the Republic of Belarus represented in Table 4.

Table 2 Analysis of indicators characterizing the use of digital services in the field of public administration by regions of the Republic of Belarus in 2024

Name of the indicator	Share of organizations using the Internet, % (D1)	Share of organizations using website, % (D2)	Multivariate average (MA)	Region rank
Republic of Belarus, total	99.3	71.0	–	–
Brest	99.8	70.0	0.964493	3
Vitebsk	99.6	67.5	0.954337	6
Gomel	99.9	67.2	0.956951	5
Grodno	98.0	75.4	0.993701	1
Minsk City	99.2	74.0	0.990809	2
Minsk	99.6	69.3	0.959102	4
Mogilev	82.9	29.1	0.757618	5

Source Authors' development based on [8]

Table 3 Analysis of indicators characterizing user satisfaction with the quality and cost of Internet services in the regions of the Republic of Belarus in 2024

Name of the indicator	Share of organizations with high data transfer speed, % (K1)	Proportion of those satisfied with the quality of Internet services, % (K2)	Share of household expenditure on communication services, % (K3)	Multivariate mean (MMM)	Region rank
Republic of Belarus, total	59.5	87.8	4.9	–	–
Brest	52.6	82.8	5.1	0.842032	7
Vitebsk	53.7	83.0	5.2	0.843025	6
Gomel	55.3	88.1	5.0	0.881057	3
Grodno	56.9	91.9	5.1	0.898835	2
Minsk city	65.9	86.8	4.2	0.98185	1
Minsk	55.2	91.5	5.7	0.863303	5
Mogilev	52.8	91.7	5.3	0.869018	4

Source Authors' development based on [8]

Table 4 Calculation of the integral indicator characterizing the level of digital potential of regional government in the Republic of Belarus in 2024

Name of the indicator	ID	II	IK	Integral Index (ICU)	Region rank
Brest	0.964493	0.773521	0.842032	0.860015	4
Vitebsk	0.954337	0.733416	0.843025	0.843593	7
Gomel	0.956951	0.761532	0.881057	0.866513	3
Grodno	0.993701	0.870848	0.898835	0.921128	2
Minsk city	0.990809	0.972805	0.98185	0.981821	1
Minsk	0.959102	0.72087	0.863303	0.847758	6
Mogilev	0.928618	0.757618	0.869018	0.851751	5

Source Authors' development

The analysis reveals a pronounced and hierarchical digital divide among the regions of the Republic of Belarus. The Minsk City stands as a clear outlier with the highest level of digitalization potential (0.9818), a result of its status as the capital and a major hub for digital infrastructure and investment. Among the other regions, a distinct tiered structure is evident: Grodno Region emerges as the regional leader (0.9211), followed by a middle group consisting of Gomel, Brest, and Mogilev regions (scores between 0.8665 and 0.8518). The Minsk and Vitebsk regions form a third, lower tier (0.8478 and 0.8436, respectively), indicating they face the most significant challenges in their digital governance capabilities.

This ranking demonstrates that digital potential is not uniformly distributed and that a region's administrative and economic profile, as seen with Minsk City, is a dominant factor. The significant gap between the capital and other regions, and the clear leaders and laggards among the regions themselves, underscores the need for targeted, region-specific digitalization strategies rather than a one-size-fits-all national approach.

As for the Vitebsk Region, it ranks 7th overall. Therefore, the weaknesses in the digital potential of the Vitebsk regional government are of significant interest. They are presented in Table 5.

The analysis of the Vitebsk Region's digital potential reveals a critical weakness: its digital infrastructure and adoption are insufficient to support effective e-government. While internet access among organizations is a strength, the region's low rates of website usage, minimal adoption of paperless e-services, inadequate internet speeds, and poor user satisfaction indicate a systemic failure to transition from basic connectivity to meaningful digital interaction. This fundamentally hinders the region's ability to digitalize public administration, as both businesses and households lack the reliable tools and positive experience necessary for engagement. The identified weaknesses point not to a lack of initial access, but to deeper institutional and infrastructural barriers that prevent the practical application of digital technologies in governance.

Table 5 Weaknesses and strengths of the digital potential of the regional government of the Vitebsk region

Strengths	Weaknesses
1. A significant proportion of organizations using the Internet	1. The low proportion of organizations using the website, which negatively impacts the interaction of the business entity with contractors and government agencies 2. The lowest share of organizations using the Internet to obtain government services electronically without the need to use paper documents by region, which slows down the development of electronic document management, including with government agencies 3. Minor share of organizations with high data transfer rates, which is an obstacle to the digitalization of public administration on the part of business entities 4. The lowest proportion of users satisfied with the quality of Internet services among the regions of the Republic of Belarus, which hinders the digitalization of public administration on the part of households

Source Authors' development

A comprehensive assessment of the official websites for all regions of the Republic of Belarus was undertaken for the year 2025 to quantitatively evaluate the maturity of their digital governance potential. This evaluation, the results of which are systematically presented in Tables 6 and 7, employed a multi-criteria framework analyzing dimensions such as service integration, usability, and information transparency.

To ensure the robustness and cross-border validity of the analytical framework, the methodology was applied to three regions of the Russian Federation. The results, detailed in Table 8, enable a comparative assessment and reveal that the challenges in transitioning from informational websites to interactive service platforms are not unique to Belarus, but a common hurdle in regional digitalization.

A comparative study of regional official web portals reveals significant heterogeneity in the level of development of e-government infrastructure. Analysis of

Table 6 Comprehensive assessment of websites of the regions of the Republic of Belarus, 2025, pt. 1

Assessment module	Criterion	Weight of the criterion (%)	Grade			
			Brest region (%)	Vitebsk region (%)	Gomel region (%)	Grodno region (%)
1. Availability and relevance of information	1.1 Data relevance	10	10	10	10	10
	1.2. Completeness of content	10	10	10	10	10
2. Functionality and electronic services	2.1. Integration with the electronic services portal	20	0	0	0	20
	2.2. Online booking and services	15	0	0	0	0
3. Technical quality and usability (UX/UI)	3.1. Adaptability and accessibility	15	15	15	15	15
	3.2. Ease of navigation	10	10	10	10	10
4. Interactivity and feedback	4.1. Communication channels	10	10	10	10	10
	4.2. Transparency of feedback	5	5	5	5	5
5. Safety and technology	5. Website security	5	5	5	5	5
	Final score	100	65	65	65	85

Source Authors' development based on [13–16]

Table 7 Comprehensive assessment of websites of the regions of the Republic of Belarus, 2025, pt. 2

Evaluation module	Criterion	Weight of the criterion (%)	Grade		
			Mogilev Region (%)	Minsk region (%)	Minsk city (%)
1. Availability and relevance of information	1.1. Data relevance	10	10	10	10
	1.2. Completeness of content	10	10	10	10
2. Functionality and electronic services	2.1. Integration with the electronic services portal	20	0	0	0
	2.2. Online booking and services	15	0	0	0
3. Technical quality and usability (UX/UI)	3.1. Adaptability and accessibility	15	15	15	15
	3.2. Ease of navigation	10	10	10	10
4. Interactivity and feedback	4.1. Communication channels	10	10	10	10
	4.2. Transparency of feedback	5	5	5	5
5. Safety and technology	5. Website security	5	5	5	5
	Final score	100	65	65	65

Source Authors' development based on [17–19]

empirical data reveals a persistent dichotomy between the universally achieved high level of basic web presence and the fragmented implementation of advanced transactional services.

The assessment results demonstrate that there is consensus among all subjects on key criteria, such as content relevance and representativeness, technical reliability, usability, and basic cybersecurity measures. This suggests that standards for the content and technical accessibility of digital resources can be considered universally adopted in administrative practice.

However, a key differentiating factor was identified in the module assessing the functionality and provision of electronic services. The qualitative gap manifests itself in the presence or absence of deep integration with government service portals and, most significantly, online booking mechanisms. Most regions of the Republic of

Table 8 Comprehensive assessment of websites of the regions of the Russian Federation, 2025

Evaluation module	Criterion	Weight of the criterion (%)	Grade		
			Leningrad region (%)	Moscow region	Smolensk region (%)
1. Availability and relevance of information	1.1. Data relevance	10	10	10	10
	1.2. Completeness of content	10	10	10	10
2. Functionality and electronic services	2.1. Integration with the electronic services portal	20	0	20	20
	2.2. Online booking and services	15	0	0	15
3. Technical quality and usability (UX/UI)	3.1. Adaptability and accessibility	15	15	15	15
	3.2. Ease of navigation	10	10	10	10
4. Interactivity and feedback	4.1. Communication channels	10	10	10	10
	4.2. Transparency of feedback	5	5	5	5
5. Safety and technology	5. Website security	5	5	5	5
	Final score	100	65	85	100

Source Authors' development based on [20–22]

Belarus and the Leningrad Region of the Russian Federation demonstrate a deficiency in this component, limiting their web platforms to passive information channels.

In contrast, a cohort of leading regions, including Grodno, Moscow, and, especially, Smolensk Region, demonstrates a transition to a more mature stage of digitalization. Their experience confirms the feasibility of transforming an official website from a publishing tool into a fully-fledged service and transaction platform. Smolensk Region, which achieved the highest results, serves as a benchmark model demonstrating the comprehensive implementation of all assessed metrics.

A comprehensive assessment of the official websites of the regions of the Republic of Belarus demonstrates the achievement of a uniform and sustainable level of development across the basic components of digitalization. All regions, including the Minsk City as a separate entity, demonstrate impeccable results in the relevance and completeness of information content, demonstrating a well-established system for supporting and updating web resources. Similarly, technical aspects, including adaptive design, ease of navigation, and basic security measures, are implemented at a high level, ensuring a standardized user experience. The availability and accessibility of feedback channels is also a universal characteristic across all Belarusian regions.

However, this apparent uniformity conceals a systemic challenge associated with the transition to a more mature stage of e-government development. The critical factor determining the overall score is the functionality and e-services module. Data analysis reveals that, with the exception of the Grodno Region, no region has implemented integration with the e-services portal, and interactive services such as online booking are absent everywhere. This results in the overwhelming majority of regions achieving a score of 65%, while the Grodno Region, thanks to integration, achieves a score of 85%.

Thus, the obtained results allow us to conclude that the main barrier to the further development of e-government at the regional level is not so much technological limitations as institutional and procedural factors. This type of barrier is not only characteristic of digitalization initiatives but is also common in other ICT-related aspects of regional development, such as advancing financial literacy [23].

Overcoming the identified gap requires targeted efforts to optimize administrative regulations, ensure interdepartmental electronic interaction, and stimulate citizen digital engagement. Replicating successful cases from leading regions appears to be the most promising vector for reducing regional asymmetries in the quality of public service delivery online.

Across Belarus's regions, we can detect an existing structural imbalance. Despite the creation of a robust web presence infrastructure, their official portals largely operate within a one-way communication paradigm, serving primarily informational rather than service functions. Bridging this gap and transforming digital platforms into tools for delivering comprehensive electronic government services requires focusing on overcoming institutional and procedural barriers, as well as disseminating the successful experience of the Grodno Region in integrating with existing e-government systems.

Mathematical methods and models are widely used both to assess the closeness of relationships between indicators and for forecasting. The forecasting object often includes gross national and regional products, as well as other indicators, such as those presented in scientific research [24, 25].

To assess the relationship between the level of e-government development and the economic performance of states and their regions, we conducted a correlation analysis between these two variables: the e-Government Development Index (EGDI) and GDP per capita based on purchasing power parity (PPP). The data for this analysis are presented in Table 9.

The calculations yielded a correlation coefficient of $r = 0.531716$. This value indicates a moderate positive relationship between the two variables. This means that there is a significant, but not perfect, relationship between the two parameters under study: when one changes, the other will likely change in the same direction, but this does not happen with absolute certainty every time. In our view, this is entirely reasonable, as GDP per capita is influenced by a large number of factors. Nevertheless, the existence of a correlation between the two indicators highlights the need to improve the digitalization of governance, including at the regional level, as

Table 9 Data for assessing the correlation between the level of e-government development and GDP per capita, 2024

Country	EGDI	GDP per capita at PPP, USD	Country	EGDI	GDP per capita at PPP, USD	Country	EGDI	GDP per capita at PPP, USD
Cambodia	0.5754	7764	Ireland	0.9138	130,460	Peru	0.8070	8452
Chile	0.8827	34,306	Italy	0.8356	40,226	Philippines	0.7621	3985
Costa Rica	0.8009	31,030	Jordan	0.6849	4618	Poland	0.8648	25,023
Croatia	0.8818	48,487	Korea	0.9679	52,204	Portugal	0.8415	28,844
Cyprus	0.8619	62,217	Latvia	0.8852	23,368	Romania	0.7636	20,072
Estonia	0.9727	48,077	Lithuania	0.9110	29,386	Saudi Arabia	0.9602	35,057
Finland	0.9575	64,070	Luxembourg	0.8466	137,517	Spain	0.9206	35,297
France	0.8744	63,639	Malaysia	0.8111	11,867	Sweden	0.9326	72,351
Germany	0.9382	71,902	Malta	0.8886	42,347	Thailand	0.8351	24,717
Greece	0.8674	42,986	Mexico	0.7850	14,158	Uruguay	0.9006	36,418
China	0.8718	27,078	Netherlands	0.9538	68,219	Russia	0.8532	47,713
Hungary	0.8043	46,657	Panama	0.7298	19,103	Peru	0.8070	17,798
Indonesia	0.7991	16,447	Paraguay	0.7251	6416	Belarus	0.7445	34,658

Source Authors' development based on [26, 27]

its positive dynamics can lead to increased economic efficiency for the state and its regions.

4 Discussion

The digitalization of regional governance is fundamentally reshaping decision-making procedures. The development of specialized automated information systems is crucial for providing management with the timely, high-quality information essential for effective governance [28]. These systems transform raw data into transparent, digitized management information, which is vital for developing effective decisions based on comprehensive and objective data. In this context, the integration of modern digital tools becomes pivotal, enabling continuous monitoring, the analysis of extensive datasets, and the formulation of well-founded decisions through information technology.

Research into digitalization confirms that IT-driven strategies can significantly improve efficiency and effectiveness in public administration. However, successful implementation is not guaranteed by technology alone; it relies critically on several foundational factors, including strong leadership, robust institutional capacity, active stakeholder engagement, and strategic resource allocation [29]. Conversely, significant challenges such as data privacy concerns, limited technical capacity, and internal resistance to change often impede progress. Thus, digital tools can be leveraged to achieve specific socio-economic goals (better governance, successful business), with a sober understanding that this requires systematic research, strategic implementation beyond mere technology adoption, and navigation of a challenging external environment [30].

Furthermore, a nuanced understanding of "digital potential" is required. Studies indicate that this potential is significantly influenced by high rates of broadband internet usage among both organizations and the population in a region [31]. Surprisingly, there is no direct correlation between the financial expenditures on digitalization and the resulting level of digitalization potential in regions. This is corroborated by analyses of the statistical relationship between regional well-being and digitalization parameters, which often reveal only a weak to moderate correlation [32]. This suggests that, with the current approach focused predominantly on funding for technology—such as devices and specialized software—it is impossible to ensure the breakthrough development of regional economic systems.

At the same time another study reveals that true regional digitalization is not just about technological readiness but about the actual use of services and the deployment of accessible, region-specific infrastructure. For developing nations, overcoming institutional and procedural barriers to bridge the gap between digital potential and tangible performance is crucial for achieving inclusive and sustainable digital governance [33].

In conclusion, while digitalization is an undeniable prerequisite for successful, competitive, and innovative regional development, simply implementing technology is insufficient. To maximize the benefits, regions must look beyond mere financial investment and focus on developing comprehensive digital strategies. These strategies must holistically address all aspects of the process, from building digital literacy and infrastructure to fostering a culture of data-driven decision-making, thereby ensuring that digital transformation translates into genuine and sustainable regional advancement [34].

5 Conclusion

In conclusion, it should be noted that the moderate positive relationship identified by the authors between the level of e-government development and the level of economic development of the state, expressed by the indicator Gross domestic product per capita at PPP, confirms the need to consider the level of digitalization of governance as one of the factors determining the state and pace of development of the national economy of the state.

Digitalization of governance should encompass all levels of government, including the regional level. Therefore, the authors proposed a methodology for assessing the digitalization potential of regional governance. This methodology is based on calculating an integrated indicator based on a system of individual indicators reflected in statistics. The indicators selected by the authors were grouped into three groups, allowing them to assess access to digital services, the use of digital services in public administration, and user satisfaction with the quality and cost of internet services. This methodology has been tested in the regions of the Republic of Belarus and is based on data from the National Statistical Committee of the Republic of Belarus.

The developed methodological approach allows us to identify the key factors hindering the digitalization of regional governance, including those dependent and independent of internet service users. The primary tool for digitalizing governance is the websites of regional government and administrative bodies. Therefore, the authors conducted a comprehensive assessment of the websites of all regions of the Republic of Belarus and three regions of the Russian Federation, which revealed that the main barrier to the further development of e-government at the regional level is not so much technological limitations as institutional and procedural factors.

The authors' findings allowed them to identify potential for accelerating governance digitalization processes based on addressing institutional and procedural challenges. Addressing these challenges will accelerate the pace of governance digitalization and, consequently, positively impact the country's economic development.

References

1. "National Strategy for Sustainable Development of the Republic of Belarus for the Period up to 2040". Ministry of Economy of the Republic of Belarus website. https://economy.gov.by/uploads/files/NSUR/natsionalnaja-strategija-ustojchivogo-razvitija-respubliki-belarus-na-period-do-2040-goda.pdf
2. "The Program of Socio-Economic Development of the Republic of Belarus for 2021–2025". Ministry of Economy of the Republic of Belarus website. https://economy.gov.by/uploads/files/macro-prognoz/Programma-2025-nov-red.pdf
3. "On approval of the Program of Activities of the Government of the Republic of Belarus for 2025–2029: Resolution of the Council of Ministers of the Republic of Belarus No. 254 of May 8, 2025". Ministry of Economy of the Republic of Belarus website. https://economy.gov.by/uploads/files/NSUR/ob-utverzhdenii-programmy-dejatelnosti-pravitelstva-respubliki-belarus-na-2025-2029-gody.pdf
4. "On the State Program "Digital Development of Belarus" for 2021–2025: Resolution of the Council of Ministers of the Republic of Belarus No. 66 of February 2, 2021". The National Legal Internet Portal of the Republic of Belarus. https://pravo.by/document/?guid=12551&p0=C22100066
5. Kudina MV, Voronov AS, Gavrilyuk AV (2023) Implementation of digital platforms for decision-making in public administration. Public administration. Electronic Bulletin no. 100. https://doi.org/10.24412/2070-1381-2023-100-166-179
6. National Center for e-Service website, https://portal.gov.by/. Last accessed 03 Nov 2025
7. Digitalization of the economy, National Statistical Committee of the Republic of Belarus. https://dataportal.belstat.gov.by/osids/thematic-sections-info/72. Last accessed 05 Nov 2025
8. National Statistical Committee of the Republic of Belarus website, www.belstat.gov.by
9. Stankevich D, Gaibnazarova Z, Naumovich O, Tuychieva M, Shadieva G (2025) The role of universities in digital transformation: Challenges and prospects. In: Skhvediani A, Kulachinskaya A, Mateeva Petrova M (eds) Digital transformation of socio-economic and technical

systems: Theory and practice. Lecture notes in networks and systems, vol 1309, pp 179–193. Springer, Cham. https://doi.org/10.1007/978-3-031-85608-2_13
10. Gaibnazarova ZT (2024) Artificial intelligence and smart railway transport as a tool to accelerate the transformation process in Uzbekistan. In: Sari M, Kulachinskaya A (eds) Digital transformation: What are the smart cities today? Lecture notes in networks and systems, vol 846. Springer, Cham. https://link.springer.com/chapter/10.1007/978-3-031-49390-4_19
11. Plotnikov VA, Maslyuk AV (2022) Prospects for digitalization of public administration processes. Management Consulting no. 3
12. Sherstneva OM (2024) Digital potential of the region: theoretical aspects and assessment in the Republic of Belarus. Bull Vitebsk State Technol Univ 49(3). https://doi.org/10.24412/2079-7958-2024-3-97-106
13. Brest region official website. www.brest-region.gov.by. Last accessed 06 Nov 2025
14. Vitebsk region official website. www.vitebsk-region.gov.by. Last accessed 06 Nov 2025
15. Gomel region official website. www.gomel-region.gov.by. Last accessed 06 Nov 2025
16. Grodno region official website. www.grodno-region.gov.by. Last accessed 07 Nov 2025
17. Mogilev region official website. www.mogilev-region.gov.by. Last accessed 07 Nov 2025
18. Minsk region official website. www.minsk-region.gov.by. Last accessed 07 Nov 2025
19. Minsk City official website. www.minsk.gov.by. Last accessed 07 Nov 2025
20. Leningrad region official website. www.lenobl.ru. Last accessed 08 Nov 2025
21. Moscow region official website. www.mosreg.ru. Last accessed 08 Nov 2025
22. Smolensk region official website, www.admin-smolensk.ru/. Last accessed 08 Nov 2025
23. Lisichonak A, Yemialyanau A, Pozdnyakova I, Abdullayeva M, Zatevakhina A (2025) Instruments for teaching financial literacy in the digital transformation age. In: Skhvediani A, Kulachinskaya A, Mateeva Petrova M (eds) Digital transformation of socio-economic and technical systems: Theory and practice. Lecture notes in networks and systems, vol 1309. Springer, Cham. https://doi.org/10.1007/978-3-031-85608-2_14
24. Gaibnazarova ZT (2023) Prospects for improving the forecasting of the gross domestic product of regions based on mathematical models. In: 9th IEEE International Conference on Information Technology and Nanotechnology (ITNT), Samara, Russian Federation. https://doi.org/10.1109/ITNT57377.2023.10139074
25. Gaibnazarova ZT, Solieva BT, Iminova NA (2024) Forecasting the risk of bankruptcy in poorly formalized processes. Sci & Tech 23(1):80–86. https://doi.org/10.21122/2227-1031-2024-23-1-80-86
26. UN Study: E-Government 2024. https://publicadministration.un.org/egovkb/en-us/Reports/UN-E-Government-Survey-2024
27. World Bank GDP per capita (current US$) data. https://data.worldbank.org/indicator/NY.GDP.PCAP.CD
28. Kovalenko Y, Zakharova O, Yakusheva O, Yakushev O, Hulak D, Kozlovska S (2021) Priorities for the development of regional information management based on system digitalization. SHS Web Conf 126. https://doi.org/10.1051/shsconf/202112604002
29. Tano I (2024) Digitalization in local governance. Pantao (International Journal of the Humanities and Social Sciences). https://doi.org/10.69651/pijhss030317
30. Ziankova I, Gaibnazarova Z, Malei A, Sharafanova E, Gasheva O (2025) Digital technologies for comprehensive promotion of a startup. In: Skhvediani A, Kulachinskaya A, Mateeva Petrova M (eds) Digital transformation of socio-economic and technical systems: Theory and practice. Lecture notes in networks and systems, vol 1309, pp 39–55. Springer, Cham. https://doi.org/10.1007/978-3-031-85608-2_4
31. Egorov NE, Kovrov GS, Tishkov SV, Volkov AD (2022) The potential of digitalization of resource regions of the Russian North. MIR (Modernization. Innovation. Research) 13(2). https://doi.org/10.18184/2079-4665.2022.13.2.238-251
32. Chistnikova IV, Antonova MV, Mikhailichenko MI (2022) Scientific approach to the study of the digitalization impact on the regional economy. E-Management (In Russ.). https://doi.org/10.26425/2658-3445-2022-5-4-72-81

33. Gupta R, Muttoo SK, Pal SK (2020) Regional e-governance development index for developing nations. Digit Govment: Res Pract 1(3). https://doi.org/10.1145/3386163
34. Shutaeva E, Pobirchenko V (2024) Digitization as a necessary condition for the development of foreign economic activity of the regions of Russia. Sci Bull: Financ, Bank, Invest 0(1). https://doi.org/10.29039/2312-5330-2024-1-183-196

State-Funded Basic Research: Stakeholder Engagement in the Digital Environment

Svetlana Soldatova, Elena Maslennikova, and Artem Turenko

Abstract We present the results of a study that provides the basis for proposed changes to the system of stakeholder engagement in basic research funded by the Russian Federation's state budget. This research was commissioned by the Ministry of Science and Higher Education of the Russian Federation with the aim of expanding and improving stakeholder participation in the planning, monitoring, and approval of the results of state-funded basic scientific research. The goal is to make basic scientific research more practical and shorten the time between the development of a basic idea and its actual implementation. As a result of the study, a new stakeholder group, qualified customers, was identified and included in the process of development and implementation of basic research. We defined the characteristics of qualified customers as parties motivated to request basic scientific results aimed at solving practical problems within the institutional conditions of the Russian Federation's science and innovation sector. A practical outcome of this study is the establishment of stakeholder engagement regulations that set out the rules and procedures for interaction within the digital environment of the 'Science and Innovations' domain. This paper also describes the technological queries service as an element of the domain's structure, designed to support effective communication among stakeholders. To test the proposed changes, an experiment was conducted, along with an awareness campaign to involve potential customers and researchers in the project. The test involved major companies-customers from various industries and public administration bodies, including regional executive authorities. The greatest interest was shown by organisations from the oil, gas, and petrochemical sectors, as well as from the engine manufacturing sector. Over 1500 responses from potential researchers were received to the queries placed in 'Science and Innovations' domain.

S. Soldatova (✉) · A. Turenko
FSBI 'Directorate of Scientific and Technical Programs', Moscow, Russia
e-mail: soldatova@fcntp.ru

E. Maslennikova
Moscow Metropolitan Governance Yury Luzhkov University, Moscow, Russia

© The Author(s), under exclusive license to Springer Nature Switzerland AG 2026
A. Kulachinskaya et al. (eds.), *Digital Transformation: What Kind of Smart Life do we have today?*, Lecture Notes in Networks and Systems 1895,
https://doi.org/10.1007/978-3-032-20582-7_8

Keywords Basic research · Stakeholders · Qualified customers · Digital environment · 'Science and Innovations' domain

1 Introduction

The role of public funding for basic research in ensuring favourable conditions and prerequisites for scientific and technological development and economic growth has been noted by many scholars. Atkinson and Moschella argue that the results of publicly funded research not only increase the productivity and focus of private sector research, but also encourage companies to increase their efforts to develop promising areas of innovation [1]. Prettner and Werner have demonstrated that there is an optimal level of government investment in basic research that maximises long-term social welfare, but even in developed countries, the actual level is below the optimal level [2]. The potential for government investment in basic research to have a positive impact on socio-economic dynamics is nowhere near its limit. Akcigit et al. [3] presented arguments in favour of government policy aimed at developing basic research and strengthening its engagement with the private sector. Experiments conducted by Parui and Prettner using a generalised model of economic growth supported the conclusion that government funding for basic research should be increased relative to its current level in OECD countries [4]. Nevertheless, researchers urge decision-makers to develop strategies aimed at strengthening cooperation among all participants in the innovation process, including science, in addition to providing a more thorough justification for the priority areas of government funding for basic research. Recent studies indicate that the fragmentation and conflicting interests of the parties involved in the creation and implementation of innovations have become an obstacle to solving pressing contemporary issues, particularly those related to sustainable development, thus raising concerns about stakeholder management [5–7]. In this chapter, we present the results of a study conducted as part of an initiative of the Ministry of Science and Higher Education of the Russian Federation aimed at developing stakeholder engagement in the process of planning, monitoring and approval of state-funded basic scientific research, aimed at strengthening the practice-oriented nature of basic science and shortening the time between the development of a basic idea and its practical implementation.

2 Institutional Conditions and Infrastructure for Stakeholder Engagement in State-Funded Basic Research

2.1 Stakeholders in Basic Science Research Projects and Rules for their Engagement

According to Miles, the variety of definitions of 'stakeholders' and 'stakeholder theory' that exist across different scientific fields can be narrowed down to a few widely agreed-upon principles. Stakeholder theory is a management theory that highlights the fact that organisations engage in relationships with many groups and focuses on these relationships in terms of processes and outcomes for the organisation and stakeholders [8]. The essence of the stakeholder approach is the establishment and management of relationships among stakeholders. As Ackermann and Eden state, the stakeholders of any organisation include groups or individuals who are affected by the organisation, as well as those who can influence it [9]. For classification purposes, researchers also apply additional criteria beyond the factor of influence [10, p. 1864].

Examining the practices of Russian organisations conducting basic scientific research with financial support from the state, we noted that the list of stakeholders in a basic scientific project is strictly limited in accordance with the established procedure for planning and financing state-funded research. The stakeholders that are actually taken into account when making decisions on scientific projects are the founders, namely the state executive bodies to which these organisations report, and the Russian Academy of Sciences (RAS). The founders define the state tasks for basic scientific research based on proposals (effectively project applications) submitted by the organisations themselves, and they also determine the amount of funding to be allocated within their available budgetary limits. When preparing applications for basic research, organisations consider previously obtained scientific results, accumulated research expertise and existing research infrastructure, along with several less significant factors. Proposed research topics must correspond to the Programme of basic scientific research for the long term, which is coordinated by RAS. As the most influential participant in the development of the Programme, the Academy of Sciences sets the framework and determines the priorities for the selection of research topics. Another significant factor influencing research organisations is the evaluation of scientific results obtained in basic research. The expert assessment conducted by the Russian Academy of Sciences determines the prospects of a particular scientific project, assessing whether it is successful and whether it should receive continued state funding.

Within the described model, stakeholders who are capable of generating demand for specific scientific results for subsequent transition to applied research, development, and pilot production with the aim of solving practical problems do not have effective leverage over decisions regarding basic research. According to the model developed by Mitchell, Agle, and Wood, such groups can be classified as latent stakeholders. Latent stakeholders do not have the right to participate in the formation of the Programme. The current version of the government decree regulating the development and implementation of the Programme states that the initiative in defining the Programme's objectives lies with 'researchers and/or research teams, scientific organisations, state academies of sciences, and foundations supporting science, technology and innovation activities based on their own logic of scientific development' [11]. Latent stakeholders do not have the opportunity to influence the planning of basic research, participate in monitoring its implementation, or evaluate the results obtained. Currently, only research organisations themselves, their founders, and the Russian Academy of Sciences hold the right to participate in these processes.

Recognising the limitations of this structure, especially its insufficient orientation towards solving pressing practical problems, the Ministry of Science and Higher Education of the Russian Federation has launched an initiative aimed at involving latent stakeholders in the formation and implementation of state-funded basic scientific research projects. The authors of this chapter have participated in the development and implementation of this initiative.

2.2 Digital Environment for Stakeholder Engagement in Basic Scientific Research Projects

The 'Science and Innovations' domain is a unified digital environment that integrates the digital services and information resources needed to provide comprehensive support for scientific activities in the interests of the state, the scientific community, and business. It includes the Unified State Information System for Accounting for Civilian Research, Development, and Technological Work (EGISU NIOKTR). The domain is intended for researchers, research managers, and representatives of government agencies and foundations.

The domain offers a range of services that provide access to up-to-date information on scientific research and development in the Russian Federation. These services include 'End-to-End Traceability,' 'Statistical Reporting,' 'Control of State Tasks,' and others. In addition, a separate service, 'Technological queries from business,' has been created to facilitate interaction between companies in the real sector of the economy, federal executive authorities, authorities of the constituent entities of the Russian Federation, and scientific and educational organisations of higher education.

3 Results: Expanding Stakeholder Engagement in State-Funded Basic Research

3.1 Proposed Changes to Stakeholder Engagement Regulations and Infrastructure

The starting point for developing the proposed changes was to identify the composition and characteristics of latent stakeholders who are motivated to request basic scientific results focused on solving practical problems. We assumed that such demand would primarily come from organisations operating in the real sector of the economy. These organisations must have deep expertise in a specific technological field and be interested in both obtaining the results of basic scientific research and their subsequent implementation in research, development and production. They must also have sufficient resources to conduct applied research, undertake development and pilot production. These organisations must have internal research and development costs and engage in innovative activities, which must be confirmed by reports sent to the Federal State Statistics Service. Such organisations must have a broad vision of scientific areas that are important for the development of the country or region. It would also make sense to include federal and regional authorities in this group. Russia does not have a state agency similar to DARPA (Defense Advanced Research Projects Agency), the US agency that acts as a customer for basic research aimed at solving practical problems. Therefore, the starting point for the project was to identify potential customers capable of creating demand for basic scientific results, as well as to establish their status legally. We used the term 'qualified customers,' which has become standard in Russian scientific and technical policies thanks to a presidential decree [12], as a way to refer to interested parties who request scientific results.

During the design phase, it was necessary to define the key rules for engagement between the expanded group of stakeholders and to formalise them in the form of regulations. Amendments also had to be made to certain legal provisions included in several existing official documents. The introduction of the right to participate in the formation of the Programme for qualified customers is a major innovation. In order to implement this, changes to the government decree [11] were proposed, resulting in an expansion of the group of individuals responsible for defining the Programme's objectives. As project developers, we proposed the introduction of multiplier coefficients for the financing of basic research projects carried out at the request of qualified customers. These coefficients are applied provided that qualified customers confirm the applicability of the research outcomes.

The regulations for stakeholder engagement developed as part of this project form the basis for creating a comprehensive framework for stakeholder engagement in the digital environment of the 'Science and Innovations' domain. The regulations define: (1) project participants; (2) the interpretation of the concept of a 'qualified customer' as it applies to the project's goals and objectives; (3) the composition of the category of qualified customers and the eligibility criteria; (4) the digital infrastructure for engagement; (5) requirements for the format and structure of a qualified customer's query for the scientific results of basic research; (6) the sequence and timing of approval of the query by other project participants; (7) requirements for the format and structure of potential executants' responses to a qualified customer's query; (8) the sequence and timing of the development, coordination and approval of the scientific research project submitted by the executant selected by the qualified customer; (9) the format of the cooperation agreement between the qualified customer and the executant; (10) the composition and timing of monitoring activities; (11) the procedure for presenting and evaluating the scientific results of basic research, as well as determining their practical application by the qualified customer. The interaction framework outlined in the regulations is implemented in the digital environment of the 'Science and Innovations' domain (Fig. 1).

As mentioned above, to support stakeholder engagement in basic research projects carried out at the query of a qualified customer, a dedicated service titled 'Technological queries from business' has been created within the 'Science and Innovations' domain. This service enables qualified customers to submit their queries to a wide range of potential researchers, receive responses, and provide feedback. The value of the service for both qualified customers and potential executants increases with the number of users involved in the service—an effect commonly referred to as the network effect. The term 'technological queries' in the service's title may be misleading regarding the nature of the queries submitted. In fact, the term simply reflects the fact that a qualified customer is turning to potential executants to conduct

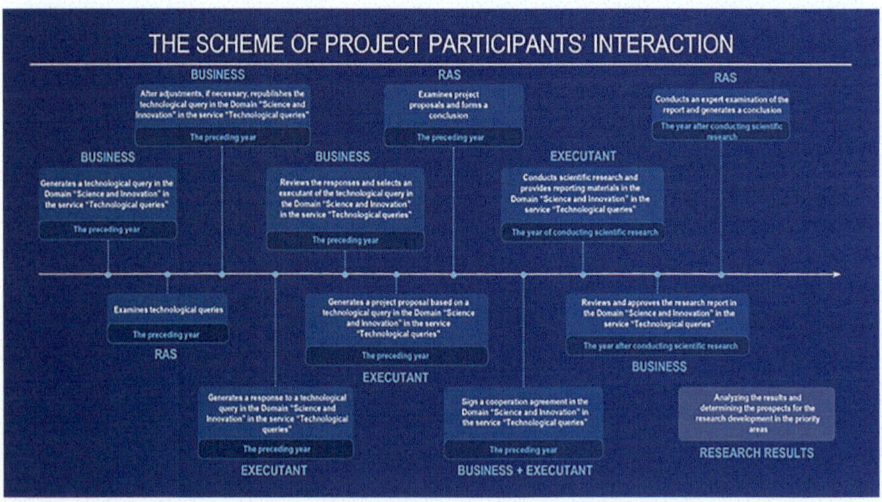

Fig. 1 The scheme of interaction of stakeholders of fundamental scientific research carried out in the interests of a qualified customer

more basic scientific research in the interests of the real sector of economy and high-tech industries in order to obtain new scientific knowledge for the creation of new technologies and solutions to practical problems.

3.2 *Testing the Proposed Changes*

An awareness campaign was conducted to engage potential qualified customers, educational and scientific organisations, and other interested parties in the experiment. Information materials were sent to five federal executive bodies (the Ministry of Industry and Trade of the Russian Federation, the Ministry of Health of the Russian Federation, the Ministry for the Development of the Russian Far East and Arctic, the Ministry of Energy of the Russian Federation, and the Ministry of Natural Resources and Environment of the Russian Federation), as well as to three state corporations: Rostec, Roscosmos, and Rosatom. Over 40 meetings and events were held with representatives of interested parties, including leading companies in the real sector of the economy, the Russian Academy of Sciences, scientific organisations, and higher education institutions. Webinars were also organised with the participation of the developer of the 'Technological queries from business' service within the 'Science and Innovations' domain, during which the implementation process of this tool was discussed. More than 1000 consultations were provided to potential project participants in response to emerging questions. An additional communication channel was subsequently created in the form of a group chat to coordinate interaction between stakeholders. This group chat included representatives of both qualified customers and executants. A wide range of issues was discussed, from operational matters related to work in the 'Science and Innovations' domain to legal consultations and announcements of thematic events dedicated to strengthening engagement between the business and scientific communities. The group chat helped to bring together various parties and identify shortcomings in the implementation of the tool and avenues for improvement.

As a result, key customer companies from a variety of industries and government agencies, in particular regional executive authorities, were involved in the experiment. The greatest interest was shown by organisations from the oil, gas, and petrochemical sectors (SIBUR Holding, Gazprom Neft), as well as from the engine manufacturing sector (United Engine Corporation and its subsidiaries). In response to the posted technological queries (more than 1500 in total), over 1500 responses were received from potential executants. Notably, universities accounted for a larger share of responses than academic institutes.

4 Discussion

Going back to the impact of government funding for basic research on innovation development indicators, it's worth noting that, according to Georgiou, Maksymenko, and Russo, government funding for scientific research and development in the US, the UK, Japan, and Spain has led to a substantial increase in private investment in Research and Development in the recent years. However, researchers consider the main effect of public investment in science and development to be the exchange of knowledge and interaction between companies and scientific organisations in joint innovation activities [13, p. 206]. Kim et al. [14] emphasise that priority should be given to the formation of public policy aimed at strengthening interaction between participants in the innovation process. Their findings are particularly relevant to our study because they are based on empirical research conducted in South Korea, a country where, similarly to Russia, the public sector plays a significant role in the national research system.

One of the main directions of state policy aimed at increasing stakeholder engagement in scientific research and development is platformisation. da Silva Neto [15] highlight the importance of network effects, which determine the high efficiency of scientific digital platforms as a tool and environment for interaction between stakeholders in a scientific project. Specifically, their study describes the history of the Lattes platform in Brazil, which was created as a government initiative and was initially used exclusively to collect and provide users with data on researchers and research projects. Over time, the government added features to encourage users to interact with each other. Kajal et al. [16] similarly describe the role of India's National Science and Technology Management Information System in the country's innovative development. According to their observations, the system's functions have long gone beyond meeting the purely informational needs of users. Todosiichuk acknowledges the significant role of the Russian EGISU NIOKTR in the interaction of participants in scientific and innovative activities [17]. The importance of the platform approach for organising scientific and technological cooperation at the inter-country level is further emphasised by Shabalina et al. [18]. Several authors have devoted their research to studying the communicative and social aspects of stakeholder engagement in innovation projects using scientific and technological platforms [19–22].

The examples above show that support for stakeholder engagement in scientific and, more broadly, innovative projects in the digital environment is being actively studied and developed. However, this issue is still rarely examined in the context of basic research. This may be due to the long-standing view that basic research is devoted exclusively to the generation of new knowledge and is initiated predominantly by scientists motivated by their own research interests. But the interpretation of basic science and its role in meeting the needs of society has broadened, leading to the emergence and promotion of various concepts reflecting the existence of a certain type of basic research carried out to solve practical problems, such as the HIBAR (Highly Integrative Basic and Responsive) concept [23]. Mechanisms for public support of basic research focused on practical needs have been created and are being developed in the USA, the EU, the UK, and Japan [24]. Russia has also joined this movement, and our study is proof of that.

5 Conclusion

The result of the study, as well as the pilot implementation of the proposed project developed on its basis to involve qualified customers in the planning, monitoring, evaluation, and approval of the results of state-funded basic research projects, is the creation of a functioning system of interaction between interested parties that has shown its first positive outcomes. The successful implementation of the system became possible because of the regulations for cooperation developed in collaboration with the authors of this chapter, as well as the expansion of the 'Science and Innovations' domain and the introduction of a dedicated service that facilitates the active participation of qualified customers in increasing the practical impact of basic research.

References

1. Atkinson RD, Moschella D (2024) Myth 38: Federal R&D crowds out private R&D. In: Technology fears and scapegoats. Palgrave Macmillan, Cham
2. Prettner K, Werner K (2016) Why it pays off to pay us well: the impact of basic research on economic growth and welfare. Res Policy 45(5):1075–1090
3. Akcigit U, Hanley D, Serrano-Velarde N (2020) Back to basics: basic research spillovers, innovation policy, and growth. Rev Econ Stud 88(1):1–43
4. Parui P, Prettner K (2025) Public provision of healthcare and basic research: what are the joint effects on economic growth and welfare? J Econ Growth 30:383–415
5. Bacq S, Aguilera RV (2022) Stakeholder governance for responsible innovation: a theory of value creation, appropriation, and distribution. J Manag Stud 59(1):29–60
6. Jia Y, Yu X, Yu X (2025) Stakeholder theory: basic theory explaining innovation-driven entrepreneurial action subjects. In: Cai L, Ge B, Li X, Yu X (eds) A theoretical foundation for innovation-driven entrepreneurship. Springer, Singapore
7. Dow S, Shi Y (2025) The end and beginning: the shift toward stakeholder capitalism. In: Corporate finance under climate crisis. Palgrave Macmillan, Cham
8. Miles S (2017) Stakeholder theory classification: a theoretical and empirical evaluation of definitions. J Bus Ethics 142:437–459
9. Ackermann F, Eden C (2011) Strategic management of stakeholders: theory and practice. Long Range Plan 44(3):179–196
10. Ribeiro SD, Wagner ME, Alves H, Raposo M (2012) A model for stakeholder classification and stakeholder relationships. Manag Decis 50(10):1861–1879
11. Order of the government of the Russian Federation No. 3684-r of 31 December 2020 'On the approval of the programme of fundamental scientific research in the Russian Federation for the long-term period (2021–2030)'. https://rulaws.ru/goverment/Rasporyazhenie-Pravitelstva-RF-ot-31.12.2020-N-3684-r/. Last accessed 27 Nov 2025

12. Decree of the president of the Russian Federation No. 145 of 28 February 2024 'On the strategy for the scientific and technological development of the Russian federation'. https://legalacts.ru/doc/ukaz-prezidenta-rf-ot-28022024-n-145-o-strategii/. Last accessed 27 Nov 2025
13. Georgiou O, Maksymenko M, Russo S (2025) R&D management and technology commercialization. Management for professionals. Springer, Cham
14. Kim ES, Choi Y, Byun J (2024) Big data and network analysis in national innovation systems: the roles of academia, industry, and government research institutes and their interactions. In: Hung JC, Yen N, Chang JW (eds) Frontier computing on industrial applications volume 4. FC 2023, Lecture notes in electrical engineering, vol 1134. Springer, Singapore
15. da Silva Neto VJ, Chiarini T (2023) The platformization of science: towards a scientific digital platform taxonomy. Minerva 61:1–29
16. Kajal M, Trikha R, Singh K (2024) A comprehensive study of the governance of India's scientific, technological, and innovative endeavors. In: Singh K, Chongtham N, Trikha R, Bhardwaj M, Kaur S (eds) Science, technology and innovation ecosystem: an Indian and global perspective. Springer, Singapore
17. Todosiichuk AV (2025) State stimulation of creation processes and the use of scientific and technical products. Sci Tech Inf Proc 52:102–109
18. Shabalina L, Shavkun G, Timko M (2025) Support for scientific and technical cooperation of the EAEU countries through the creation of a digital technological platform. In: Tao F, Gadekallu TR, Kumar V, Akberdina V, Kuzmin E (eds) Artificial intelligence and digital transformation, Lecture notes in information systems and organisation, vol 78. Springer, Cham
19. Fecher B, Kahn R, Sokolovska N, Völker T, Nebe P (2021) Making a research infrastructure: conditions and strategies to transform a service into an infrastructure. Sci Public Policy 48(4):499–507
20. Smolina ES, Mantulenko AV (2025) Architecture of digital transformation of science. In: Mantulenko VV, Horák J, Kučera J, Ayyubov M (eds) Proceedings of the international scientific conference "digital future: science, education, and innovative development of socio-economic systems". DFSED 2025. Lecture notes in networks and systems, vol 1552. Springer, Cham
21. Bracci E, Bruni ME (2025) Stakeholder engagement and the mindset open to innovate in the digital era. In: Hazır Ö, Bruni ME (eds) Tomorrow's data empowered project management. Future of business and finance. Springer, Cham
22. Loureiro SMC, Romero J, Bilro RG (2020) Stakeholder engagement in co-creation processes for innovation: a systematic literature review and case study. J Bus Res 119:388–409
23. Whitehead LA, Slovic SH, Nelson JE (2020) Re-invigorating HIBAR research for the 21st century: enhancing fundamental research excellence in service to society. Technol Innov 21(2):153–167
24. Dezhina IG (2020) Transformational research: new priority of the state after the pandemic. E.T. Gaidar Institute for Economic Policy Foundation, Moscow
25. Mitchell RK, Agle BR, Wood DJ (1997) Toward a theory of stakeholder identification and salience: defining the principle of who and what really counts. Acad Manag Rev 22(4):853–886

Artificial Intelligence in Russian Healthcare: Current Status and Application Prospects

L. V. Larchenko, A. I. Demina, M. Gabibova, and A. N. Paranina

Abstract Russia is undergoing a digital transformation across various sectors of the economy. One of the key areas is the implementation of artificial intelligence in the healthcare system. Particularly significant progress has been observed in the implementation of artificial intelligence in medical image interpretation. The aim of this study is to analyse the challenges arising from the implementation of artificial intelligence in the Russian healthcare system, including the preparedness of medical professionals to use artificial intelligence in their practice. The study utilized a content analysis of scientific literature to identify key concepts, trends, and key areas of AI technologies successfully applied in medicine. To analyse perceptions of and readiness for the application of artificial intelligence technologies in clinical practice, a sociological study was conducted using in-depth interviews. Physicians from various specialties across Russia participated in the survey. The study assessed physicians' attitudes toward AI technologies, their willingness to work with such technologies, and their level of understanding of their operating principles. It also provides insight into the key organizational aspects of using artificial intelligence in medicine. The use of artificial intelligence systems in clinical practice offers potential for improving the quality of diagnostics and preventive medicine.

Keywords Artificial intelligence · Digital technologies · Healthcare · Digital skills of healthcare workers · Telemedicine

L. V. Larchenko (✉) · M. Gabibova · A. N. Paranina
Herzen State Pedagogical University of Russia, St Petersburg, Russia
e-mail: lubalar@mail.ru

A. I. Demina
HSE University, St Petersburg, Russia

© The Author(s), under exclusive license to Springer Nature Switzerland AG 2026

A. Kulachinskaya et al. (eds.), *Digital Transformation: What Kind of Smart Life do we have today?*, Lecture Notes in Networks and Systems 1895,
https://doi.org/10.1007/978-3-032-20582-7_9

1 Introduction

With the rapid digital transformation of modern society, the integration of artificial intelligence (AI) technologies into all sectors of the economy, including healthcare, is becoming increasingly important. The Fourth Industrial Revolution, characterized by new technologies of recent decades such as the Internet of Things, big data technologies, robotics, and artificial intelligence, is creating fundamentally new opportunities for improvement of medical diagnostics and treatment.

Modern academic circles demonstrate significant differences of opinion regarding the precise definition of artificial intelligence. John McCarthy, a pioneer in the field, believes: "It is the science and technology of creating intelligent machines, especially intelligent computer programs. Artificial intelligence is concerned with the task of using computers to understand the workings of human intelligence, but is not limited to using methods observed in biology" [1]. One of the key goals of creating artificial intelligence technologies has become the imitation of human intelligence. In the development of artificial intelligence and its use in non-IT sectors, the COVID-19 pandemic has become a key turning point, catalysing the development of digital medicine and the implementation of cutting-edge technologies, particularly artificial intelligence. Healthcare in Russia and globally faces similar challenges due to the specific nature of the industry, including a shortage of qualified specialists, a high workload for medical professionals due to maintaining patient records, significant budgetary costs for maintaining the industry, and the need to improve the quality of medical services. A study by Burkovskaya et al. revealed that over 70% of surveyed medical workers report high levels of workload associated with functions not directly related to the treatment and diagnostic process [2]. AI technologies can become a healthcare problems solving tool, as they can reduce medical costs, improve the quality of diagnosis and treatment, optimize medical professionals' work by automating routine tasks, and expand medical coverage in remote regions.

The use of artificial intelligence in healthcare can lead to a transformation of the labour market. In this regard, French researcher Moustapha Zouinard studied the unprecedented performance of AI, leading him to conclude that AI technologies will inevitably spread across all socioeconomic sectors of public life [3]. As part of his research, Zouinard developed a model for synergistic collaboration between human and artificial intelligence. His concept is based on the principle of complementarity, where artificial intelligence acts not as a substitute, but as a tool to expand human capabilities. The key thesis of the work is the assertion that humans must retain strategic management functions, while artificial intelligence should act as a booster of cognitive and analytical abilities.

The objective of the study is to analyse the challenges arising from the implementation of AI in the Russian healthcare system, including the preparedness of medical professionals to use artificial intelligence in their practice.

2 Research Materials and Methods

The theoretical and methodological basis of the study is based on the fundamental works of Russian scientists devoted to the digital transformation of medicine. In particular, the work of Alikperova on the prospects and risks of implementing artificial intelligence technologies in healthcare [4] and Sharova's research on the ethical aspects of interactions between medical professionals and artificial intelligence systems [5]. The economics of artificial intelligence, particularly its application in healthcare, were examined in the research of Lukichev [6].

As part of the study of international experience, the works of foreign researchers were analysed, including Trakhtenberg's concepts on the professional communities adaptation to technological innovations [7] and Zouinard's theoretical models revealing the principles of optimal human interaction with artificial intelligence in the professional environment [3].

The empirical basis was a content analysis of the scientific literature: international and Russian experience in the application of artificial intelligence in healthcare in public and private clinics. Content analysis allowed us to identify key concepts, trends, and priority vectors for the development of AI technologies. Qualitative research methods were also used: in-depth interviews with 23 physicians of various specialties from across the Russian Federation were conducted to analyse the perceptions and readiness of medical professionals to use artificial intelligence technologies in clinical practice.

3 Research Results

3.1 *Key Applications of Artificial Intelligence in Healthcare*

Analysis of international and Russian experience with artificial intelligence in healthcare has identified four categories of artificial intelligence solutions in medicine: digital assistants that simplify routine procedures for medical professionals; diagnostic platforms that operate based on medical image analysis; robotic systems used in surgical interventions; and telemedicine technologies.

Digital Assistants. Modern digital assistants are presented in the form of chatbots and voice assistants powered by artificial intelligence algorithms. The primary objective of such systems is to optimize the work of medical professionals by automating standard processes, including medical history collection and data recording. An important area of focus is accelerating the rapid allocation of patients based on the severity of their condition to ensure timely assistance. Furthermore, these solutions provide synchronization with electronic medical records, allowing for the transfer and systematization of information in a standardized format. The most popular functions of digital assistants both abroad and in Russia include: automatic filling of electronic medical records, preliminary diagnosis; differential diagnosis of patients using clarifying questions; processing and transcribing voice input into text. While digital assistants aimed at optimizing and automating physician workflows were presented above, chatbots are also being developed to assist patients. A virtual chatbot cannot replace a full consultation with a physician, but it can perform the following tasks: analysing patient complaints; generating recommendations for next steps—seeking medical attention or treating at home; and monitoring patient conditions, which is especially important during treatment for mental illnesses.

Diagnostic Platforms. Platform solutions based on artificial intelligence for diagnosing and predicting diseases have been developed and are being used in various regions of the world, including Asia, Europe, and North America. Modern artificial intelligence algorithms demonstrate high efficiency in identifying abnormalities and predicting the development of pathologies in such fields of medicine as endocrinology, neurology, oncology, cardiology, physical medicine, gynecology, urology, and dentistry. These solutions primarily work with medical imaging, including CT, MRI, ultrasound, and X-ray data. Using computer vision methods and neural network models, the systems analyse images and generate preliminary diagnostic reports.

Robotic Systems. These systems operate on the basis of artificial intelligence. These systems combine high-precision mechanics, advanced data processing algorithms, and AI-based surgical procedure management systems, enabling surgical procedures to be performed with minimal patient intervention. The main advantage of such systems is their ability to provide unprecedented precision of surgical manipulations, unattainable with traditional surgical methods. The functionality of modern robotic systems includes several key capabilities. Firstly, they are equipped with AI algorithms that help surgeons plan optimal access to the surgical site, taking into account the patient's anatomical features. Secondly, special stabilization systems eliminate even the slightest involuntary movement of instruments, which is especially important when working with microscopic structures. Many robotic systems are capable of performing complex surgeries through minimal incisions, significantly reducing the risk of postoperative complications and shortening the recovery period.

Telemedicine Technologies. These innovative solutions combine advanced data analysis technologies with remote interaction between doctor and patient. The key difference between AI-based telemedicine technologies and remote consultations with a doctor is that the former involves the interpretation of laboratory test results and medical images, and the generation of a preliminary diagnosis by AI, followed by a physician review of the results. The main advantage of such systems is their ability to significantly reduce the time required to establish a preliminary diagnosis, while simultaneously reducing the workload of medical specialists. It is important to emphasize that these technologies do not replace the work of specialists, but rather act as a powerful tool to support medical decision-making. This type of technology expands a physician's capacity to handle the number of patients they can see, thereby increasing medical coverage for the largest population. This approach is particularly valuable. This provides residents of remote regions with access to high-quality diagnostics without the need for long trips to medical centers.

Thus, based on the content analysis, it can be concluded that AI solutions for diagnosing and identifying patient pathologies, as well as AI-powered digital assistants, have the greatest potential in healthcare. Robotic AI systems for surgical interventions also hold great promise, but there are already cases of patient injuries during surgery. Telemedicine technologies are at the peak of their development, and many private clinics in Russia are already implementing telemedicine consultations with specialists.

3.2 A Sociological Study of Medical Professionals' Opinions on the Possibilities of Using Artificial Intelligence in Healthcare

In order to explore the potential of using artificial intelligence in healthcare, a sociological study of medical professionals' opinions was conducted. This qualitative study identified socio-psychological barriers that may hinder and slow the adoption of artificial intelligence technologies in the medical sector down. Twenty-three physicians from eight regions of the country participated in the survey. Their professional experience ranged from 5 to 25 years, ensuring coverage of different generations of medical experience. Participants were selected with gender and age balance in mind, and represented both public and private medical institutions across the Russian Federation. Data was collected through in-depth interviews conducted in person and online. The interview methodology utilized D. Silverman's qualitative research approach [8]. The final sampling criterion was reaching "thematic saturation"—the point at which new interviews ceased to yield significantly different information and responses began to repeat themselves. As a result, 23 respondents were interviewed, which corresponds to generally accepted standards for qualitative research in medical sociology.

Medical specialists from various fields identified specific areas in which artificial intelligence could significantly optimize their professional activities. In particular, a physiotherapist noted the need to automate the process of preparing legally significant documents, especially patient consent forms. Dentists unanimously highlighted the importance of implementing automated medical data recording systems, citing the existing Voice2Med solution, which converts doctor dictation into structured medical records [9]. Of particular interest are the proposals of psychotherapists, who see the potential of AI in automating the creation of individualized cognitive behavioural therapy protocols. Therapists, in turn, are willing to delegate a whole range of organizational and diagnostic functions to artificial intelligence systems:

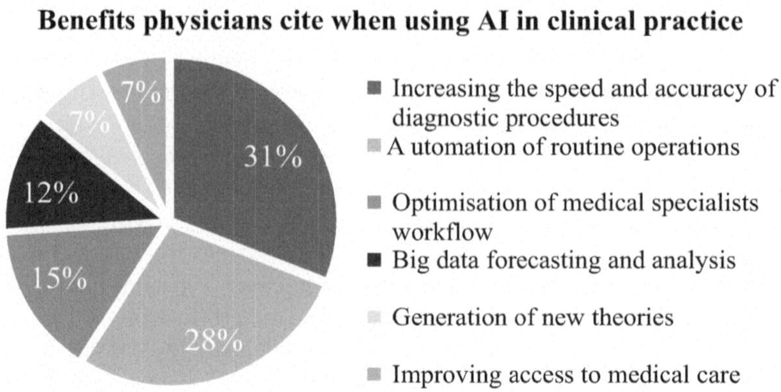

Fig. 1 Benefits identified by physicians when using AI in clinical practice. *Source* Compiled based on in-depth interviews

from automated patient monitoring and medical image analysis to the initial systematization of symptoms. An analysis of medical opinions allows us to classify the potential applications of AI in medicine into three key areas: automation of routine administrative processes, support for medical decision-making (including medical image interpretation), and comprehensive analysis of large medical data sets for epidemiological forecasting and personalized treatment approaches.

The data presented in Figs. 1 and 2 were obtained using qualitative content analysis methods. The initial transcribed in-depth interviews were segmented into semantic units, followed by the identification of key thematic areas. Thematic analysis structured the empirical data to identify recurring semantic patterns. Key conceptual propositions (key points) identified during the analysis were systematized into thematic clusters. The quantitative data representation in diagrams reflects the frequency of each key idea mentioned by respondents as a percentage.

Analysis of the survey results revealed the benefits of using artificial intelligence in clinical practice, as identified by study participants. A clear hierarchy of importance was identified. The dominant factors were increased speed and accuracy of diagnostic procedures (31%) and automation of routine procedures (28%). The smallest group (7%) included aspects such as generating new hypotheses and theories and adjusting treatment regimens, as well as a similar group (7%) that noted increased access to medical care. The obtained results reflect not only the current priorities of practicing physicians regarding digital technologies but also the potential development vectors for medical AI solutions.

Let's consider the risks identified by medical professionals (Fig. 2).

The most frequently cited risk was the fear that AI errors could lead to negative consequences for patients (33%). The next most important risk was the devaluation of medical expertise due to the widespread use of AI technologies in healthcare (17%). Interviewees also identified the risk of a violation of patient confidentiality due to potential data leakage (14%). Next, they highlighted the risk of a decline in physician

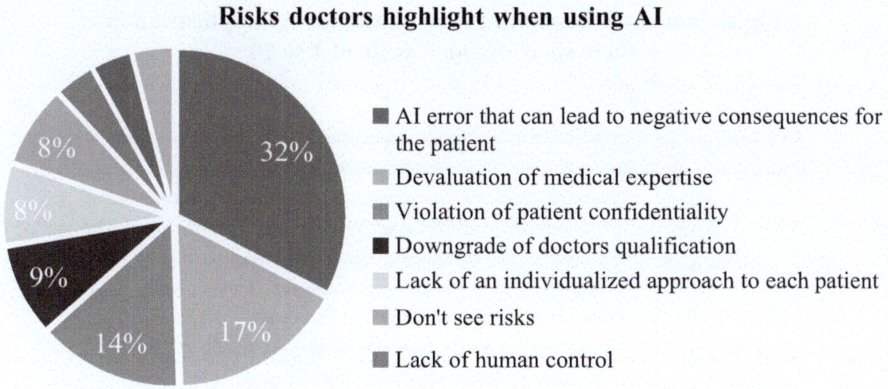

Fig. 2 Risks identified by doctors when using artificial intelligence. *Source* Compiled based on in-depth interviews

qualifications, including a decline in the quality of clinical judgment among medical professionals (9%). Additionally, a lack of an individualized approach to each patient was noted (8%), while similarly, no risks were present (8%). 4% each cited the lack of human oversight, lack of accountability to patients, and the fear of patients self-medicating without consulting a doctor based on their independent use of AI.

At the end of the interview, participants were asked to answer the question: on a scale of 1–10, how much do they favour or oppose the use of artificial intelligence in healthcare (where 1 is the most opposed, 10 is the most in favour). The survey results are presented in Fig. 3.

As Fig. 1 shows, the most positive outlook was shared by physicians in the following specialties: rehabilitation therapist, epidemiologist, surgeon, pediatrician, endocrinologist, psychotherapist, hygienist, and dentist. This is explained by the fact that these specialties involve a significant amount of work related to the interpretation of medical images. During interviews, physicians noted that this type of technology could be highly sought after, as it would help eliminate human error. Meanwhile, the lowest assessments of the potential for using artificial intelligence in their work were given by psychiatrists, physical therapists, intensive care specialists, neurologists, and clinical psychologists. This result can be explained by the fact that these specialties require a physician's subjective assessment of the patient's condition, as well as skills such as intuition, empathy, and clinical reasoning. At the current stage of technological development, these skills cannot be described in protocols and transferred to artificial intelligence.

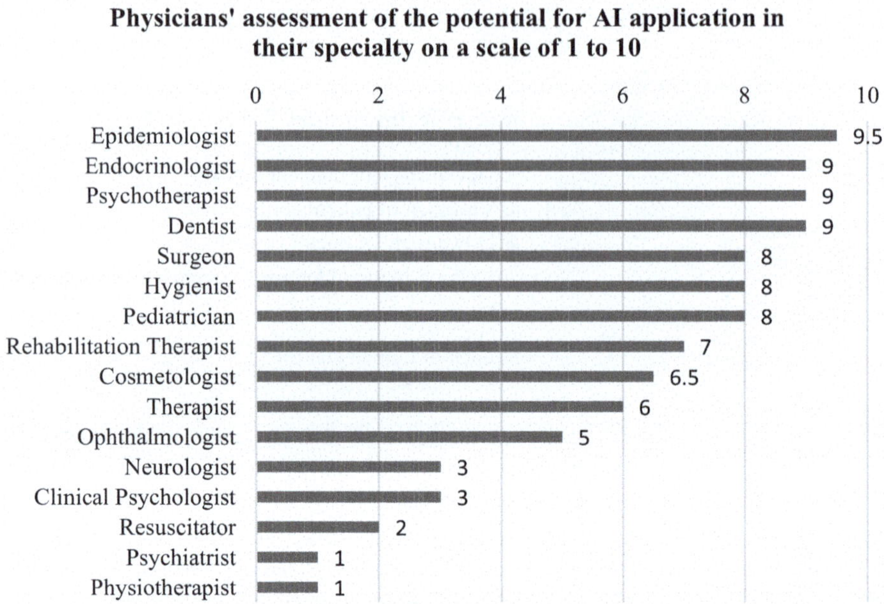

Fig. 3 Physicians' assessment of the potential of AI usage in their specialty on a scale of 1–10. *Source* Compiled based on in-depth interviews

4 Discussion and Conclusion

Despite the limited sample size, which prevents the study's findings from being generalized to all physicians in Russia, the collected data provides insight into key organizational aspects of using artificial intelligence in medicine. Specifically, the study assessed physicians' attitudes toward AI, their willingness to work with such technologies, and their level of understanding of their operating principles.

The results showed that most medical professionals positively perceive the implementation of AI in clinical practice. However, an important nuance emerged: many physicians confuse the concepts of automated systems operating according to a predetermined algorithm with full-fledged artificial intelligence, indicating the need for additional training aimed at developing the digital skills of medical professionals. During the interviews, a list of medical tasks was compiled where the use of AI could significantly optimize workflows. Of particular interest were the physicians' innovative ideas for using AI, as well as their clear vision of which medical data could be useful for training algorithms.

The main conclusion of the study is the need for an active dialogue between the developers of AI-based systems and the medical community. Physicians, as direct users of technology, can not only identify the most in-demand features but also warn of potential risks. Among the main concerns, respondents cited:

1. The risk of AI errors with serious consequences for patients;

2. A decrease in the value of medical expertise due to excessive reliance on algorithms;
3. Issues with protecting the privacy of medical data.

These results confirm that when creating and implementing AI solutions in healthcare, it is critical to establish transparent communication between technology experts and medical professionals [10]. Furthermore, the development of AI-based medical technologies is stimulated by related industries—from IT development to the production of specialized equipment—creating new areas of economic growth.

Modern research confirms that artificial intelligence technologies are actively integrating into global medical practice, shaping a new paradigm for medical care. Rational and effective use of artificial intelligence systems in clinical practice opens up prospects for improving the quality of diagnostics and preventive medicine. Systemic digitalization of healthcare, including the implementation of AI solutions, creates the preconditions for a qualitative improvement in medical services: accelerating diagnostic processes, developing individualized treatment protocols, and ensuring the accessibility of high-tech care for various population groups. In the long term, such transformations can not only modernize the work of medical institutions but also have a significant impact on socio-demographic trends, specifically, contributing to increased life expectancy.

References

1. McCarthy J (2007) What is artificial intelligence. https://www-formal.stanford.edu/jmc/whatisai/whatisai.html. Last accessed 27 Nov 2025
2. Burkovskaya YV et al (2022) Attracting specialists with non-medical education to work in a medical organization. Curr Probl Health Care Med Stat 5:437–461
3. Zouinar M (2020) Évolutions de l'Intelligence Artificielle: quels enjeux pour l'activité humaine et la relation Humain Machine au travail? Activités 17–11. http://journals.openedition.org/activites/4941. Last accessed 27 Nov 2025
4. Alikperova NV (2023) Artificial intelligence in healthcare: risks and opportunities. City Healthc 4(3):41–49
5. Sharova DE et al (2021) On the ethical aspects of implementing artificial intelligence systems in healthcare. Digit Diagn 2(3):356–368
6. Lukichev PM, Chekmarev OP (2022) Economics of artificial intelligence: possibilities and problems of use in healthcare. Issues Innov Econ 12(2):1111–1130
7. Trajtenberg M (2018) Artificial intelligence as the next GPT: a political-economy perspective. In: Agrawal A, Gans J, Goldfarb A (eds) The economics of artificial intelligence: an agenda. University of Chicago Press, Chicago and London, pp 175–186
8. Silverman D (2000) Doing qualitative research. In: Sage Publications Ltd, Thousand Oaks, CA
9. Kryuchkova ES, Egorov EN (2023) Speech and handwriting recognition technologies. In: Young people and science results ants perspectives. Collection of materials from the all-Russian scientific and practical conference of students and young scientists with international participation, Saratov, pp 216–217. Saratov State Medical University named after V.I. Razumovsky, Saratov
10. Baker A et al (2020) A comparison of artificial intelligence and human doctors for the purpose of triage and diagnosis. Front Artif Intell 3:543405

Co-Evolution of Social and Financial Technologies: The Present Stage

Viktoriya Razletovskaia and Igor Stepnov

Abstract The advent of modern financial technologies must be contemplated beyond the confines of the financial sector, but rather within the overarching paradigm of a global system. This macro-level analysis unveils the fundamental principles that govern the evolution and interconnectedness of the network system. On a global scale, they exhibit a convergent interconnectedness with social technologies. Utilizing an evolutionary approach to the development of financial innovations and their interconnection with scientific and technological progress, as well as a civilizational approach to the dynamics of the development of social technologies and management theories, the chapter elucidates the specifics of the modern stage of their co-evolution: the general direction and continuity of their development. Utilizing the paradigm of the present stage of their practical implementation, a taxonomy of categories of Social financial technologies is conducted, which predetermines the necessity to regulate novel actors of convergent technologies and their roles and interaction processes.

Keywords The coevolution of financial and social systems · Social Fintech · Interdisciplinary approaches · Use of AI and neurotechnology in fintech

1 Introduction

The increasing mutual influence of social and financial systems and their technologies has been confirmed both in theoretical studies by many authors over the centuries and in practical terms at the current stage of international practice in the development of financial and technological processes. The authors reveal the interdisciplinary nature of financial technologies and confirm the evolutionary perspective on development—the formation and spread of financial innovations and social technologies. Financial innovations are considered as key components of developing social technologies—a set of institutions, norms, legal frameworks that organize human interactions.

V. Razletovskaia (✉) · I. Stepnov
MGIMO University, Moscow, Russian Federation
e-mail: dlglab@inno.mgimo.ru

© The Author(s), under exclusive license to Springer Nature Switzerland AG 2026

A. Kulachinskaya et al. (eds.), *Digital Transformation: What Kind of Smart Life do we have today?*, Lecture Notes in Networks and Systems 1895,
https://doi.org/10.1007/978-3-032-20582-7_10

The geneology of this approach originates in the fundamental schools of economic thought: German historicism Hildebrand [1], rejecting the ideas of universal non-historical economic laws, without taking into account national and religious characteristics, and American institutionalism Veblen [2], who coined the term "evolutionary economy," considering the reasons for the development of institutions before the legal innovations of Hugo Grotius [3]. Razletovskaya [4], constructing transnational corporations as a subject of international law. Going beyond natural law and legal design, he demonstrated how fundamental legal doctrines are critical social technologies that make possible and shape specific regimes of financial organization and global coverage.

Modern researchers Wallerstein [5] and Arrighi [6] synthesized historical and institutional ideas for the analysis of global capitalism and the modern global financial system as part of a historical social system with age-old trends. So Arrigi [6] confirms and highlights the patterns about the increased impact of social factors on the accumulation regime with the course of evolution.

Complementing the study of historical social systems with cyclical rhythms, Perez [7] argues that each technological surge requires a deep restructuring of the institutional and social structure—the "socio-institutional sphere." The latest works of sociologists, philosophers, historians and anthropologists Harari [8], Prokhorov [9], Chuprov [10], Ferguson [11], Bjerg [12], Zizek [13], continue the integrative evolutionary tradition, exploring the nature of money as a certain order, the organization of socio-economic relations and approaches to understanding it.

The patterns of development of financial and social technologies have been studied taking into account cluster-network theory [14] and behavioral theories of German, American, and Soviet sociology (behavioral and irrational finance, cognitive psychology—Goldobina et al. [15], Parsons [16], Goldner [17], Kirdina et al. [18]). Some of the latest works published on the subject of research are noteworthy: "Other People's Money: The Real Business of Finance" by Kay [19], "The origin of wealth: The Radical Remaking of Economics and What it means for business and society" by Beinhocker [20], and "FinTech: Finance, Technology, and Regulation" by Buckley et al. [21].

The evolutionary perspective of financial and social technologies confirms that financial instruments from credit systems and joint-stock companies to digital assets are designed inside, influenced and shaped by social technologies, legal doctrines, political institutions, cultural norms and values.

The current stage of convergence of social and financial technologies actualizes theoretical studies of their interconnectedness and determines the continuation of the study of co-evolution, the diversity of their types in order to develop adequate mechanisms for regulating their development both at the global and national levels.

2 Methodology

The study comprehensively and systematically examines the evolutionary dynamics of convergent technologies, taking into account the civilizational approach, as well as the managerial and institutional approaches at both the national and international levels. The influence of convergent technologies on social interaction processes and their forms, as well as processes of managing fintech development at the global level, are examined. Assessing the influence of the emerging financial and technological infrastructure on social interaction processes and forms requires additional study and detailed research beyond the scope of this article.

Updating the evolutionary approach originally outlined by Arrighi [6] in systemic cycles of capital accumulation and I. Wallerstein's world-system approach [5] as applied to the development of financial technologies, and expanding them with an interdisciplinary, managerial, and civilizational approach, the study examines, alongside capitalist ones, financial and economic relations, systems, and their technologies determined by religious orders, social engineering, and socialist regimes.

At the same time, it is necessary to take into account the limited possibilities of using full-fledged network analysis in the field of financial and social technologies (based on the index of centrality, interconnectedness, and other indicators), due to the lack of reliable and complete databases.

Using evolutionary and civilizational approaches, the stages of development and classification of social technologies as an element of economic sociology have been identified as a rethinking of dominant models of development. The financial aspects of social technologies and forms of their concentration have been studied in accordance with the identified stages. The processes of co-evolution of social and financial technologies, manifested in common principles of direction and continuity, as well as in the development of financial and technological infrastructure, have been studied. This allowed us to conclude that the current stage of convergence of social and financial technologies represents the coordination of organized financial redistribution systems and processes, using financial and social technologies, and its technological consolidation at both the global and national levels. The classification of types of financial technologies from the point of view of social aspects shows that the regime of capital accumulation, due to the strengthening of the influence of social factors, is evolving into an intellectual and technological industry, a cognitive economy of attention, and the coordination of the creation and consolidation of socio-economic relations of resource redistribution. This will require appropriate directions and mechanisms for regulating these processes.

3 Results

Financial and social systems are forms of organization of social and financial relations between all subjects of the reproduction process for the redistribution of resources. The main mechanisms and instruments for their formation and functioning are financial and social technologies [4].

The history of the use of innovations and technologies in the field of financial and economic relations, from the innovative technology of money as an equivalent of exchange to financial engineering, financial management, the institutional sphere, the use of neurotechnologies and artificial intelligence (see Table 1), shows that fintech, by its nature as a model, algorithm, is aimed at forming new interrelationships [22]. It does not mediate social interactions in the understanding of classical finance, but constructs and forms them [23]. In this study, fintech is considered not as a financial innovative means of payment, but primarily as a way of organizing socio-economic relations in a certain order and as a form of new institutionalism in sociology [23].

The essential characteristics of fintech are constructivism and the creation of value through redistributive processes that use networked forms of power [14]. The movement of value and managerial content of fintech—the need for control over socioeconomic relations and planning through the creation of interdependencies—have led to the concentration of finance and power, development of financialization, and its institutionalization.

Forms of concentration of financial capital and power developed:

- From religious orders and wars to the creation of coordinating supra-religious and supranational organizations and settlements;
- From the consolidation in space and time of local patterns of capital accumulation cycles in the dynamics of Arrighi [6] to the processes of institutionalization;
- Through the development of imperialism and hegemony;
- The convergence of social and financial technologies, socialist finance with the state planning system;
- Corporatism, concentration of finance at the state-corporate level and the level of transnational corporations;
- Globalization and digitalization processes;
- The development of modern forms of fintech and processes of financial and technological integration.

Their development determined the tasks of social and management technologies.

Global markets produce results when they are managed (historical examples: formally organized and regulated Venetian capitalism, informally organized and regulated Genoese diaspora capitalism, transnational organized structures, the Dutch revival of regulated capital through registered joint-stock companies and institutions [6]).

Cosmopolitan organized structures play a special role in their creation—transnational nations, local civilizations, diasporas.

Table 1 Leading financial innovations and technologies, stages of financialization and capital concentration, and accumulation cycles according to G. Arrighi

	Leading financial innovations and technologies	Financialization and concentration of capital	Cycles of accumulation Giovanni Arrighi
Mobilization model before the fifteenth century	Financial innovation in government and religious governance, birth of banking fourteenth century emergence of cross-border nations	Technologies of concentration—Wars and political and religious conflicts. Money means of exchange, control and mobilization of flows. Religions as supranational mechanisms of organization of relations. sixteenth-seventeenth century creation of supra-religious organizations based on personnel management creation of non-national settler states	Genoese-berian (15 early 18 century)
Redistribution-accumulation model of the fifteenth-nineteenth centuries (investment and marketing model) Capital accumulation cycles	1. Use of mathematical methods in finance Development of banking—institutionalization 2. Investment financial technologies of capital concentration, Financing wars,-leading bond technology applied to ensure development and accumulation processes 3. Financial technologies began to be applied in symbiosis with social technologies. financial technologies appear, providing regulation and direction of investment processes in the right direction, stimulating demand from customers and supply from producers	Concentration technologies—financial crises, devaluation inflation, speculation, credit bubbles and market oversaturation, loans, spending for military purposes Late 19th-early twentieth century globalization, interdependence international specialization Institutionalization—financial markets, exchange floors, banks, stock markets, national banking systems	Dutch mid-seventeenth century-late eighteenth century) British early XIX century-early XX century
		Concentration of finance at the state-corporate level (Germany) and corporatism of transnational corporations. creation of socialist finance USSR, China. Technologies of concentration—global world level of division and specialization in the technological cycle, social engineering	American—Early XX century
		Concentration of finance at the state-corporate level (Germany) and corporatism of transnational corporations. creation of socialist finance USSR, China. Technologies of concentration—global world level of division and specialization in the technological cycle, social engineering	

(continued)

Table 1 (continued)

	Leading financial innovations and technologies	Financialization and concentration of capital	Cycles of accumulation Giovanni Arrighi
Model globalization model financial engineering, financial wars, financial digitalization model	Financial engineering[1] tools and technologies based on stock market instruments, Unification of international financial instruments, global standards of financial behavior. Infrastructure for their servicing—international financial corporations operating in global markets	Mid twentieth century Bretton Woods system, financial engineering, global governance, creation of socialist camp **Concentration technologies**-Global financial crises- banking, mortgage, sanctions wars, scenarios, etc.	
Cognitive-social-emerging model	Cognitive financial technologies, Digital financial technologies Orientation on redistribution processes The next stage is conditioned by a qualitatively new level of service using neuroassistants and neurocomunications in fintech The future of fintech is AI, neurotechnologies and databases. Analytics	Digitalization stage, digital currencies, Bitcoin, AI Neurotechnology, Big Data new business models-fragmentation of international financial system, development of fintech, Multi-agent systems	

Source Compiled by the authors using sources [1–29]

[1] *Leveling cross-border risks and providing insurance and long-term planning of financial flows.*

Social technologies in research are considered from the perspective of a managerial and institutional approach within the framework of economic sociology as a means of social transformation of socio-economic relations in social systems and structures, as well as socio-engineering activities—the organized, sequential, and step-by-step implementation of actions that ensure the specified dynamics of the development process and are aimed at achieving goals.

With the convergence of social technologies, scientific and technological progress (AI, neurotechnologies, bio-technologies), financial innovations, and technologies are increasingly focused on redistribution processes. Finance not only concentrates resources to ensure control over power, but also shows signs of convergent redistributive technology, which are the financial aspects of social technologies. This confirms the patterns identified by Arrighi [6] regarding the strengthening of the influence of social factors on the regime of capital accumulation as evolution progresses.

Thus, at the current stage of development, the convergence of social and financial technologies is changing the priorities of national financial regulators and international organizations. Central banks, along with focusing on risk and inflation management, under the coordination of international financial organizations, are supplementing their activities with the promotion of the financial aspects of social technologies: green fintech, ESG investment agenda, and blended finance. The priorities of international organizations are to concentrate resources on the environmental and social agenda.

German sociology initially occupied a special place in world sociology, followed by American sociology. Since the beginning of the twentieth century, the main directions of social technologies have been theories and models: Taylor's management theory, Elton Mayo's human relations, statistics, and structural functionalism. The Frankfurt School, criticizing capitalism, pays special attention to authoritarianism and totalitarianism. It explores issues of power and knowledge, network society, colonial problems, racism, feminism, the environmental agenda, gender and digital sociology networks. In management theory, a modern bioalgorithmic model is being formed, based on strategic management and a systematic approach, social engineering, human resource management, geopolitics, and the intellectual industry.

By concentrating public, corporate, and private financial resources on social technology priorities, including those coordinated at the level of international organizations as representatives of institutionalism, the differences in the processes of resource concentration and redistribution at the level of capitalist and socialist systems are gradually being leveled out, and specialization in the global economic structure is intensifying. Since its inception, capitalism has been aimed at eliminating competition, establishing interdependence and control, while technological rationality in a broad sense determines and transforms into requirements for strict planning and control of the process in order to eliminate disruptions in supply and sales, otherwise the risks would increase. The history of capitalism is a struggle against this uncertainty through market manipulation [24, 25]. Historical examples: In social technologies, since the mid-twentieth century, there has been a transition to a planned economy at the level of capitalist states, with the emergence of social programs such as vaccination, school education reforms, and socialist finance. At the end of the twentieth century, digital platforms, electronic budgets, and social networks appeared, and control and coordination intensified and became more concentrated (Table 2).

The common features of co-evolution and the patterns of development of financial and social technologies are manifested in the unified principles of direction and continuity, the integrity of their interconnection with the concentration of capital and financialization, and financial and technological integration.

Due to the strengthening influence of social factors, the regime of capital accumulation [6] is evolving into an intellectual and technological industry that coordinates the creation and consolidation of socio-economic relations and the redistribution of resources [4]. Social technologies demonstrate that education and the intellectual industry influence the consumption of information and materials and resemble religious orders.

The identified dynamics of coevolution, based on a process- and actor-network-oriented approach, show how forms of resource concentration and control have changed due to the convergence of social and financial technologies into organized financial redistribution systems at the global level (see Fig. 1).

The methods and technologies for managing the development of financial technologies are changing: from individual methods to the creation of convergent management technologies, organized financial redistribution systems and processes, including scenarios, sequences of transformations of methods and technologies for organizing financial relations, processes at the global level [4]: financialization and its institutionalization, concentration of resources at the state-corporate level and the level of transnational corporations, creation of global management mechanisms—networks of international organizations—characterized by convergence of technologies and an inter-sectoral approach, scale of activity, and population coverage.

Cosmopolitan organized structures—transnational nations, local civilizations, and diasporas—play a special role in the creation of the global system of financial and economic relations, forming national models for the development of financial and social systems.

The leading procedures and methods of concentration and redistribution of resources, including the use of finance in historical retrospect, are presented from religious orders and principles of redistributive processes to capital accumulation cycles (Genoa-Iberian, Dutch, British, American), and to corporate-organized national accumulation structures (German states, Austria-Hungary, and Russia. German state corporatism. Creation of corporate-organized national structures of accumulation on the basis of concentration and redistribution of resource-socialist finances Soviet Union and China, creation of trans-state nations and cosmopolitan organized structures, international organizations, financial systems of socialist states—socialist camp).

Thus, levels are distinguished—from diasporas and local civilizations to the creation of cosmopolitan organized structures—transnational nations.

At the current stage of development, historical trends in the co-evolution of financial and social technologies are confirmed by processes of international integration,

Table 2 Co-evolution of financial and social technologies

Social Technology and Economic Sociology	Traditional methods until the twentieth century- customs, religious norms and practices, community governance, family upbringing, traditions, knowledge transmission. sixteenth-seventeenth century Creation of supranational and supra-religious organization and Non-national continental settler democracy USA (first Spaniards, English colonies 17th c) concentration of wealth and intellectual potential of Britain and Spain and France—eighteenth century-US constitution	End of eighteenth century classical social theories: corporatism (Pope 1891), class struggle K. Marx 1867, sociology as a science of norms, values and institutions, Max Weber role of religion and bureaucracy	Nineteenth century mass production technologies, health and welfare system, insurance, pensions. Early twentieth century theories and models, Taylor management theory, Elton Mayo human relations, statistics, structural functionalism, Frankfurt School—authoritarianism and totalitarianism, Power and knowledge. Castells network society, colony problems, racism, feminism, environmental agenda, gender and digital sociology-networks. Mid twentieth century-transition to planned economy at state level, socialist programs-vaccination reforms school education, socialist finance Late twentieth century. Digital platforms, e-budget, social networks, control and coordination	Constructivism of social models and processes, Intellectual industry, AI and big data, neurotechnologies, Blockchain, virtual reality
Historical stages Financial technologies	Mobilization model up to the fifteenth century	Redistribution-accumulation model of the fifteenth-nineteenth centuries (investment and marketing)	Globalization model financial engineering	Cognitive-social model—emerging model, financial digitalization

Source Compiled by the author using sources [1–29]

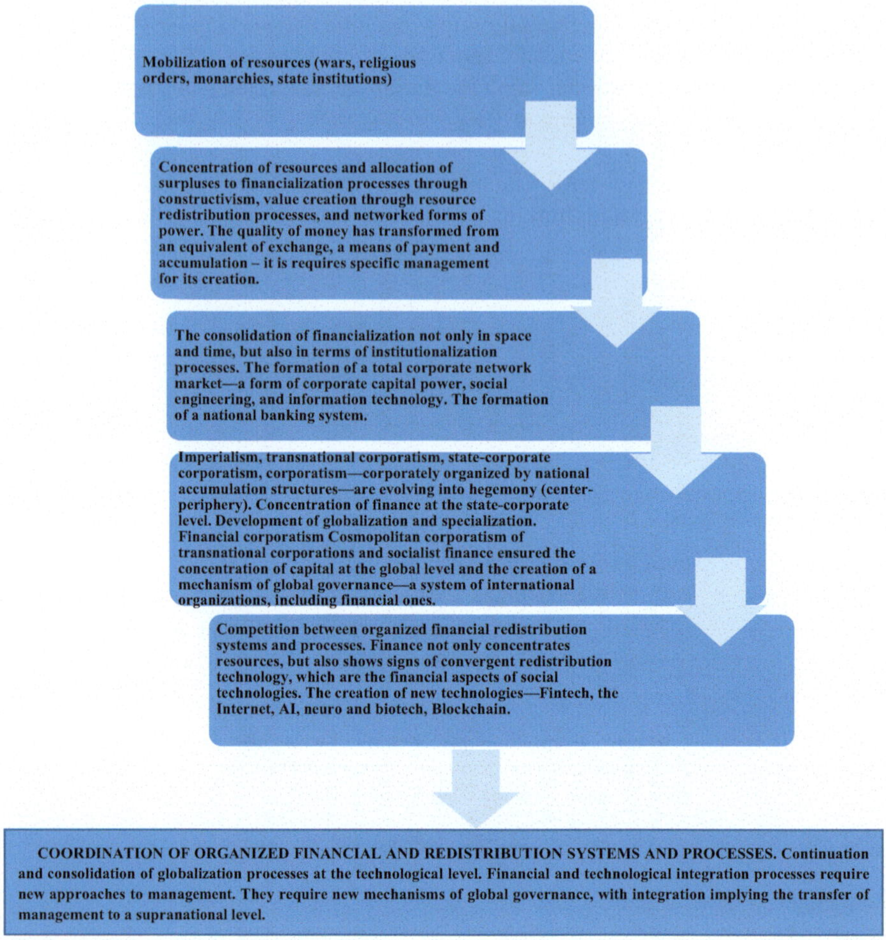

Fig. 1 Stages and forms of integration of socio-economic relations

stages, forms, and directions of coordination in the development of fintech at the international level. The strengthening influence of social factors is evident in modern types of financial technologies.

At the present stage, evidence of international integration in the field of fintech is the regional nature of the spatial scale of these processes, the formation of clusters and alliances in the field of financial and technological development [27] based on the principle of specialization in the new technological cycle.

Thus, the priorities of integration cooperation in the cluster (EC, USA Singapore UK and Japan) are testing and approving blockchain projects [26], agreeing on international standards, and ensuring interoperability. In the emerging cluster (BRICS,

EAEU, SCO, APEC, IBSA)—the creation of a common financial market, participation in alliances, the creation of payment infrastructure, ensuring interoperability [4].

At the same time, countries are at different stages of financial and technological integration (see Table 3).

At the industry level, partnerships play a key role in fintech companies' strategies. Overall, 84% of fintech companies surveyed collaborate with traditional financial institutions, most often through integrations using application programming interfaces (APIs) (52%), technology providers (41%), and financing agreements (36%) [29].

Table 3 Countries and participants by stages of financial integration in the development of FinTech at the international level

Stages of financial integration	Countries and participants in this stage in the field of fintech development at the international level
1. Processes of development and consolidation of financial markets	At the level of emerging alliances, participants in the EAEU, SCO, APEC
2. Digitalization of the financial sector	Participants in the EAEU, SCO, APECA coordinated process at the level of international organizations and global institutions Alliances for the coordination of international standards in fintech
3. Testing and approval of international blockchain technology projects in the financial sector as a continuation of the processes of development and consolidation of financial markets in technological form	The EU, the US, China, Singapore, Israel, the UK, Japan, and BRICS + members International projects on regulatory cooperation in the development of central bank digital currencies (Israel, Norway, Sweden): (Tunisia and France); BIS and Switzerland, (Singapore, Malaysia, South Africa, Australia), (Hong Kong, China, United Arab Emirates, Thailand). Intergovernmental supervisory body and council of central banks. Permanent secretariat (EU, China and its provinces). Development of the technological infrastructure of decentralized financial technologies of transnational corporations (IBSA participants)
4. Creation of a global cooperation mechanism based on a decentralized financial system, cross-sector integration, and the merging of financial and non-financial services	United States, China, United Kingdom, Japan Technological infrastructure and social coordination. Regulation of integrated high-tech clusters and the development of financial information technologies. Building an effective system of inter-sectoral coordination and regulating the development of the crypto ecosystem at the global level

Source Compiled by the author using sources [1–29]

Fintech trends signal an integrated and coordinated sustainable financial future.

An important trend is the inclusion of value components in financial products, rather than just economic benefits. Millennials and Generation Z prefer socially responsible businesses. In this regard, the Social fintech sector is actively developing, focused not only on making a profit, but above all on transforming socio-economic relations, social systems and structures, as well as socio-engineering activities, ensuring the specified dynamics of the development process to achieve goals. The Social fintech sector includes both specialized companies focused on charity and support for vulnerable populations, inclusiveness, and accessibility of financial services, as well as sustainable funds, green fintech, blended finance, ESG investment agendas, and community-based models: Fintech platforms that use the sharing economy and models built around communities and lifestyles: sports, music, youth subcultures that promote collaboration and collective financial sustainability, fintech focused on generations (the Silver Economy), which allow individuals to raise funds for personal and community projects.

According to fintech accelerator Finnovista, Social fintech companies could capture 30% of Mexico's financial market in just ten years. China, India, and Brazil, three countries with exceptionally high percentages of unbanked populations, are among the world's top five countries in terms of fintech adoption [29].

There is a regional imbalance in the directions of social fintech development. Fintech companies in the Global North tend to focus on traditional banking services such as money transfers, insurance, and online banking, which are already offered by conventional financial institutions. Fintech in the US and Europe is focused on increasing the use of AI, products for metaverses, and neurotechnologies. Here, fintech built around communities is focused on image, lifestyle, and loyalty.

On the other hand, in the Global South, fintech focuses on serving those who lack sufficient funds for financing. They typically provide microfinance, payment, and crowdfunding services. This makes them particularly attractive to countries in Africa, Asia, and Latin America. Fintech in China and Russia is focused on building ecosystems, scoring and assessment, developing alternative payment services, developing digital currencies for cross-border payments, and strengthening information security.

However, it is also necessary to highlight qualitative changes in the development of fintech. 80% of fintech companies surveyed are implementing artificial intelligence (AI) in several business areas. Customer service and process automation are leading the way in AI adoption: 91% of fintech companies are already implementing AI or plan to do so in the near future in these areas. AI, regional interconnection, open banking, and an open financial system are the main areas of fintech development for the next five years.

These processes indicate that what is happening is not simply the cross-border expansion and integration of fintech with traditional financial infrastructure, but rather a joint transformation of digital and financial services, moving beyond its boundaries through the use of AI and neurotechnologies.

The degree of independent decision-making by economic actors is being delegated to fintech or reduced.

From the point of view of the use of social technologies, AI, and neurotechnologies in fintech, the following types can be identified that influence socio-economic relations, economic sociology, and the sociology of digitalization.

- Fintech using algorithms strictly according to regulations—chatbots without AI.
- Chat GPT, LLM neural network, performs prompts in dialogue with the user (CHATGPT, PERPLEXITY, DEEPSEEK).
- AI investment financial assistant performs prompts, works under the guidance of the user Chatbot with AI personal assistant.
- AI agent that autonomously performs tasks and makes decisions independently (performing professional tasks: finance, accounting, legal, marketing, insurance, investment).
- AI operator independently performs tasks with access to data and a controlled device, a full-fledged robot with AI.
- Automation of social interaction and fintech business processes. The Social fintech technology and company sector: specialized, focused on charity and support for vulnerable populations, inclusivity and accessibility of financial services, Sustainable Funds, Green Fintech, mixed finance, ESG investment agenda.
- Fintech platforms—models based on lifestyle communities or generation-oriented (Silver Economy).
- Platform solutions, e-commerce, embedded finance with gamification and monetization, based on stratification. Monetization and gamification lead to addiction and dependence on algorithmic interactions. Developers monetize the use of gamification and the integration of games into business applications. New gaming platforms are expected to revolutionize the way people communicate and search for information. Search engines that use AI to respond to user requests not only provide sources of information and content, but also process it.
- Connection to medicine and diagnoses, customizable expense control for people with dementia, mental health issues, or special needs (example: True Link offers debit cards with customizable controls).
- Assessment of borrower risk based on consumer behavior—inflows and outflows from consumer bank accounts. Stratification.
- Fintech linking current consumption and lifestyle with investments and savings—stratification issues. Fintech tasks—simplifying the customer journey: the process of interacting with a fintech product should be simpler and more understandable for young people. Expanding the toolkit by including family products, savings for children, and creating a "persuasion architecture" that will encourage the right long-term decisions by involving children and employers.
- Growth of avatars with artificial intelligence—Monetization of social and emotional interactions as a service. Provides unlimited passive income through a multi-level subscription model: the intensity and clarity of interaction is ranked, and the connection is monetized. (For example, 80% to the avatar and 20% to the platform) Users are vulnerable to forming emotional attachments to simulations, and the platform profits from interactions that seem real but remain one-sided. AI creates infinitely personalized content for subscribers by imitating fame, fantasy,

and intimacy. Emotional dependence on AI chatbots arises. Synthetic intimacy replaces biological interaction and communication. Consumption of such services causes addiction.

Under the influence of AI, neurotechnology, and social technology, the function and role of FinTech extends beyond financial innovation.

These tools are moving beyond economic relations. The main focus is on social interactions. They go beyond the boundaries of "loyalty" based on consumer choice, emphasizing behavioral responses, attachment, reinforcement of dependence, emotional dependence, and imitation of synthetic human relationships and interactions. They do not simply mediate financial and economic relations based on the decisions of their subjects, but act as moderators and organizers, full-fledged participants in these relations, determining their interactions and possibilities. Fintech is increasingly focused on modeling and social simulation of economic and financial behavior.

Behavioral patterns and forms of interaction are being formed on the basis of algorithms. Procedures are emerging that are interrelated with values and rules governing the behavior of individuals and social groups.

Economic relations are closely linked to social relations. New practices in financial and economic relations, driven by the evolution of fintech, are shaping new processes and their participants, new institutions—formal and informal rules of conduct that affect society as a whole.

Qualitative identification of processes and their participants (based on new types of fintech) allows us to conclude that a new socio-technological institution is being formed. At the current stage of its institutionalization, corresponding to financial and technological integration, the infrastructure of social technologies is being formed and built: from the emergence of new directions and types of Social fintech to the activities of international organizations in working groups and national central bank regulators in the areas of fintech and social technology regulation.

Economic institutions influence the functions of social institutions: regulatory, developing common socially significant patterns of behavior and ideals, and integrative, ensuring the preservation of the integrity of the institution through consolidation, compatibility of personal goals with the goals of the group of participants, and mobilization of private resources of group members to achieve goals. And they require appropriate forms of social control. As a result, a clear new, technologically entrenched, status-role structure of social relations is established.

Thus, the changes caused by the evolution of financial technologies—interaction processes and technologies, new institutions—raise questions about the study of changes in economic sociology, the transformation of the economic system, and the adaptation of the psychology of participants in economic relations, the impact on their value systems, and changes in economic institutions (property, economic mechanisms).

Given the importance of social aspects in the development of fintech, it is advisable to extend its regulation beyond financial legislation and innovation regulation. Fintech should also be considered as an element of social infrastructure and social policy, which requires appropriate regulatory approaches. It is advisable to introduce neurotesting of products in the field of fintech, licensing of developers, standardization and control of neural network training in fintech, control and testing of AI bias in stratification and scoring, and to determine medical parameters and regulate dependencies and psychological effects of technologies. Monitor and evaluate the impact of social financial technologies on social interaction processes, economic sociology, and social policy. These interdisciplinary approaches and directions for regulating the development of Social fintech require reform of the existing structure of international organizations and government bodies at the national level, as well as appropriate forms of coordination between them.

4 Discussion

The strengthening of the impact of social factors on the regime of capital accumulation with the course of evolution is reflected in the modern processes of development of financial technologies, concentration of capital and power, the consequences occurring in the global economic and political context, and their influence on geopolitics and economic sociology. Thus, in the process of convergence of financial and social technologies, the focus on ensuring control over capital accumulation processes [6] is transformed into the coordination of organized financial redistribution systems and processes, using financial and social technologies, and its technological consolidation at both the global and national levels. The regime of capital accumulation, due to the strengthening of the influence of social factors, is evolving into an intellectual and technological industry of coordination for the creation and consolidation of socio-economic relations and the redistribution of resources.

The common features of the co-evolution of financial and social technologies are manifested in the principles of directionality and continuity.

The principle of directionality—According to historians [11] and sociologists [28], the evolution of money leads to an inexorable movement toward an ever-stronger concentration of financial capital—financialization.

If we consistently compare and apply the evolutionary approach, originally outlined in the world systems analysis by Wallerstein [5] and Arrighi [6] in the systemic cycles of capital accumulation, to the development of financial technologies and the civilizational approach to the development of financial and economic relations determined by religious orders, social engineering, and socialist regimes, the following directions of co-evolution of financial and social technologies can be traced.

The co-evolution of financial and social technologies shows that there is indeed a focus on the concentration of resources, but financialization is not the goal, but only another step towards a highly organized level of integration of socio-economic relations and redistribution systems at the global level, in increasing the degree of integrity, concentration, and expediency of these processes.

The main direction of evolution is from low-organized forms of human relations to more highly organized and complex redistribution systems, in order to increase their degree of integrity and expediency, to consolidate and preserve them in space and time and in terms of processes.

The principle of continuity.

- Current trends in the development of financial and social technologies are interdependent with the cycles of capital accumulation described by Arrighi [6]. This is confirmed on the basis of an actor-network and process approach: financialization, fragmentation of the global financial system, financial aspects of social technologies and priorities of international organizations, trends in fintech development, and processes of financial and technological integration.
- Cross-sectoral aspects and convergence of technologies are leading to new qualitative connections and changes in business processes, participants, and priorities. Consolidation and concentration processes indicate that the future global financial and technological architecture has been formed and the next stage of its development is beginning within the framework of a new global technological cycle in accordance with the specialization of countries. The convergence of technologies and new structures requires a new global mechanism for cooperation and a transition to financial and technological regulation of social processes based on decentralized financial systems, new forms of interaction and interoperability of digital financial systems, as well as between several multipolar reserve digital currencies.

5 Conclusion

Financial technologies are the financial aspects of social technologies and represent a way of organizing financial and economic relations in a certain order.

Fintech shows signs of convergent redistributive technology.

The processes of financial and technological integration at the global level and the formation of future components of the international financial and technological system structure indicate the next stage of technological globalization and an organized transition to a coordinated global financial and technological policy.

There are observable trends toward the transformation of the global financial and institutional infrastructure into a financial and technological infrastructure.

Cosmopolitan organized structures—transnational nations, local civilizations, and diasporas—play a special role in the processes of financial and technological integration. The main directions and processes of regulation at the current stage of development are digital geopolitics and geoeconomics, the intellectual industry, and tools such as big data, AI, neurotechnology, and biotechnology. This requires special attention and approaches in the processes of regulating the development of fintech at the national and supranational levels and in international cooperation.

At the current stage of development, the use of AI and neurotechnology in fintech is expanding its impact on social interactions. The development of the Social fintech sector necessitates viewing it as an element of social technological infrastructure and social policy, which will require appropriate approaches to its regulation.

References

1. Hildebrand B (2011) Political economy of the present and the future [Politicheskaya ekonomiya nastoyashchego i budushchego]. (In Russian)
2. Veblen T (2006) Why is economics not an evolutionary science? Terra Economicus. 4(2):99–111
3. Jus Talionis in Hugo Grotius's treatise "On the Law of War and Peace." Ethical Thought 3:139–165. https://et.iphras.ru/article/view/2491
4. Razletovskaya VV (2024) Management of the processes of development of modern financial technologies: evolution of national models [Upravlenie processami razvitiya sovremennyh finansovyh tekhnologij: jevoljucija nacional'nyh modelej]: Dissertation of Doctor of Economic Sciences, Moscow State Institute of International Relations (University) of the Ministry of Foreign Affairs of the Russian Federation. (In Russian)
5. Wallerstein I (1974) The modern world-system, vol 1. Capitalist agriculture and the origin of the European World-Economy. Academic Press, New York
6. Arrighi G (2006) The long twentieth century: Money, power and the origins of our time [Dlinnyj dvadcatyj vek: Den'gi, vlast' i istoki nashego vremeni]. Izdatel'skij dom "Territoriya budushchego", Moscow. (In Russian)
7. Perez C (2013) Technological revolutions and financial capital. Bubble dynamics and the period of prosperity [Tekhnologicheskie revolyucii i finansovyj kapital. Dinamika puzyrej i periody procvetaniya]. Delo Publishing House RANEPA, Moscow. (In Russian)
8. Harari YN (2016) Sapiens. A brief history of mankind [Sapiens. Kratkaya istoriya chelovechestva]. Sindbad, Moscow. (In Russian)
9. Prokhorov AP (2002) Russian model of management [Rossijskaya model' upravleniya]. Art. Lebedev Studio, Moscow (In Russian)
10. Elistratov AS (2008) The origin of socio-historical direction of economic thought in Russia (on the example of the economic doctrine of A.I. Chuprov) [Stanovlenie social'no-istoricheskogo napravleniya ekonomicheskoj mysli v Rossii (na primere ekonomicheskogo ucheniya A.I. Chuprova)]: Avtoref. dis.... kand. ekon. nauk, Moscow. (In Russian)
11. Ferguson N (2023) The rise of money. Financial history of the world [Vozvyshenie deneg. Finansovaya istoriya mira]. AST, Moscow. (In Russian)
12. Bjerg O (2018) How is money made? Philosophy of post-credit capitalism [Kak delayutsya den'gi? Filosofiya postkreditnogo kapitalizma]. Ad Marginem Press, Moscow (In Russian)
13. Dubyansky AN (2020) Philosophical view of money (About the book by U. Bjerg "How is money made? The philosophy of post-credit capitalism"). Vopr Ekon (3):129–140. https://doi.org/10.32609/0042-8736-2020-3-129-140
14. Castells M (2011) A network theory of power. Int J Commun 5:773–787
15. Goldobina SY (2012) Modern interpretation of basic social institutions of G. Spencer [Sovremennaya interpretaciya bazovyh social'nyh institutov G. Spensera]. Psihol, Sociol i Pedagog (9). https://psychology.snauka.ru/2012/09/986. Accessed 03 Oct 2023. (In Russian)
16. Batygin GS (2003) Structural functionalism of Talcott Parsons [Strukturnyj funkcionalizm Tolkotta Parsonsa]. Vestnik RUDN, seriya Sociologiya 4–5:6–34 (In Russian)
17. Gvishiani DM (2007) Selected works on philosophy, sociology and system analysis [Izbrannye trudy po filosofii, sociologii i sistemnomu analizu]. In: Popkov YS, Sadovsky VN, Soitov AA (eds) Seriya "Filosofy Rossii XX veka". "Канон+" ROOI "Reabilitaciya", Moscow. (In Russian)
18. Kirdina SG (2014) Institutional matrices and the development of Russia: Introduction to X-Y theory [Institucional'nye matricy i razvitie Rossii: Vvedenie v X-Y-teoriyu]. 3rd edn. Nestor-Istoriya, Saint Petersburg. (In Russian)

19. Kay J (2015) Other people's money: The real business of finance. Public Affairs, New York
20. Beinhocker ED (2007) The origin of wealth: The radical remaking of economics and what it means for business and society. Harvard Business Review Press, Boston
21. Buckley RP, Arner DW, Zetzsche DA (2023) FinTech: Finance, technology and regulation. Cambridge University Press, Cambridge
22. Razletovskaya VV (2023) Systematization of approaches to regulating the development of financial technologies [Sistematizaciya podhodov k regulirovaniyu razvitiya finansovyh tekhnologij]. Probl Ekon i Yuridicheskoj Prakt 19(3):288–294 (In Russian)
23. Razletovskaya VV (2000) Managerial content, concept and essence of financial technologies [Upravlencheskoe soderzhanie, ponyatie i sushchnost' finansovyh tekhnologij]. Regional'naya Ekon: Teor i Prakt 21(10):1985–2000. https://doi.org/10.24891/re.21.10.1985. (In Russian)
24. Arrighi G, Silver BJ (1999) Chaos and governance in the modern world system. University of Minnesota Press, Minneapolis, MN
25. Harvey D (2003) The new imperialism. Oxford University Press, Oxford
26. Maslov AV, Shvandar KV (2023) New international projects on the use of central bank digital currencies in the transformation of cross-border settlements. Financial Journal 15(2):47–58. https://doi.org/10.31107/2075-1990-2023-2-47-58
27. Razletovskaya VV (2023) Strategic alternatives of integration of national fintech development models into the world infrastructure [Strategicheskie al'ternativy integracii nacional'nyh modelej razvitiya finteha v mirovuyu infrastrukturu]. Izvestiya Sankt-Peterburgskogo Gos Ekon Univ (5):32 (2023). (In Russian)
28. Hilferding R (1969) Financial capital. A study of the newest phase in the development of capitalism [Finansovyj kapital. Novyshaya stadiya v razvitii kapitalizma]. Izdatel'stvo social'no-ekonomicheskoj literatury, Moscow. (In Russian)
29. World Economic Forum (2025) Cambridge centre for alternative finance: The future of global fintech: From rapid expansion to sustainable growth. Insight Report, 2 edn. https://www3.weforum.org/docs/WEF_Future_of_Global_Fintech_Second_Edition_2025.pdf

Research on Smart Government Policies Aligned with Green Finance Policy Instruments to Achieve Sustainable Economic Development Goals

Inha Vladimirovna Ziankova, Gulnora Mardievna Shadieva, Zumrat Talatovna Gaibnazarova, Alena Borisovna Maley, and Olga Vasilevna Gasheva

Abstract The chapter discusses the theoretical foundations and practical cases for the implementation of smart government policy instruments working in a coordinated implementation with transparent, trusting, effective green finance policies for sustainable development. The purpose of the research is to consider the formation of strategic recommendations to support green finance for sustainable development in a coherent development with smart government policy instruments. The following tasks were set and solved: monitoring of policy documents, the works of scientists to identify the global conditions in which the coordinated policy of smart government and the policy of green finance are developing; consideration of the content of green finance for sustainable development, lessons learned from key components and instruments of green finance; rationale as smart government and green finance policy tools for sustainable development: working with data, digital tools for process management and monitoring, risk management and strategic planning, intersectoral and interregional coordination. The research methodology is based on the principles of systems analysis, formal logic, and an interdisciplinary scientific approach. In the formation of the theoretical and practical basis for the formation of strategic

I. V. Ziankova · A. B. Maley · O. V. Gasheva
Euphrosyne Polotskaya State University of Polotsk, Novopolotsk, Belarus
e-mail: i.zenkova@psu.by

A. B. Maley
e-mail: e.malei@psu.by

O. V. Gasheva
e-mail: ogasheva@psu.by

G. M. Shadieva
Samarkand Institute of Economics and Service, Samarkand, Uzbekistan
e-mail: shodiyeva12345@gmail.com

Z. T. Gaibnazarova (✉)
University of Science and Technologies, Tashkent, Uzbekistan
e-mail: zumrat59@rambler.ru; z.gaibnazarova@usat.uz

© The Author(s), under exclusive license to Springer Nature Switzerland AG 2026

A. Kulachinskaya et al. (eds.), *Digital Transformation: What Kind of Smart Life do we have today?*, Lecture Notes in Networks and Systems 1895,
https://doi.org/10.1007/978-3-032-20582-7_11

recommendations, monographic and descriptive methods, a method of analysis and generalization, and a method of economic analysis were used. The novelty of the study lies in the compilation of unique experience in forming strategic recommendations for the coordinated implementation of smart government policies with green finance policies for sustainable development. A practical approach has been developed to harmonize smart and green finance policies for sustainable development, which is characterized by a system of institutional measures at the micro and macro levels.

Keywords Smart government · Smart contracts and digital registries · Smart city · Smart and green finance policy · Sustainability · Green finance

1 Introduction

In the digital age, the link between green finance and smart government is fundamental to achieving sustainable development goals.

The link between green finance policy and smart government policy is seen through specific macro-tools: the creation of digital platforms for collecting and aggregating environmental data; work on data and reporting transparency, indicators for environmental impact verification; systematization of digital tools for management and monitoring through smart contracts and digital registries; organizing digital auctions and tenders for emission quotas to regulate processes through digital channels; risk management and strategic planning through climate scenario modelling; creation of climate risk maps.

At the head of the Decree of the President of the Republic of Belarus No. 381 "On Digital Development" is the introduction of information technologies, state platforms, the introduction of pilot projects in state statistics and ecology, as well as the exchange of information within the Eurasian Economic Union, which strengthens the coherence and efficiency of decision-making in state policy and in the policy of green finance for the sustainable development of the national economic system [1].

In Belarus, there is a study of the introduction of smart government tools in the creation of "smart cities." For example, the introduction of a regional state standard digital platform "Smart City (Region)" in Minsk, regional centers, cities and districts with a population of 80 thousand people or more, providing services on their basis in various spheres of life and management [2].

In Belarus, the digital state model is built through the transition from the e-government model to the digital state model; creation of the Smart City platform aimed at providing a wide range of digital services to ensure a comfortable living environment for the population. At the first stage—in 17 cities of the country; construction of 5G networks on a fundamentally new architecture of wireless broadband access for the introduction of the latest world technologies in various sectors of the economy and social sphere; translation of more than 400 administrative procedures into electronic form; ensuring traceability of goods through the unification of digital logistics services on a single platform; trade and cooperation cooperation through the introduction of electronic document management in trade [3].

Belarus has world-class forums—EcoForum 2025 with UNDP partnership and regulatory documents National Action Plans for a Green Economy and the National Strategy for the Development of a Circular Economy [4].

The nature of smart government and green finance policies is reflected in a combination of global and local trends.

These include: problems of the development of the electronic state in the Republic of Belarus and development prospects differentiation of e-government models, PRC practice in the development of modern digital law implementation of the national environmental monitoring system of China; specifics of environmental supervision of the Russian Federation; EU Innovation Fund study [5–8].

The concept initiative "Creating a green partnership for sustainable development" was put forward at the forum of the Shanghai Cooperation Organization countries on green development [9].

Forum on Civil Society and the Green Economy in China focuses on the green economy [9].

Republican Environmental Forum focuses on regional adaptation to climate change, principles of a cyclical economy, preservation of ecosystems, the relationship between environmental protection and sustainable development [10].

Green finance is key to achieving the United Nations Sustainable Development Goals (SDGs), particularly SDG 7, which promotes access to affordable and clean energy [11]. Its importance goes beyond energy to include climate action and conservation [11, 12].

Global clean energy investment reached $1.7 trillion in 2022, with more than $500 billion raised from green bonds alone, according to the International Energy Agency, bridging funding gaps outlined in the Paris Agreement [13].

Eastern China Green Growth Analysis: Factors and Policies Tentatively Outlined in Research by Chinese Scientists [14, 15].

Belarusian researchers demonstrate the principles of a circular economy in their analysis: "The efficient use of energy efficiency resources through the principles of a circular economy creates synergistic effects between economic growth" [16].

As smart government and green finance policy tools, the studies summarize: the practice of implementing the Russian public procurement portal, where eco-criteria are introduced, international reviews of Green Public Procurement practices, reviews of the world's ETS systems, cap-and-trade analytics, country indicators and eco-ratings, open data for sustainability ratings, approaches to financing sustainable projects, green project implementation practice, Climate Data Store tools for modeling, Roshydromet climate scenarios, open sources of risk maps (European system), international risk assessment and mapping methodologies [17–26].

The following studies have found practical applications in the world:

(1) How investments in renewable energy, environmental regulations and financial development contribute to the transition to renewable energy [27];

(2) Optimal planning of an automated energy Internet taking into account uncertainty [28].

However, it should be noted that all the studied tools are described separately, either relative to smart government or to green finance policies, therefore, the purpose of this study is to summarize smart policy tools working in a coordinated implementation with transparent, trusting, effective green finance policies for sustainable development.

The research topic corresponds to the priority areas of fundamental and applied scientific research of the Republic of Belarus, is consistent with the concept of the National Strategy for Sustainable Socio-Economic Development of the Republic of Belarus for the period up to 2040.

The genesis of economic thought has shown that research has failed to capture the impact of green finance, examining how innovation in green finance disrupts established accounting systems and affects the overall performance of factors; quantitative analysis of policy interventions through the use of systems dynamic modelling to examine synergistic effects of green finance policies across sectors, including threshold effects associated with green bond issuance and reduced carbon intensity; regional impact assessment through the use of financial geography analysis to understand how green capital flows can mitigate regional development imbalances, particularly through targeted green credit policies that enhance sustainability in environmentally sensitive areas; developing a mapping framework for green finance and SDGs through the creation of a dynamic model that links green finance practices to specific SDG targets, supported by empirical evidence from the green finance reform pilot zone; digital technologies through research on how blockchain and artificial intelligence increase transparency and accountability in green finance, including systems for tracking environmental benefits. The lack of resolution of these tasks actualizes the research issue under consideration.

The relevance of the indicated problems, the insufficient degree of their theoretical and practical development predetermined the choice of the research topic, its goals and objectives.

2 Methods

The research methodology is based on the principles of systems analysis, formal logic and an interdisciplinary scientific approach.

When researching issues on the close relationship between the implementation of smart government policies and green finance policies for sustainable development, a research methodology was used from monitoring policy documents, the works of scientists to identify the global conditions in which policies are developing. Further through the analysis of digital tools for management and monitoring; stimulation and regulation through digital channels; risk management and strategic planning; analysis of intersectoral and interregional coordination; analysis of models of mixed financing and artificial intelligence to formulate strategic recommendations for strengthening the tightness of the coordinated development of green finance policies for sustainable development with the coordinated development of smart government policies.

Research methods: systematic approach, analysis, synthesis, induction, deduction, scientific abstraction, logical generalization, modeling, comparison, inference by analogy.

Results Obtained

Developing a theoretical framework for a close link between smart government policies and green finance policies for sustainable development: reviewed the content of the green finance policy category for sustainable development; digital process management tools are systematized at various levels of the economic system: data management, transparency in reporting; digital tools for management and monitoring; stimulation and regulation through digital channels; risk management and strategic planning; intersectoral and interregional coordination; the following are substantiated and considered as digital tools for managing and monitoring processes: models of mixed financing; blockchain-based environmental asset tokenization; biodiversity-related derivative financial instruments; models using artificial intelligence to evaluate results, practical cases are considered; machine learning algorithms, time series analysis, statistical methods, dimensionality reduction methods, cryptocurrency initiative, green bonds and private investment are justified as risk management and strategic planning tools.

Strategic recommendations were formed to support green finance for sustainable development in a coordinated development with smart government policy instruments.

The novelty of the study lies in the compilation of unique experience in forming strategic recommendations for the coordinated implementation of smart government policies with green finance policies for sustainable development.

3 Research Results

The target indicators of the digital transformation of the Belarusian society are indicated (NSUR, 2040) "the growth rate of the GVA of the digital economy—at least 3 times in 2026–2040; joining the number of 50 states in the world ranking on the index of readiness for the use of advanced technologies in 2040 (55th place out of 166 countries in 2023) and 80 states in the world ranking on the index of readiness for artificial intelligence (107th place out of 193 states in 2023); increase in the export of computer, telecommunications and information services by at least 1.7 times in 2026–2040" [29].

We learned to study a number of high-level research results around optimizing investments in renewable energy sources, a mechanism for coupling financial development and environmental policy:

(1) Machine Learning Applications to Optimize Photovoltaic Systems in Zero Green Energy Buildings [30].
(2) Hybrid molecular fuzzy recommendation systems to assess the impact of climate change on renewable energy efficiency [31].
(3) Molecular fuzzy decision model to optimize renewable energy investments for carbon neutrality [32].

Studies of Belarusian researchers focused on the "green economy" and regional sustainable development strategies, and seeks to introduce green concepts in regional economic management:

(1) Sustainable socio-economic development model: implementation through assessment of regional ICT capacity [33].
(2) Efficient use of energy efficiency resources based on circular economy principles [16].
(3) Developing an inclusive society with green economy and education tools [34].
(4) E-Commerce Web Page Conversion Optimization [35].

Green finance for sustainable development considered

Green financing for sustainable development refers to a dynamic financial ecosystem that is consistently implemented with the digital tools of the state's smart policy. This ecosystem creates bidirectional feedback loops between capital flows and sustainable development outcomes through which financial instruments not only finance projects that benefit the environment, but also generate cycles of economic efficiency, social inclusion and environmental protection.

Smart government creates a technology, regulatory and infrastructure environment where green finance can function effectively, scale and deliver measurable benefits. The study summarizes the tools through which the implementation of smart government policies is implemented in close connection with green finance policies for sustainable development.

Smart government is addressed through digital process management tools at various levels of the economic system: data management, transparency in reporting; digital tools for management and monitoring; stimulation and regulation through digital channels; risk management and strategic planning; intersectoral and interregional coordination.

The study summarizes the experience that green finance has a profound impact on the sustainable development of enterprises in three critical areas:

Financing Mechanism—By providing targeted, low-cost financial resources, green finance enables businesses to invest in sustainable technology and infrastructure. This approach reduces capital barriers to green projects and drives green transformation.

Risk Management—Green Finance introduces comprehensive risk assessment frameworks that integrate environmental, social and governance (ESG) criteria. This approach helps businesses identify, assess and mitigate potential risks associated with sustainable development, contributing to long-term strategic planning.

Innovating. Financial instruments such as green bonds and sustainability-related loans create powerful incentives for technological innovation. Businesses are interested in developing cleaner technologies, improving resource efficiency and reducing carbon emissions to access preferential financing terms.

The multidimensional impact of green finance is fundamentally changing corporate strategies, aligning economic performance with environmental sustainability [36].

Considered are smart government tools and green finance policies for sustainable development (data, transparency and reporting).

Smart government creates centralized digital platforms for collecting, aggregating and analyzing environmental data: CO_2 emissions, air and water quality, energy consumption, and the state of green spaces. Investors and financial institutions receive verified, standardized data for risk assessment. Further, climate risks, physical and transitional, become measurable. Thus, Impact Verification allows you to prove that the green bond project actually reduced emissions by the amount declared. Research addresses critical gaps in standardized exposure reporting by developing a cross-jurisdictional compatibility matrix and proposing new approaches to regulating the base for testing climate-resilient financial instruments.

In addition, the studies look at a three-tier verification system that combines satellite monitoring, IoT sensor networks and decentralized ledger technology to significantly increase confidence in green investment outcomes.

The chapter summarizes the experiences of the key components and instruments of green finance (Table 1).

The study describes such instruments as: green funds, bonds and insurance, bonds with sustainable development principles, bonds of carbon assets, transition finance and bonds of transition, the Green Index investment fund with a detailed description of both their functions and practical examples of implementation, taking into account the principles of smart government: equity financing of individual "green" industries, providing recommendations for transparency of reporting and disclosure.

These tools provide practical channels for mobilizing capital to achieve environmentally beneficial outcomes and serve as a bridge between economic growth and environmental responsibility [9].

For example, green bonds specifically issued for renewable energy projects such as wind and solar tend to have more substantial and long-lasting effects than green general-purpose bonds [39].

Thus, the effectiveness of green finance depends not only on the volume of investments, but also on the coordination of financial instruments with clearly defined environmental goals, with the implementation of the principles of smart government. Green bond investments have measurable impact on both carbon reduction and renewable energy supply [11].

For example, cluster analyses in these regions also reveal a positive correlation between economic performance and overall ESG risk estimates [40].

It was revealed that, a significant risk of "green flushing" remains due to the lack of uniform classification standards and consistent disclosure rules for green bonds,

Table 1 Key components and tools of green finance

Green financial instruments/ mechanisms	Descriptions	Cases/data	Source of information
Green fund	Equity financing for selected green industries	China National Green Development Fund	Green finance development centre [37]
Green Bond Principles (GBP)	Providing recommendations on transparency and information disclosure	Around 200 global issuers, underwriters and investors follow the GBP	APEC Asia–Pacific economic cooperation [38]
Sustainability Linked Bonds (SLB)	Developed in accordance with the sustainability linked bond principles (SLBP) issued by the international capital market association (ICMA)	China issued its first batch of SLBs	Green finance development centre [14]
Carbon asset bonds	First issue	In august 2022, shanghai pudong development bank [15] underwrote the first batch of carbon asset bonds	Green finance development centre [14]
Transition finance and transition bonds Финансирование		In the first half of 2022, 23 transformation bonds totaling $2.1 billion were issued globally	Green finance development centre
Green index investment fund		By the end of the third quarter of 2022, the size of domestic green index investment funds exceeds RMB 130 billion	Green finance development centre

Source Based on data[14, 15, 37, 38]

which requires strengthening the principles of data openness as a smart government policy tool [12].

The experience of linking green finance policies and smart government policies is summarized.

By the first half of 2017, for example, the company issued five green bonds totalling €4.5bn, all of which are earmarked for renewable energy development. These projects have been externally evaluated and certified by independent third parties, ensuring transparent and responsible allocation of funds. This approach not only strengthens confidence among investors, but also increases the efficiency and effectiveness of capital use in environmentally oriented initiatives.

Looking closely at smart government policy tools—the openness and transparency of working with data with the practical implementation of green funds—attention is drawn to how this connection is supported by specific green industries through shared financing.

For example, China's National Green Development Fund has a registered capital of 88 billion yuan (roughly $14 billion) and provides strong capital support for clean energy and environmental protection. Green funds first raise capital from investors and then invest it in projects that meet environmental standards, thereby ensuring an accurate match between capital and green projects.

In addition, green funds tend to have characteristics of long-term investments, which is useful to alleviate the problem of initial investment of large amounts of capital in green projects, the return cycle is relatively long and thus can contribute to the sustainable and healthy development of green industry.

Digital tools for process management and monitoring. Smart government tools such as mixed finance models involving public–private cooperation; blockchain-based environmental asset tokenization; biodiversity-related derivative financial instruments; models using artificial intelligence to assess sustainability results contribute to the digital transformation of the accounting system and affect the overall performance of factors, determine the channels of green investments.

Green credit is a strategic financing instrument used by banks and financial institutions to support environmentally responsible businesses and projects. Rather than simply promoting growth, it channels capital into sectors that meet environmental goals. Credits are usually issued after a thorough evaluation process that ensures applicants meet certain sustainability standards. Once approved, financial institutions closely monitor the placement of funds to ensure they are being used for intended green purposes.

In China, green lending policies have affected market dynamics, penalizing high-carbon businesses and rewarding those who align with ESG values. For example, carbon-intensive firms face declining valuations, while ESG-compliant businesses enjoy stronger investor interest. This market differentiation has contributed to increased investment in green R&D and created a virtuous cycle that simultaneously promotes innovation and environmental responsibility.

Green insurance as a risk management tool. When implemented effectively, green insurance helps reduce corporate exposure to environmental sanctions and encourages more sustainable business practices. By shifting some of the risk burden onto insurers, firms are more likely to invest in cleaner technologies and adopt green production practices, leading to wider benefits for both the economy and the environment.

For example, green stock indices provide investors with benchmarks for environmentally oriented companies, helping guide capital flows to sustainable business models.

In 2022, the value of China's green index investment funds topped 130 billion yuan, reflecting growing market confidence in the sector. Carbon finance tools—such as emissions trading schemes and carbon asset bonds—are also gaining traction.

Sustainability-related bonds (SLBs) are another emerging instrument. Tied to firms "environmental performance, these bonds adjust financing costs based on sustainability metrics".

In May 2021, China issued SLBs worth 7.3 billion yuan, creating powerful incentives for companies to improve their environmental footprint. Globally, transition bonds are also gaining popularity, with $2.1 billion issued in the first half of 2022, highlighting their potential in broader decarbonization strategies.

Together, these tools—from green bonds to insurance arrangements—form a diversified financial ecosystem aimed at driving the transition to sustainability. By redistributing capital to environmentally beneficial sectors and discouraging harmful practices, green finance serves not only as a policy lever, but also as a catalyst for industrial transformation. Companies are increasingly shifting their asset allocation towards innovation, with each playing a specific but interconnected role in building a low-carbon economic future.

Risk management and strategic planning. The data set includes a certain number of observations covering macro indicators, which provides a reliable basis for analyzing non-linear relationships within eco-economic systems.

The integration of several economic indicators into a single analytical structure allows you to more fully understand the dynamics of green finance. For example, variance inflation factor (VIF) analysis is used to examine potential multicollinearity among key resilience variables. The correlation between green taxation (TNR) and clean energy investment (GS) with VIF signals a level of interdependence that needs further investigation [10, 41–43].

Among these variables, economic policy uncertainty (EPU) plays a key deterrent role. If the practical case shows substantial variance, then it highlights policy volatility that can significantly affect financial flows and investment behavior in green sectors.

Concluded to address cross-country comparability, the study applies machine learning algorithms to construct composite indices that normalize resilience metrics at vastly different scales.

This provides a better understanding of the relationship between economic growth and environmental impact. For example, while the average GDP across the entire dataset is 8.606, CO_2 emissions average 30.609, indicating the carbon intensity associated with growth, especially in developing countries undergoing industrial transformation [44].

The implementation of smart government tools in terms of transparency of indicators is important, since, for example, the theory of environmental potential emphasizes that economic development must remain within the boundaries of the ecosystem, which requires a thorough assessment of the intensity of the exploitation of natural resources.

Artificial intelligence, especially natural language processing, is increasingly being used to standardize ESG reporting based on unstructured corporate information.

Quantum computing and geospatial analytics offer new tools to optimize green investment portfolios and assess environmental risks with unprecedented accuracy.

Hybrid models combine machine learning with causal methods and offer new opportunities to quantify directional relationships between green finance flows and ecosystem restoration outcomes.

AI-based pattern recognition is used to develop dynamic materiality matrices that prioritize industry sustainability metrics, allowing more targeted allocation of green investments.

Real-time geospatial analytics now enable project-level verification of environmental requirements, thereby reducing the risk of pollution in climate-focused investments.

Experimental applications of quantum computing in portfolio optimization demonstrate the potential to solve complex, multi-purpose problems that balance financial returns with minimizing carbon footprint across time horizons.

These technological synergies create new pathways to realize the triple bottom line concept at scale, especially through the development of interoperable sustainability accounting platforms that integrate IoT sensor data from intelligent infrastructure with blockchain-based carbon credit markets.

The following summarizes the practical experiences of smart government tools and green finance for sustainable development [45, 46] (Table 2).

Analysis of Table 2 showed that machine learning algorithms allow to take into account the features of ranking the importance of random forest models in financial modeling and stability modeling help to determine important variables, which positively affects real-time decision making and the promotion of sustainability by improving the quality of data.

In turn, cryptocurrency-funded green infrastructure projects, decentralized finance platforms (DeFi) support sustainable investments, as well as promote transparency, reduce transaction costs and support innovative mechanisms for financing green projects through decentralized solutions.

Time series analysis: Methods such as ARIMA modeling and exponential smoothing help predict productivity trends in sustainability-related industries. This analysis allows us to assess the potential long-term impact of improving data quality on key industry and sustainability indicators.

Statistical methods: Regression analysis is used to examine the relationship between improved data quality and measures of resilience. This approach supports informed decision-making by identifying important correlations.

Dimensionality reduction method: Principal component analysis (PCA) simplifies large data sets by reducing less significant measurements while retaining the most informative features, facilitating clearer analysis.

Cryptocurrency initiatives: This includes green infrastructure projects funded by cryptocurrency and decentralized financial (DeFi) platforms that promote sustainable investment. These initiatives increase transparency, reduce transaction costs and offer innovative financial solutions for green projects.

Table 2 Systematization of smart and green finance policy instruments for sustainable development

Smart and green finance policy tools	Specific cases/data	Ways to advance the sustainable development goals
Machine learning algorithms	Importance ranking features of random forest models in financial and sustainability modelling help identify important variables	Influence real-time decision-making and promote sustainability through improved data quality
Time series analysis	ARIMA modelling and exponential smoothing for predicting performance in industry and sustainability	Estimating the long-term impact of data quality development and improvement on core industry and sustainability aspects
Statistical methods	Regression analyses to calculate correlations between data quality improvements and individual sustainability indicators	Supporting decision-making through correlation analysis
Dimensionality reduction method	Principal Component Analysis (PCA) for data compression	Compress large datasets by eliminating the least important dimensions and retaining the most diagnostic features
Cryptocurrency initiative	Cryptocurrency-funded green infrastructure projects, decentralised finance (DeFi) platforms support sustainable investment	Promote transparency, reduce transaction costs and support innovative financing mechanisms for green projects through decentralised solutions
Green bond	Green bonds attract investment in projects that promote environmental sustainability	Promoting environmental sustainability through financing renewable energy development, energy efficiency upgrading and sustainable infrastructure projects
Private investment	Private investment through diversification of funding sources and diversification of systemic risk	Supporting innovative approaches to addressing environmental challenges and promoting the transition to a sustainable economy

Source Based on data [45, 46]

Green Bonds: Designed specifically to attract investment in projects that support environmental sustainability, green bonds fund renewable energy initiatives, energy efficiency and sustainable infrastructure development.

Private investment: Diversifying sources of financing through private investment helps reduce systemic risks. These investments contribute to the development of innovative strategies to address environmental challenges and promote the transition to a sustainable economy [47].

Green finance not only funds existing projects, but also accelerates innovation. Preferential financing conditions encourage companies to invest in clean technologies and energy-saving upgrades.

For example, big data tools help financial institutions identify potentially green technology businesses, ensuring efficient capital allocation. This targeted support fosters innovative ecosystems that are critical to both environmental progress and long-term competitiveness.

Support from cryptocurrency initiatives and decentralized financial platforms is also introducing innovative financing mechanisms for green projects. These new technologies promote green infrastructure by increasing transparency and reducing transaction costs, facilitating a broader transition to a low-carbon and environmentally friendly approach.

The practical implementation of smart government through risk management and strategic planning through the use of machine learning algorithms, time series analysis, statistical methods, dimensionality reduction techniques, cryptocurrency initiative, green bonds and private investment has an impact on changes in green finance policies and on indicators of sustainable development.

Intersectoral and Interregional Coordination is Essential for the Effective and Sustainable Development of Green Finance.

Their regional ICT assessment study shows that the maturity of digital infrastructure directly correlates with the pace of green finance adoption, suggesting that techno-logical readiness serves as a critical enabler of sustainable financial transformation. This finding is particularly relevant for developing regions, where gaps in ICT infrastructure can constrain green finance initiatives.

International cooperation is undoubtedly an extremely important way to promote the integration of green finance and real economic development.

Thus, enhanced international cooperation is a necessary way to promote the global development of green finance through the internationalization of green finance standards and the facilitation of cross-border capital flows.

4 Discussion

At the head of the NSUR, 2040 in the Republic, "the task is to form a multi-level management system for the development of a closed cycle economy, including the creation of a state information and analytical management system in the field of waste management; creation of an electronic trading platform—a waste exchange; a single digital platform of a closed cycle economy to popularize the principles of a circular economy among business entities." Further, the task was set "to continue work on the development of the legislative framework and improving the tools for the implementation of environmental policy to form a national system of carbon regulation taking into account the convergence of the climate agenda within the EAEU" [48].

It was concluded that in Belarus, smart policy and green finance policy in the future until 2040 will have a vector of rapprochement and coherence relative to each other for sustainable development through smart policy tools (NSUR, 2040)—creating openness and transparency of data, digital platforms, electronic trading platforms.

Thus, the close relationship between decision-making tools (machine learning algorithms, artificial intelligence, quantum computing, and geospatial analytics) within smart government and green finance policies enables the measurement and evaluation of environmental externalities in investment decision-making processes [49–51].

So, emerging tools such as decentralized finance platforms (DeFi) and cryptocurrency-funded infrastructure initiatives promote transparency and increase access to finance, especially in regions underserved by traditional banking. These mechanisms illustrate how traditional and digital finance can be aligned with the UN Sustainable Development Goals (SDGs).

When shaping smart policies, it is important to take into account the experience of green financing practices in different regions. Differences in regional development significantly affect the implementation of green finance [42].

By further deepening cooperation with BRICS countries and countries along the Belt and Road and actively participating in the development of a multilateral sustainable financing catalog, China can play a more influential role in the international arena.

In the UK and Singapore, regulatory tools allow fintech start-ups to test green financial products, such as tokenised renewable energy assets, in a controlled environment.

China's green finance pilot zones provide favorable policies and credit structures aimed at expanding the adoption of clean technologies.

Such initiatives are an example of a shift to more adaptive regulation that empowers market-driven green innovation.

Active participation in global green financial governance, including sharing of China's practical experiences and technological innovations, can contribute to a fairer and more rational international system of green finance. Through multilateral cooperation mechanisms such as the Belt and Road Initiative, the space for international cooperation in green finance can be expanded.

The study of best practices and technologies accumulated by other countries in the field of smart government, green finance is of great importance for the implementation of sustainable development goals. The government can reduce the cost of green financial products through financial subsidies and tax incentives to increase their competitiveness in the market. In addition, the government can create a scientific and accurate method for assessing the environmental benefits of green finance products and improve the internal governance mechanism to ensure healthy and orderly business development of green finance.

5 Conclusion

Monitoring of global policy documents, the work of scientists, made it possible to summarize that global documents are considered separately by smart government policy instruments, green finance policies, sustainable development policies, which made it possible to formulate as the purpose of this study a synthesis of smart policy tools working in a coordinated implementation with transparent, a credible, effective green finance policy for sustainable development and the formulation of practical and theoretical recommendations for their tight implementation.

The tools through which the implementation of smart government policies is implemented in close connection with green finance policies for sustainable development are summarized. Smart government is addressed through digital process management tools at various levels of the economic system: data management, transparency in reporting; digital tools for management and monitoring; stimulation and regulation through digital channels; risk management and strategic planning; intersectoral and interregional coordination.

Digital tools for process control and monitoring have been studied. Smart government tools such as mixed finance models involving public–private cooperation; blockchain-based environmental asset tokenization; biodiversity-related derivative financial instruments; models using artificial intelligence to assess sustainability results contribute to the digital transformation of the accounting system and affect the overall performance of factors, determine the channels of green investments.

Considered smart government tool—risk management and strategic planning. Concluded to address cross-country comparability, the study applies machine learning algorithms to construct composite indices that normalize resilience metrics at vastly different scales. This provides a better understanding of the relationship between economic growth and environmental impact.

It is summarized that the close connection between decision-making tools (machine learning algorithms, artificial intelligence, quantum computing and geospatial analytics) within smart government and green finance policies makes it possible to measure and evaluate environmental externalities in investment decision-making processes.

It is shown that the practical implementation of smart government through risk management and strategic planning through the use of machine learning algorithms, time series analysis, statistical methods, dimensionality reduction methods, cryptocurrency initiative, green bonds and private investment affects changes in green finance policies and indicators of sustainable development.

It was concluded that enhanced international cooperation was a necessary way to promote the global development of green finance through the internationalization of green finance standards and the facilitation of cross-border capital flows.

Strategic recommendations were developed to support green finance for sustainable development in a coherent manner with smart government policy instruments. The recommendations include tools: first, the creation of a comprehensive information-sharing platform to centralize information on financing needs and environmental benefits of green projects to facilitate access to information for potential investors. Second, promoting the trade flow of green financial products to improve market liquidity, which is key to attracting long-term capital to a green financial market. Blockchain technology can also be used to track the flow of green bond funds to ensure funds are actually being used for environmental projects. This transparent mode of operation improves market efficiency and gives regulators an effective monitoring tool. Third, the introduction of an independent assessment agency could strengthen scrutiny of the environmental benefits of green finance projects, thereby reducing the risk of greenwashing. The creation of a special fund is also an effective way to stabilize the source of financing for green financial innovation. These measures can not only attract more financial institutions to participate in green financial practices, but also contribute to the diversification of green financial products. Fourth, to strengthen incentives for green finance, governments should implement pre-designed policy measures: financial subsidies and tax incentives can directly reduce the cost of financing green projects and increase their competitiveness in the market. Fifth, policy frameworks must be regional and sectoral to ensure maximum effectiveness and contextual relevance.

The novelty of the study lies in the synthesis of the unique experience of implementing smart government policies with the tools of green finance policies in close connection to achieve the goals of sustainable development of the national economic system.

References

1. Decree of the President of the Republic of Belarus No. 381 "On Digital Development". https://president.gov.by/fp/v1/910/document-thumb__51910__original/51910.1701783426.b012284ff2.pdf. Last accessed 15 Sep 2025
2. About the project smart cities of Belarus. https://www.mpt.gov.by/ru/o-proekte-umnye-goroda-belarusi. Last accessed 15 Oct 2025
3. Digital state model, smart cities. https://www.government.by/news/model-cifrovogo-gosudarstva-umnye-goroda-seti-5g-roman-golovchenko-nazval-glavnye-zadachi-v. Last accessed 15 Sep 2025
4. XVIII Republican Ecological Forum. https://www.undp.org/belarus/press-releases/xviii-republican-ecological-forum-brings-together-over-400-participants-advance-climate-action-and-circular-economy-and?utm_source=chatgpt.com. Last accessed 15 Oct 2025
5. Electronic state in the Republic of Belarus. Development prospects. https://apni.ru/article/13544-elektronnoe-gosudarstvo-v-respublike-belarus-perspektivy-razvitiya. Last accessed 15 Oct 2025
6. Burdenko EV (2023) E-government models. Issues Innov Econ 1:59–76
7. Shaoxue J (2024) Review of PRC legislation and practice in the development of modern digital law. RUDN J Law 28(3):528–545
8. Xu Y, Guo S (2024) Do green finance policies inhibit the financialisation of manufacturing enterprises? Sustain 16(15)
9. Deng WY (2023) A bibliometric review of domestic green finance research. J Shanghai Univ (Soc Sci Ed) 40(2):87–100

10. Zhang S, Liang BB, Xu SZ, Hou JL (2023) Empirical analysis of green finance and high-quality economic development in the Yangtze River Delta based on VAR and coupling coordination model. Front Environ Sci 11
11. Mavlutova I, Spilbergs A, Verdenhofs A, Kuzmina J, Arefjevs I, Natrins A (2023) The role of green finance in fostering the sustainability of the economy and renewable energy supply: Recent issues and challenges. Energies 16(23)
12. Mudalige HMNK (2023) Emerging new themes in green finance: a systematic literature review. Futur Bus J 9(1):108
13. Sadiq M, Le-Dinh T, Tran TK, Chien F, Phan TTH, Quang Huy P (2023) The role of green finance, eco-innovation, and creativity in the sustainable development goals of ASEAN countries. Econ Res-Ekon Istraživanja 36(2)
14. Nan L, Shinan Q, Ying C (2021) China green bond report
15. Shanghai Pudong Development Bank (2022). https://sustainabilityreports.com/reports/shanghai-pudong-development-bank-spdb-2022-csr-report-pdf/. Last accessed 15 Oct 2025
16. Ziankova IV (2021) Efficient use of resources in the field of energy efficiency through the principles of the circular economy. In: Topical issues 2021: Web of conferences, vol 266
17. Unified information system in the field of procurement. https://zakupki.gov.ru. Last accessed 25 Sep 2025
18. IISD. https://www.iisd.org. Last accessed 25 Sep 2025
19. ICAP. https://www.icapcarbonaction.com/en/. Last accessed 24 Sep 2025
20. World Bank Group. https://www.worldbank.org/en/topic/climatechange/publication/state-and-trends-of-carbon-pricing. Last accessed 24 Sep 2025
21. OECD. https://www.oecd.org/environment/indicators-modelling-outlooks. Last accessed 25 Sep 2025
22. The Sustainable Development Goals Report 2023. https://unstats.un.org/sdgs/report/2023. Last accessed 25 Sep 2025
23. Kickstarter Environmental Resources Center. https://www.kickstarter.com/environment. Last accessed 25 Sep 2025
24. Climate Change Service. https://climate.copernicus.eu/climate-data-store. Last accessed 2025/09/25
25. Roshydromet. https://www.meteorf.ru. Last accessed 25 Sep 2025
26. IPCC. https://www.ipcc.ch. Last accessed 25 Sep 2025
27. Wei L, Yedan S et al (2023) How renewable energy investment, environmental regulations, and financial development derive renewable energy transition: Evidence from G7 countries. Renew Energy 206:1188–1197
28. Wei L, Huainian Z, Haoliang Y (2022) Uncertainty-guided optimal scheduling of automated energy internet. IEEE Trans Industr Inf 18(8):5662–5669
29. Ministry of Economy of the Republic of Belarus. https://economy.gov.by/uploads/files/NSUR/natsionalnaja-strategija-ustojchivogo-razvitija-respubliki-belarus-na-period-do-2040-goda.pdf. Last accessed 25 Sep 2025
30. Wei L, Yedan S et al (2023) Machine learning applications for photovoltaic system optimization in zero green energy buildings. Energy Rep 9:2787–2796
31. Shen Y, Liu W, Dinçer H et al (2025) Hybrid molecular fuzzy recommender systems for impact of climate change on renewable energy performance. Renew Energy
32. Shen Y, Liu W, Yüksel S et al.: A molecular fuzzy decision-making model for optimizing renewable energy investments towards carbon neutrality. Renew Energy
33. Ziankova IU, Lisichonak AP, Yemialyanau AA, Pozdnyakova IA, Gasheva OV Sustainable socio-economic development model: Implementation through regional ICT potential assessment. https://doi.org/10.1051/bioconf/202412001029. Last accessed 15 Sep 2025
34. Ziankova I, Lisichonak A, Yemialyanau A (2019) Development of inclusive society with the instruments of "green" economy and education. In: Environment, technology, resources, 12th international scientific and practical conference, pp 353–358
35. Li Z, Zhang Z (2022) Optimizing web page conversion in e-commerce. Management and marketing: Experience and challenges: Sat. scientific. Works/A.N. Varaksin; under the general ed. V.S. Golik. Minsk, pp 244–249

36. Li Y, Sun Y, Li L (2024) Can green finance promote the development of new quality productivity? Finance and Economy 12:1–11
37. Wang CN (2023) Publications—green finance & development center
38. Liu J, Qi X (2024) Transition finance helps low-carbon transition. China Finance 18:95–96
39. Alharbi SS, Al Mamun M, Boubaker S, Rizvi SKA (2023) Green finance and renewable energy: a worldwide evidence. Energy Econ 118
40. Kuerbis B, Mueller M, Hapuarachchi K (2023) Exploring the role of data enclosure in the digital political economy. Telecommun Policy 47(8)
41. Liu D, Chang Y, Yao H, Kang Y (2023) The impact of green finance on green economy development efficiency: Based on panel data of 30 provinces in China. Front Environ Sci 11
42. Yale School of Management. The importance of climate finance research in addressing climate change
43. Alharbi SS, Al Mamun M, Boubaker S, Rizvi SKA (2023) Green finance and renewable energy: A worldwide evidence. Energy Econ 118
44. Gaibnazarova ZT (2023) Prospects for improving the forecasting of the gross domestic product of regions based on mathematical models. In: 9th IEEE International Conference on Information Technology and Nanotechnology (ITNT), Samara, Russian Federation, pp 1–4. https://doi.org/10.1109/ITNT57377.2023.10139074
45. United Nations (2023) Report of the inter-agency task force on financing for development
46. Ziolo M, Bak I, Cheba K (2020) The role of sustainable finance in achieving sustainable development goals: Does it work? Technol Econ Dev Econ 27(1):45–70
47. Gaibnazarova ZT, Solieva BT, Iminova NA (2024) Forecasting the risk of bankruptcy in poorly formalized processes. Sci & Tech 23(1):80–86. https://doi.org/10.21122/2227-1031-2024-23-1-80-86
48. NSUR—2040. https://economy.gov.by/uploads/files/NSUR/natsionalnaja-strategija-ustojchivogo-razvitija-respubliki-belarus-na-period-do-2040-goda.pdf. Last accessed 15 Sep 2025
49. Gaibnazarova ZT (2024) Artificial intelligence and smart railway transport as a tool to accelerate the transformation process in Uzbekistan. In: Sari M, Kulachinskaya A (eds) Digital transformation: What are the smart cities today? Lecture notes in networks and systems, vol 846. Springer, Cham. https://link.springer.com/chapter/10.1007/978-3-031-49390-4_19
50. Ziankova I, Gaibnazarova Z, Malei A, Sharafanova E, Gasheva O (2025) Digital technologies for comprehensive promotion of a startup. In: Skhvediani A, Kulachinskaya A, Mateeva Petrova M (eds) Digital transformation of socio-economic and technical systems: theory and practice. Lecture notes in networks and systems, vol 1309. Springer, Cham, pp 39–55. https://doi.org/10.1007/978-3-031-85608-2_4
51. Stankevich D, Gaibnazarova Z, Naumovich O, Tuychieva M, Shadieva G (2025) The role of universities in digital transformation: challenges and prospects. In: Skhvediani A, Kulachinskaya A, Mateeva Petrova M (eds) Digital transformation of socio-economic and technical systems: theory and practice. Lecture notes in networks and systems, vol 1309. Springer, Cham, pp 179–193. https://doi.org/10.1007/978-3-031-85608-2_13

Functional Features of Integrated Thinking in Digital Transformation

Natalia Mushketova, Victoria Telyatnikova, Elena Marusinina, Flor Zakharchenko, and Anzhelika Syrbu

Abstract Digitalization in global expansion has simplified and shortened the time for labor operations, but the main attribute has been the publicity of business activities in all sectors of the economy. Thanks to information sites, all areas of activity should be disclosed in a statement of compliance with digital security requirements for the financial and non-financial, public and government sectors, and the household sector. That is why the relevance of the digital economy and sustainable development paradigm is evident, justifying an integrated way of thinking to analyze the feasibility of using tools to transform resources into capital and generate income. The study analyzes the principles of digital and sustainable development in the context of integrated thinking, including: openness; research approach; creativity; empathy; collaboration; consistency; compromise. The study made it possible to clarify the general and specific functions of integrated thinking in digital and sustainable development. The synthesis of general and specific functions of integrated thinking allowed us to develop a map of the functional implementation of integrated thinking (FRIM) in digital and sustainable development.

Keywords Digital transformation management · Functions of integrated thinking · Digital economy · Integrated thinking · Sustainable development · ESG factors

V. Telyatnikova (✉)
Volgograd State Socio-Pedagogical University, Volgograd, Russia
e-mail: shokhnekh@yandex.ru

N. Mushketova · E. Marusinina · F. Zakharchenko
Volgograd State University, Volgograd, Russia

A. Syrbu
The Russian Presidential Academy of National Economy and Public Administration, Volgograd, Russia

© The Author(s), under exclusive license to Springer Nature Switzerland AG 2026
A. Kulachinskaya et al. (eds.), *Digital Transformation: What Kind of Smart Life do we have today?*, Lecture Notes in Networks and Systems 1895,
https://doi.org/10.1007/978-3-032-20582-7_12

1 Introduction

All the attributes of the inventive digital economy at the end of the twentieth century were aimed at chasing a steady increase in corporate profits. However, the information transparency of the digital economy allowed us to analyze that investments aimed at solving social and environmental problems were minimized. That is why, based on public opinion about the need to curb the scale of digital fraud and environmental disasters, a vector of "integrated thinking" is emerging. It is also important to trace the close connection between digitalization processes and sustainable development. In digital transformation, there are opportunities to provide transparent and complete information about the activities of all business sizes [1]. Integrated reports of large corporations make it possible to analyze the processes of transformation of all types of resources into capital, as well as to assess the social value and the level of safety of production [2]. That is why the vector of digital and sustainable development is determined in digital transformation.

Integrated thinking as a managerial approach to holistic thinking about the resources and relationships that a corporation uses or influences, as well as the dependencies and trade-offs between them as value is created. Therefore, the routes of successful business in terms of two-way benefits are determined.

(1) on the one hand, the cost to the company is increasing;
(2) on the other hand, at the same time, value is created for stakeholders [3, 4].

Therefore, it is possible to clarify the principles of digital and sustainable development in the context of integrated thinking.

1. The principle of openness as a willingness to consider new ideas and points of view.
2. The principle of research as a desire to understand complex situations and identify hidden relationships.
3. The principle of creativity as the ability to create new and original solutions with the support of intelligent neural systems.
4. The principle of empathy as the willingness and ability to understand and consider the interests of other people.
5. The principle of cooperation as a willingness to work in a team and share knowledge in a digital environment.
6. The principle of universal consistency presupposes the ability to perceive a problem as part of a broader system.
7. The principle of compromise as a willingness to find solutions that take into account the interests of all stakeholders in the digital environment.

In terms of applying the principles of integrated thinking, it is possible to combine user needs, technological capabilities and regional business objectives to form competitive regional markets [5]. Integrated thinking as a skill for solving complex problems in the modern world allows you to take into account different points of view, identify the relationships between the elements and create completely new ideas that satisfy all stakeholders. Developing an integrated mindset of digital and sustainable development requires openness, research skills, creativity, empathy, and willingness to collaborate.

2 Methods

To study the general and specific functions of integrated thinking in digital transformation, complex and systematic approaches were used, including various methods of collecting, analyzing and synthesizing information. Source materials: articles in scientific and professional economic journals devoted to management technologies; materials of international conferences. Research methods: a systematic review of integrated thinking in digital and sustainable development; meta-analysis of the results of several studies to obtain more general conclusions; content analysis to identify key concepts and relationships; synthesis and interpretation of results.

In modern conditions of digital and sustainable development, many interrelated problems are being addressed, such as: limited access to public information; climate change; resource scarcity; social and digital inequality that generates poverty; concealment of information about environmental pollution. Externalities, both positive and negative effects, are manifested in production processes [6]. Such externalities are difficult to trace in the limited access to information. However, production effects involuntarily affect the interests of various stakeholders, including government, business, the public, public organizations and scientific institutions [7].

It is possible to identify the general functions of integrated thinking in digital and sustainable development, which are: (1) in building a link between the present and the future in the digital environment; (2) in minimizing the risks of damage in the information, social and environmental spheres; (3) in identifying prospects and opportunities for mastering digital skills; (4) in developing and implementing flexible strategies for digital and sustainable development; (5) in forming a willingness to adapt to new paradigms and challenges (Fig. 1).

Integrated thinking combines the economic sustainable success of a corporation, the strategic orientation and risk-based approach, the needs and interests of the corporation's stakeholders into a single system, creating a complex multifactorial goal-setting model and requirements for measuring its effectiveness and efficiency [8, 9].

Of course, the increasing trends aimed at protecting national information, economic interests, social responsibility and the environment affect the ontology of economic activity, which in the traditional context was synonymous only with income and profit [10].

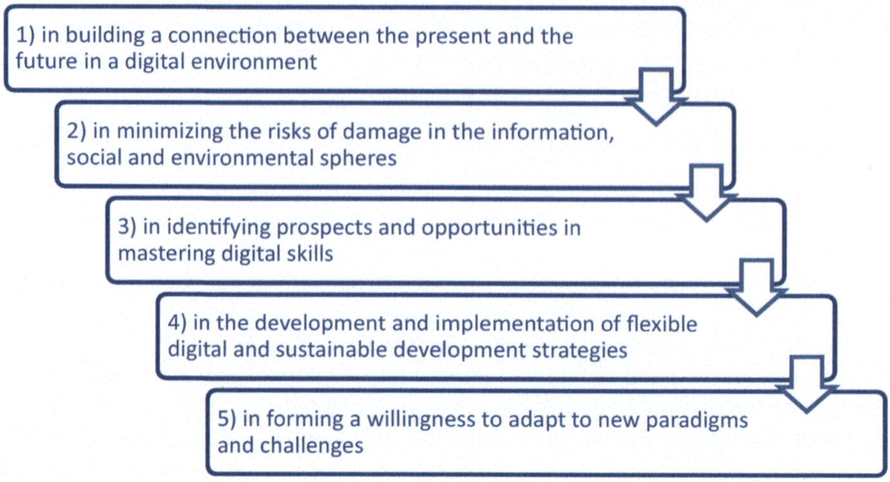

Fig. 1 General functions of integrated thinking in digital and sustainable development

3 Effective Part

The specific functions of integrated thinking in digital and sustainable development are manifested in 10 procedures that affect strategic management (Fig. 2).

The study shows that the role of integrated thinking is manifested in 13 fundamental functions: (1) substantiating the values of economic processes and forming ideas about value creation and information disclosure; (2) ensuring the institutionalism of processes, implementing in the digital environment the presentation of financial and non-financial data on ESG factors; (3) providing evidence-based analysis in the digital environment of the expediency of economic processes, optimizing management decisions in terms of value creation in the medium and long term perspective, avoiding short-term gains; (4) ensuring minimization of risks of economic processes in the digital environment, leveling risks and values for the holistic disclosure of public information; (5) providing mechanisms for planning economic value creation processes taking into account ESG factors [11]; (6) providing an institutional approach to economic processes, implementing cross-functional cooperation between corporations and institutions in partnership; (7) providing an innovative approach to economic processes in the digital environment, implementing principles and methods of strategizing new innovative business models; (8) ensuring the fulfillment of the requirements of ESG factors in economic processes, which determines a coordinated integrated information space in digital and sustainable development, taking into account ESG factors [11]; (9) ensuring the development and implementation of a corporate culture in digital and sustainable development; (10) ensuring the implementation of the principles of positive reputation, forming a positive image of an economic entity contributing to contribution to

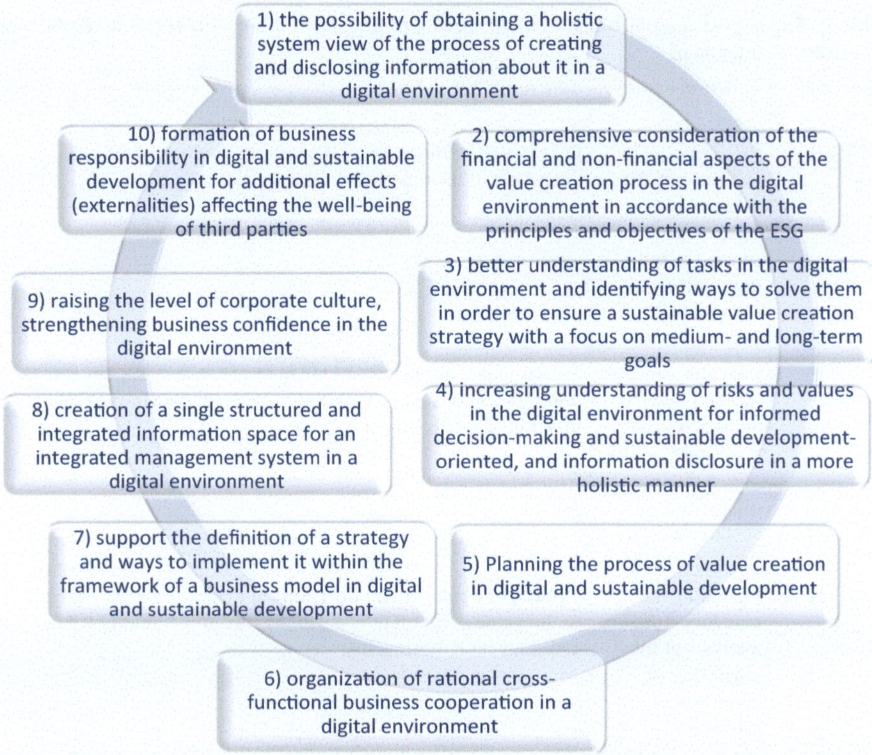

Fig. 2 Specific functions of integrated thinking in digital and sustainable development

digital and sustainable development; (11) providing support for digital implementation in management, coordinating digital transformation in all areas of activity at the micro, meso, and macro levels; (12) ensuring a balance of interests in managing economic processes, integrating psychological, sociological, strategic, and environmental involvement of human resources at the micro, meso, and macro levels in order to ensure consistency between employers, employees and society ("Model De Prince") [12]; (13) providing a circular economic approach in digital and sustainable development, leading to the introduction of the principles of circular economic closed cycles in human resource management, implementing: (1) continuing education; (2) disclosure of internal mobile giftedness; (3) launching mechanisms for preserving organizational knowledge and improving competencies; (4) algorithm for researching and unlocking the potential of human resources; (5) preventive mechanisms for professional burnout of human resources [13].

The synthesis of general and specific functions of integrated thinking allowed us to develop a map of the functional implementation of integrated thinking (FIITH) in digital and sustainable development (Table 1).

Table 1 The map of the functional implementation of integrated thinking (FIITH) in digital and sustainable development

Function code	Function name	Functional procedures
1-FIITH	Providing evidence of the value of economic processes in digital and sustainable development	Provides the formation of ideas about the process of value creation and information disclosure
2-FIITH	Providing an institutional approach to economic processes in digital and sustainable development	Substantiates the institutionalism of processes, implements the presentation of financial and non-financial data on ESG factors in a digital environment.;
3-FIITH	Providing an evidence-based analysis of the feasibility of economic processes	Optimizes management decisions in evidence-based analysis to the value creation process in the medium and long term, avoiding short-term gains
4-FIITH	Ensuring minimization of risks of economic processes	Eliminates risks and values in the digital environment for the holistic disclosure of public information
5-FIITH	Providing mechanisms for planning economic processes	Implements a mechanism for planning the value creation process, taking into account ESG factors
6-FIITH	Ensuring a coordinated partnership approach to economic processes	Implements cross-functional cooperation of corporations and institutions in a partnership environment
7-FIITH	Providing an innovative approach to economic processes	Implements principles and methods of strategizing new innovative business models in the digital environment

(continued)

Table 1 (continued)

Function code	Function name	Functional procedures
8-FIITH	Ensuring compliance with the requirements of ESG factors in economic processes	It determines a coordinated integrated information space in the management of sustainable development and the application of ESG-factor requirements
9-FIITH	Ensuring the development and execution of corporate culture	Forms and ensures the implementation of corporate culture
10-FIITH	Ensuring the implementation of the principles of positive reputation	Creates a positive image of an economic entity contributing to sustainable development
11-FIITH	Providing support for digital implementation in management	Coordinates digital transformation in all areas of activity at micro, meso, and macro levels
12-FIITH	Ensuring a balance of interests in managing economic processes	Ensures the integration of psychological, sociological, strategic and environmental engagement in order to establish a balance of interests between employers, employees and society (the "De Prince Model")
13-FIITH	Providing a circular economic approach to managing economic processes	Determines the implementation of circular economically closed cycles in digital and sustainable development by implementing: continuing education; unlocking the potential and inner giftedness of human resources; launching mechanisms for preserving organizational knowledge; a mechanism for prevention of human resource burnout

Of course, for the analysis and evaluation of FIITH in digital and sustainable development, it is important to study the indicators. The application of the integrated thinking implementation function map can be used to form a monitoring working document that allows identifying the conformity of the modern concept of digital and sustainable development.

Obviously, in digital and sustainable development, it is assumed that security conditions are provided taking into account the requirements of ESG standards. ESG standards, which define a comprehensive rating assessment of a variety of indicators, dictate the security requirements of economic systems, which are measured by non-financial indicators of information, environmental, social and corporate content.

Such non-financial indicators are determined by factors identified in sustainable development as ESG factors.

4 Discussion

The results show that it is important to develop and apply integrated thinking among management personnel in the context of digital transformation. Many authors in their research focus on the development of integrated thinking only towards sustainable development and the formation of integrated reporting [4, 9, 11]. According to Ignatva I. and Shmarova L., integrated thinking is aimed at providing stakeholders with transparent information about the activities of an economic entity, in the integration of financial and non-financial information. The availability of integrated information opens up the possibility of entering international markets. Reading information about the requirements of international business and forming your own information about the state of affairs is an important condition for business functioning [13–15].

However, integrated thinking has a more advanced functionality that allows taking into account the paradigms of the digital economy, sustainable development [15], and behavioral economics. The authors of this study believe that the functional features of integrated thinking in digital transformation are manifested in the willingness and ability to apply the tools of foresight development (35–50 years into the future) [5, 16], in the process of producing any goods, assessing the level of safety, usefulness and long-term value of the transformation of dwindling resources [17].

5 Conclusion

Modern society is aimed at addressing issues of long-term value creation in all sectors of the economy, taking into account all the consequences of decisions taken for safe production. It is obvious that factors of uncertainty and high risks in the political, legal, economic, social, and environmental fields require adaptation to changing conditions. The relevance of the study of the specific functions of integrated thinking is determined by the desire to: (1) obtain in the digital environment a holistic systematic understanding of the process of creating and disclosing information; (2) comprehensive consideration of the financial and non-financial aspects of the value creation process in the digital environment in the requirements of ESG standards; (3) enhanced understanding of tasks in the digital environment and identification of ways to solve them in order to ensure a sustainable value creation strategy with a focus on medium- and long-term goals; (4) justification of the possible emergence of risks and value in the digital environment to make informed decisions and focus on sustainable development, and to make information disclosure more holistic; (5) planning the value creation process in digital and sustainable development; (6)

organization of rational cross-functional business cooperation in the digital environment; (7) substantiation of the strategy and ways of its implementation within the framework of the business model in digital and sustainable development; (8) creation of a single structured and integrated information space in the digital environment; (9) raising the level of corporate culture, strengthening business confidence in the digital environment; (10) responsible attitude of senior personnel for the manifestation of additional effects (externalities) affecting the welfare of third parties.

References

1. Glinskaya O, Shokhnekh A, Prokopova O, Skorikova I (2019) The forecasting approach to strategic development of economic subjects. Book: Lecture notes in networks and systems. The future of the global financial system: Downfall or Harmony, vol 57. Springer Nature, Cham, Switzerland, pp 829–835. https://doi.org/10.1007/978-3-030-00102-5_89. EDN ZIJMUP
2. Kail Y, Shokhnekh A, Samsonova M, Lamzin R, Elsukova Y (2019) Prospects of GR management development in the context of forming a digital economy. Espacios 40(22):30. EDN MTRZRG
3. Abu Ezza H, Shokhnekh A, Telyatnikova V, Mushketova N (2020) Quality parameters of information systems for business in the context of digital transformations. In: E3S web of conferences, Yekaterinburg, vol 208, p 03059. https://doi.org/10.1051/e3sconf/202020803059. EDN FQXXQW
4. Vakhrushina M, Katasonova M, Vakhrushina S (2023) Integrated thinking: issues of theory and practical application. Int Account 26(1)(499):28–51. https://doi.org/10.24891/ia.26.1.28. EDN KWPRWT
5. Sidunova G, Shokhnekh A (2019) Balance-cognitive approach to foresight-development of management audit efficiency digital economic systems in the context of strategic drift. The future of the global financial system: Downfall or Harmony, vol 57. Springer Nature, Cham, Switzerland, pp 657–665. https://doi.org/10.1007/978-3-030-00102-5_68. EDN IIVQAN
6. Shokhnekh A, Skiter N, Rogachev A, Pleschenko T (2017) Features of optimal modeling of tax mechanisms in the leveling system of environmental and food security taking into account inter-industry externalities. J Environ Manag Tour 8(1)(17):100–104. https://doi.org/10.14505/jemt.v8.1(17).10. EDN XNNZJP
7. Oudah AlMMY, Shokhnekh A, Glinskaya O, Shokhnekh M, Chusov I (2019) Regional mechanisms of modernization of infrastructure of regions and the country in the system of innovational development. Book: Specifics of decision making in modern business systems: Regularities and tendencies. Emerald Publishing Limited, pp 209–227. EDN BENXZB
8. Bryukhanova N, Makolov V (2023) A stakeholder approach to the formation and use of key indicators in the management of organizations and in assessing the financial benefits of their activities. Siberian Financ School 4(152):162–174. https://doi.org/10.34020/1993-4386-2023-4-162-174.-EDNBAFKHM
9. Safonova IV (2023) Development of the corporate reporting model in the concept of an ecosystem approach. Accounting and Statistics 20(2):107–116. https://doi.org/10.54220/1994-0874.2023.42.91.010. EDN XWMFFX
10. Kayl II, Shokhnekh AV, Lamzin RM, Samsonova MV, Epinina VS (2020) Effective technologies and methods of GR management that stimulate increase of the level of socio-economic development of territories and quality of life. In: Conferense 2020, vol 91. Nizhny Novgorod. Springer, Cham, pp 191–207. https://doi.org/10.1007/978-3-030-32015-7_23. EDN SAWZZK
11. Posulikhina N (2022) ESG standards and assessment (expertise) of integrated project security. Actual Probl Russ Law 8:104–110

12. De Prins P., Van Beirendonck L., De Vos A, Segers J (2014) Sustainable HRM: Bridging theory and practice through the' Respect Openness Continuity (ROC)'-model. Management revue, pp 263–284
13. Lukashevich NS (2025) Sustainable human resource management: models, principles and paradoxes. In: Conference 2025. Minsk, vol 1, pp 96–97. EDN LDBBGY
14. Ignatova I, Shmarova L (2024) Challenges of integrated reporting implementation and their solutions. Izv Vyss Uchebnykh Zavedenii Teknol Tekstil' noi Promyshlennosti 6(414):125–134. https://doi.org/10.47367/0021-3497_2024_6_125. EDN GICPEK
15. Skiter N, Aljumaili Q, Almuslehi R, Shokhnekh A (2023) Accounting and analytical information support of financial strategy in the paradigm of sustainable development. In: E3S web of conferences 2023, vol 371. EDP Sciences, Tashkent, Uzbekistan. p 05029., https://doi.org/10.1051/e3sconf/202337105029. EDN GMVTJJ
16. Shokhnekh A, Skiter N, Boriskina T, Ketko N, Avdeeva IN (2024) Genesis standardization of integrated reporting in the context of sustainable development and climate change risks. In: BIO web of conferences, vol 141, p 04040. https://doi.org/10.1051/bioconf/202414104040. EDN RHCLAZ
17. Kayl I, Shokhnekh A, Lamzin R, Syrbu A, Elsukova Y (2020) Foresight study of the limits of development and application of the modern tools of GR management in view of the socio-economic specifics of territories' development. In: Conferense 2020. Nizhny Novgorod, vol 91. Springer, Cham, pp 179–190. https://doi.org/10.1007/978-3-030-32015-7_22. EDN CDIUGK

Smart Built Environment

Digital Transformation of Information Support for the Development and Visualization of Architectural Projects

Anna Shokhnekh, Natalya Skiter, Natalia Ketko, Adel Shokhnekh, Yuri Kolyshev, and Tatyana Boriskina

Abstract The study reveals the effects of the digital transformation of information support for the development and visualization of architectural projects: (1) the concept of digital culture in the architectural environment is studied; (2) the evolution of digital architecture is traced in parallel with the development of computer technology and software; (3) the concept of parameterization through the architectural epic and architectural styles is studied; (4) the relationship of the genesis of parameterization with the architectural classification of styles is traced; (5) the directions of work in the application of digital and virtual tools are investigated; (6) the advantages of working in a digital architectural environment are analyzed; (7) the stages of parametric design work in an architectural digital environment are classified. Of course, the software and information support for the development and visualization of architectural projects in a digital environment has enormous advantages, allowing: (1) quickly create 2D and 3D models, drawings, diagrams in AutoCAD, ArchiCAD, AutoCAD Architecture, Revit, Allplan, Vectorworks; (2) carry out accurate geometry modeling; (3) develop construction documentation (plans, sections, facades); (4) create and expand libraries of elements and materials; (5) integrate into intelligent tools. Digital transformation improves architectural information support, including: parameterization of an architectural object; adaptability of an architectural object to the external environment and internal environment; development of a library on the specifics of an architectural object; machine learning in the design of an architectural object; analysis and evaluation of technological and economic efficiency.

Keywords Digital transformation · Software and information support · Architectural digital environment · Visualization · Architectural object

A. Shokhnekh (✉)
Volgograd State Socio-Pedagogical University, Volgograd, Russia
e-mail: shokhnekh@yandex.ru

N. Skiter · N. Ketko · A. Shokhnekh · Y. Kolyshev · T. Boriskina
Volgograd State Technisch University, Volgograd, Russia

© The Author(s), under exclusive license to Springer Nature Switzerland AG 2026

A. Kulachinskaya et al. (eds.), *Digital Transformation: What Kind of Smart Life do we have today?*, Lecture Notes in Networks and Systems 1895,
https://doi.org/10.1007/978-3-032-20582-7_13

1 Introduction

In the information and technological context, modern society is assessed by the level of digital culture, which provides a variety of human needs. The cultural environment of digital technologies manifests itself in an abundance of information flows, many of which lose their relevance quite quickly. Aspects of the formation of software and information support in the digital environment is the most pressing issue today [1, 2]. The issues of using intelligent systems are also subject to development and reflection from the perspective of digital culture in all industries [3, 4]. Architecture as an art and technology focuses on socio-cultural and scientific-technical discoveries. Digitalization, robotization and virtualization of many processes in architectural design require the expansion of technological attributes. In such conditions, the transition from traditional to digital design becomes an indicator of digital culture in the architectural environment. In the digital architectural space, design is classified from the position of: (1) concepts of the architect, which is made on a machine; (2) concepts of machines and algorithms according to the architect's terms of reference.

The aesthetic and artistic component highlighted by M. A. Shuklina in digital architecture manifests itself as a design paradigm in the creative approach and creative process. In digital architecture, creativity is consistent with related research, analytical work, and the creation of new algorithms for programs and applications. Computer technology is used as a tool for creative search and research, allowing you to quickly perform mathematical calculations [5].

Research shows that many authors associate the emergence of new trends in architecture with digital culture, including:

(1) Parametric modeling and generative design technologies;
(2) The introduction of artificial intelligence in design.

Obviously, the use of digital architecture makes it possible to quickly decode information about the environment [6].

It is in digital architecture that modern information technologies are integrated into the design and construction process. In digital architecture, approaches to the creation of buildings and spaces are being transformed, revealing new opportunities and solutions [7].

2 Methods

To study the software and information support for the development and visualization of architectural projects in a digital environment, integrated and systematic approaches were used, including various methods of collecting, analyzing and synthesizing information. Source materials: articles in scientific and professional architectural journals devoted to CAD, BIM, visualization, VR/AR in architecture, project management technologies and engineering calculations; materials from international conferences on CAD, BIM and architectural informatics (for example, eCAADe, CAAD futures, ASCAAD). The objects of the study were real architectural projects developed using various software. Research methods: systematic review of architectural design; meta-analysis of the results of several studies to obtain more general conclusions; content analysis of the content of texts (articles, documents, websites) to identify basic concepts and relationships; synthesis and interpretation of the results.

The evolution of digital architecture can be traced in parallel with the development of computer technology and software.

According to the research of T. A. Gevorgyan and B. L. Valkin, it was found that computer technologies in the mid-1950s did not have sufficient computing power and the capabilities of graphic editors to output high-quality images. During this period, symbolic graphics were used, which was called ASCII graphics and ASCII Art. It wasn't until 1959 that the Massachusetts Institute of Technology used a graphics console for the first time on a new model of the TX-2 computer. In 1961, it became possible to create the simplest geometric shapes, after the invention of the computer program "Sketchpad" (Sketchpad). However, it was the advent of graphic editors that made it possible to assemble digital architecture objects on the first electronic computers (mid–60–70 s of the XX century). In the 70 s of the twentieth century, French scientist, mathematician and engineer Pierre Bezier invented a curve that allows describing any complex shapes. T. A. Gevorgyan and B. L. Valkin defines the 80 and 90 s of the twentieth century as the period of the breakthrough of digital graphics, which was actively used in the film industry, and then in the techno-utopian aesthetics of Marcos Novak (1990 years) [8].

A scientific analysis of the history of digital architecture formation conducted by N. E. Sharipbek led to the conclusion that Frank Gehry was among the first architects to use computer programs. The widespread use of the CATIA computer-aided design system allowed Frank Gehry to create many projects in the digital architectural environment, including the sculpture "Swimming Fish" for the Barcelona Olympics (1992) [6].

An architectural epic designed and presented in constructed objects embodies grandiose, monumental structures or entire ensembles reflecting the features of history, myths and culture of entire peoples or epochs. It is obvious that the idea in the architectural epic is embodied through parametric harmony, combining many architectural styles. It is through parametric settings that styles in architecture are defined, reflecting approaches to aesthetics. Architectural styles have been formed since ancient history, reflecting the imprint of various styles in certain features.

(1) Epoch and culture, collecting and revealing values, peculiarities of time, place of origin;
(2) The materials of architectural objects reflect the commitment and tradition of history (stone, wood, brick, concrete, glass, metal);
(3) Types of structures used that reflect the appearance and functionality of structures (arches, arches, domes, beams, columns);

(4) Types of decor that realize individuality, aesthetic ideals of style through decorative elements of ornaments, sculptures, frescoes;
(5) Functionality that allows you to implement the purpose and role of architectural objects.

Architectural parameterization made it possible to identify and classify the most famous architectural styles, including: ancient Greek, embodying columns, porticos, pediments, proportions, order system; ancient Roman, characterized by arches, vaults, domes, aqueducts, amphitheater; the Middle Ages, represented in Romanesque style (massive walls, semicircular arches, small windows) and Gothic (pointed arches, high vaults, stained glass windows, buttresses, pinnacles).

Of course, for a long time, there has not been a great variety of parametrization of ancient styles and styles of the Middle Ages. However, the post-Medieval and modern world are characterized by many forms, the parameters of which are classified and include such as:

- Renaissance (symmetry, proportions, classical orders, domes, pilasters);
- Baroque (splendor, dynamics, curvilinear forms, abundance of decor, theatricality);
- Rococo (grace, lightness, asymmetry, pastel colors, ornaments with shells);
- Classicism (rigor, proportions, symmetry, columns, porticos, Roman motifs);
- Empire (imperial style, monumentality, triumphal arches, military symbols);
- Art Nouveau (smooth lines, floral motifs, asymmetry, decorative art);
- Art Deco (geometric shapes, luxury, exotic materials, zigzags);
- Constructivism (functionality, practicality, geometric shapes, rejection of decor);
- Functionalism (practicality, convenience, lack of unnecessary details);
- Brutalism (raw concrete, massive shapes, functionality);
- Postmodernism (rejection of strict rules, mixing of styles, humor, irony);
- High-tech (the use of new technologies, open communications, metal structures, glass);
- Minimalism (simplicity of forms, functionality, lack of decor);
- Eco-architecture (the use of environmentally friendly materials, energy efficiency, integration with nature).

It should be noted that in reality, it is often eclectic buildings that combine elements of different styles and unique architectural traditions.

Currently, compliance with the parametric features of styles and forms of architectural images is becoming more accurate, with minimal time costs in the context of an architectural digital environment.

The transition to parametric software has changed the architectural language of communication, with its traditional plans, sections and sections [9].

As the newest architectural style, "parametrism", according to the research of K. Kasulu, O. V. Volichenko, represents a continuous differentiation and adaptability of forms supported by computational processes. The aesthetic potential of architectural practice significantly enhances parametric design, allowing you to create unique projects. Control over complex geometric shapes is carried out using computational design. The accuracy of calculations, taking into account the peculiarities of external conditions, firstly, enriches architectural expressiveness, and secondly, increases the functional adaptability and stability of architectural objects. Aesthetics are fully achieved through the application of mathematical rigor, environmental sensitivity, and structural characteristics [10].

In the work of A. V. Golyatin, E. A. Yerysheva and T. A. Demidova, an essential function is attributed to parametric modeling based on the creation of digital algorithms that control the geometry of facade elements taking into account the specified parameters. The functional features of digital techniques are implemented through the automation of facade solution generation processes using digital parametric and algorithmic methods [11].

The implementation of methods in the digital architectural environment is based on information modeling of the BIM construction model [12].

Digital architecture, according to the classification of the work carried out by A. B. Mottaeva, V. L. Kashintseva and I. A. Kubrak, includes two main areas of work: (1) the use of digital tools; (2) the use of virtual tools [13].

1. Application of digital tools to object modeling (BIM) in conditions of: (1) the application of a digital model of an architectural object; (2) the implementation of management of the entire life cycle of the object; (3) the integration of BIM data on design, construction and operation in various areas to increase the effectiveness of coordinated work on the project; 4) coordination and reduction of errors, conflicts [13].
2. Application of virtual tools to design and construction (VDC) in conditions of: (1) forecasting and analysis of construction stages using 3D, 4D, 5D and other models; (2) consistency of efficiency and quality of the construction industry; (3) the possibility of developing the design of traditional approaches to construction in various industries [13].

Models in digital architecture are distinguished by their functionality, environmental friendliness, and dynamism. Architect's drawing (Fig. 1, authored by Prof. Kolyshev Yu.B.) replaced digital tools (Fig. 2, authored by Prof. Kolyshev Yu.B.) with parametric 3D design, taking into account the interrelationships between objects in a digital architectural environment.

In the conditions of the natural and ecological situation in the world, the resource potential of digital architecture makes it possible to increase the functionality, environmental friendliness and dynamism of buildings. The architect's work in the digital design environment is aimed at enhancing the value, usefulness and suitability of the created objects, while at the same time ensuring the brightness of the historical and local history idea.

The study shows that currently in the digital environment, software and information support for the development and visualization of architectural projects offers the ability to quickly create 2D and 3D models, drawings, diagrams in AutoCAD, ArchiCAD, AutoCAD Architecture, Revit, Allplan, Vectorworks.

Fig. 1 Drawing sketch of the architect (author professor Kolyshev Yu.B.)

Fig. 2 Parametric 3D design in the digital architectural environment of an architectural artist (author Professor Kolyshev Yu.B.)

3 Effective Part

The use of digital architecture tools makes it possible to introduce a variety of forms and states into projects through: dynamism; asymmetry; unevenness; non-linearity [6].

In the digital design environment, the automation and optimization of the design of structures of an architectural object takes place from the point of view of the choice of materials, rhythm and composition.

The study allowed us to determine the stages of parametric work in a digital environment, including: (1) highlighting the design features of the parameterization of an architectural object in a digital environment; (2) identifying the features of the adaptability of an architectural object to the external environment and internal environment; (3) substantiating the features of developing a library based on the specifics of an architectural object; (4) identifying approaches to machine learning in the design of an architectural object; (5) developing approaches to assessing technological and economic efficiency. In general, the stages of project work in an architectural digital environment and their characteristics are presented in Table 1.

As can be seen from the study, all stages of architectural work in the digital environment towards the formation of information support involve such procedures as: analysis of the patterns of architectural elements; analysis of the modular division of an architectural object; analysis of the generation of an integral architectural object and its social value; analysis of expressive architectural appearance; analysis of the internal organization of architectural appearance; analysis of harmonization between the internal organization and external appearance; analysis of the flexibility of design solutions; analysis of arched and window modules; analysis of angular and variable elements; analysis of patterns in architectural solutions; analysis of successful projects; analysis of suggestive architectural solutions; analysis of the balance between automation and creativity; analysis of project cost reduction; analysis of the quality level of the project in parametric modeling; analysis of the harmonization of economy, functionality in architectural expression.

4 Discussion

The study shows that digital transformation affects all stages of work in the architectural environment and is an urgent topic for discussion. Scientists and practitioners discuss digital culture and ethics in creative work [6, 7], and explore new styles that become a continuous differentiation and adaptability of forms supported by computational processes. They discuss the aesthetic potential of architectural practice, which is complemented by parametric design, which makes it possible to create unique projects in a digital environment [10, 11]. They study the norms of digital environment control over complex geometric forms of design [11, 12]. The approaches to classifying the work carried out are being investigated, including both the use of

Table 1 Stages of parametric design work in an architectural digital environment

Stage	Procedure	Characteristics of the procedure
1. Design features of parameterization of an architectural object in a digital environment	1.1 Analysis of patterns of architectural elements	Geometry features Features of the parameters Design features Mounting features
	1.2 Analysis of the modular division of an architectural object	Features of individual functional modules Features of individual visual modules Features of adaptability of individual modules into a complete object design
	1.3 Analysis of the generation of an integral architectural object and its social value	Features of design constraints Features of aesthetic limitations Features of economic restrictions
2. Features of the adaptability of an architectural object to the external environment and the internal environment	2.1 Analysis of expressive architectural appearance	Features of facade elements Features of the location of the external zones adjacent to the object Features of external functional areas
	2.2 Analysis of the internal organization of the architectural appearance	Features of the orientation of the premises Features of window openings and natural lighting Features of communication placement
	2.3 Analysis of the harmonization between internal organization and external appearance	Ventilation features Features of stairwells and elevators Acoustic Comfort Features Features of thermoregulation
3. The specifics of developing a library based on the specifics of an architectural object	3.1 Analysis of the flexibility of design solutions	Features of the modular library of individual elements Features of adaptability of individual elements to different conditions Features of variability and unification
	3.2 Analysis of arched and window modules	Features of geometric parameters Features of mounted parameters

(continued)

Table 1 (continued)

Stage	Procedure	Characteristics of the procedure
	3.3 Analysis of angular and variable elements	Features of corner modules Features of random offsets Features of the compositions Features of creating expressiveness
4. Features of machine learning in the design of an architectural object	4.1 Analysis of patterns in architectural solutions	Features of analysis algorithms Features of automatic offers Features of collecting and processing data on completed projects
	4.2 Analysis of successful projects	Features of the analysis of aesthetically successful compositions Features of the analysis of functionally successful compositions
	4.3 Analysis of suggestive architectural solutions	Features of the output of suggestive solutions Features of the output of the most suitable combinations Features of the selection of energy-efficient combinations Features of selecting economical combinations
	4.4 Analysis of the balance between automation and creativity	Features of quality assessment of architectural solutions Features of estimating the reduction of time costs Features of increasing the level of individuality of architectural objects
5. Features of the assessment of technological and economic efficiency Stage 1. Design features of parameterization of an architectural object in a digital environment	5.1 Analysis of project cost reduction	Features of comparative analysis of time expenditures between manual and automated modeling of an architectural project Features of using digital variation of unique elements

(continued)

Table 1 (continued)

Stage	Procedure	Characteristics of the procedure
	5.2 Analysis of the project quality level in parametric modeling	Features of reducing testing procedures and adapting to the external conditions of the facility location Features of the selection of environmental materials Features of the selection of the style of cultural and local history values
	5.3 Analysis of the harmonization of economy and functionality in architectural expression	Features of cost estimation of individual modules Features of the selection of modern and economical materials Features of the selection of environmentally friendly and safe materials Features of extending the functionality of an architectural object

Source Author's development

digital tools and the use of virtual tools. The study reveals the positive impact of digital transformation on architectural models that are functional, environmentally friendly, and dynamic [12, 13].

The study structures the stages of parametric work in a digital environment using intelligent systems: highlighting the technical specifications for the parameters of an architectural object; identifying the conditions for the adaptability of an architectural object to the external environment and internal environment; expanding the library on the specifics of an architectural object for intelligent systems; providing conditions for machine learning in the design of an architectural object; (5) developing approaches to assessment of technological and economic efficiency.

5 Conclusion

The impact of the digital transformation of information support for the development and visualization of architectural projects involves the use of a range of tools and technologies that allow architects and designers to create, visualize and collaborate on projects in a digital format. Digital transformation covers a wide range of tasks, from the creation of conceptual sketches to the detailed development of construction documentation. Improve the accuracy and quality of design: CAD and BIM provide more accurate modeling and error detection at an early stage. Of course, the transformation process is accompanied by positive effects.: (1) automation of many tasks allows you to develop projects faster and optimize costs; (2) BIM and collaboration tools simplify communication and coordination; (3) 3D modeling and rendering allow you to create realistic images and animations for presentation to customers; (4) engineering calculation tools allow you to optimize the project in terms of energy consumption and environmental impact Wednesday; (5) BIM provides information at all stages of the building's life cycle, from design to operation.

References

1. Abu Ezza H, Shokhnekh A, Telyatnikova V, Mushketova N (2020) Quality parameters of information systems for business in the context of digital transformations. In: E3S web of conferences, Yekaterinburg, vol 208, p 03059. https://doi.org/10.1051/e3sconf/202020803059. EDN FQXXQW
2. Shokhnekh A, Skiter N, Boriskina T, Ketko N, Avdeeva IN (2024) Genesis standardization of integrated reporting in the context of sustainable development and climate change risks. In: BIO web of conferences, vol 141, p 04040. https://doi.org/10.1051/bioconf/202414104040., EDN RHCLAZ
3. Telyatnikova V, Shokhnekh A, Kolyshev O, Bespyatykh V, Azarskaya M (2020) Approaches to formation of an innovative platform of university complexes in the strategy of human resources development as small business entrepreneurs: Growth poles of the global economy: Emergence, changes and future perspectives, vol 1, pp 1363–1372. Springer, Cham (2020). https://doi.org/10.1007/978-3-030-15160-7_139. EDN JHUHTO
4. Lomakin N, Sazonov S, Polianskaia A, Shokhnekh A, Lukyanov G, Gorbunova A (2020) Hadoop and deductor based digital AI system for predicting cost of innovative products in conditions of digitalization of economy. Conference, 2020, Saint-Petersburg, vol 1, p 3373810. https://doi.org/10.1145/3372177.3373810. EDN WIVAYK
5. Shuklina M (2014) Aesthetic and artistic features of digital architecture. Book: Science, education and experimental design. In: Proceedings of the MARCHI: Proceedings of the international scientific and practical conference. Collection of articles, Moscow Architectural Institute (State Academy), vol 1, pp 198–201. EDN TSTNSR
6. Sharipbek N (2021) The history of the formation of digital culture and its impact on modern architecture. Bull Sci Educ 17–1(120):78–82. EDN MBPEKO
7. Allakulyeva D, Atayeva A, Amanov A (2024) Digital architecture: the influence of technology on design. Bull Sci 3, 10(79):966–969. EDN SNRYAA
8. Gevorgyan T, Valkin B Aesthetics of virtual digital architecture. Architecture and modern information technologies, vol 2, no 51, pp 362–372. https://doi.org/10.24411/1998-4839-2020-15120., DN ELTDDZ
9. Kiseleva M (2022) Digital architecture in modern society 5. In: Conference 2022, Yekaterinburg: Yekaterinburg Academy of Contemporary Art (Institute), vol 1, pp 365–368. EDN NRQGFK
10. Kasulu K, Volichenko O (2025) Parametric design as a catalyst for modern architectural aesthetics. In: Conference 2025, Ivanovo, Ivanovo State Polytechnic University, vol 1, pp 31–33. EDN ZGHBXY
11. Golyatin A, Yerysheva E, Demidova T (2025) Architectural and artistic organization of facades by means of parametric modeling. Architecture and design: history, theory, innovations. EDN BHLITH 9:196–201

12. Dubynin N (2022) Digital architecture. Academia. Arch Constr 1:19–28
13. Mottaeva A, Kashintseva V, Kubrak I (2023) Actual trends in the digital transformation of the Russian construction industry. Bull Sib Inst Bus Inf Technol 12(4):98–104. https://doi.org/10.24412/2225-8264-2023-4-98-104.EDNZWXWCW

Smart Business and Workplace

Methodical Approach to Assessing the Readiness of Organizations in the Construction Industry for Innovative Development

Valeriya Glazkova

Abstract Priorities in construction innovations and their relevance are determined by the government policy for the industry development. The organization's readiness for innovative development depends both on the external environment and on the multidimensional nature and status of internal organizational processes. Thus, organizational readiness for implementing innovations is to be determined already at the stage of planning and preparing for innovative development. Having analyzed current methodological approaches to assessing organizational readiness for innovative development described in economic and managerial literature, author developed a new complex methodological approach to this assessment. While considering specifics of construction industry organizations, our approach involves the use of a system of indicators and criteria (technical and technological, financial and organizational) for assessing organizational readiness in construction industry to implement innovations.

Keywords Innovations · Innovative development · Innovative potential · Readiness for innovations · Construction organizations

1 Introduction

Innovative development of the national construction industry has currently a particular relevance and, therefore, is defined by a number of government documents [1–4]. New methods and technologies used in the construction design, organization of production and operation of facilities stimulate labor productivity, boost competitiveness in the industry and, as a result, improve financial performance of organizations.

V. Glazkova (✉)
National Research Moscow State University of Civil Engineering (NRU MGSU), Moscow, Russia
e-mail: leram86@mail.ru

© The Author(s), under exclusive license to Springer Nature Switzerland AG 2026

A. Kulachinskaya et al. (eds.), *Digital Transformation: What Kind of Smart Life do we have today?*, Lecture Notes in Networks and Systems 1895,
https://doi.org/10.1007/978-3-032-20582-7_14

Due to the governmental support for innovative development that the Russia's construction sector enjoyed over the past few years its innovativeness has considerably increased. Thus, for the period 2019–2024 the level of innovation activity increased by more than 30%. Over the past 7 years the number of advanced production technologies developed in the industry has also increased considerably: in the processes related to design, project management, construction control and supervision the growth of this indicator was 2.5 times over the period 2018–2024, in the building materials industry the increase was 1.5 times over the same period. However, in 2019–2024 dynamics of construction organizations that implement technological innovations was rather unstable: since 2019, the share of such organizations has increased by an average of 12% compared to 2021, then until 2023 there was a decrease by an average of 12.5% per year, and in 2024 the number of innovative-friendly organizations again increased—by 17.5% compared to the previous year. This instability is caused by the impact of the external environment shaping the innovation climate in the sector, and by the poor readiness of construction organizations to create and incorporate innovations because sufficient production, financial, and organizational resources are required to implement the decisions made.

Since factors of internal environment affecting results of the organization's innovative activities include production, technological, economic, financial, organizational, managerial and personnel components, the issue of assessing organizational readiness to implement innovations is rather multidimensional. Therefore, the aim of this study is to develop a methodological approach to the assessment of organizational readiness for innovative development in construction industry. The tasks of the study to be solved are the following:

1. to study methodological basics for assessing organizational readiness for innovative development;
2. to develop a system of indicators for assessing organizational readiness for innovative development in construction industry;
3. to develop a methodological approach to assessing the readiness of construction industry organizations for innovative development.

2 Methodology and Methods

As a rule, any innovation requires three stages—preparation, implementation and the final phase (Fig. 1). Preparatory phase has an outstanding importance while providing management of organizations with the data reflecting organizational innovative development readiness. In case of favorable results, the preparatory phase will include planning organizational innovative development, preparing subdivisions and departments to implement innovations and adjust to the related changes, and identifying indicators which characterize the efficiency and effectiveness of innovation in organizations. At the preparatory phase it is decided which kind of innovation to implement and how to get it—either to buy in the market or to develop by themselves.

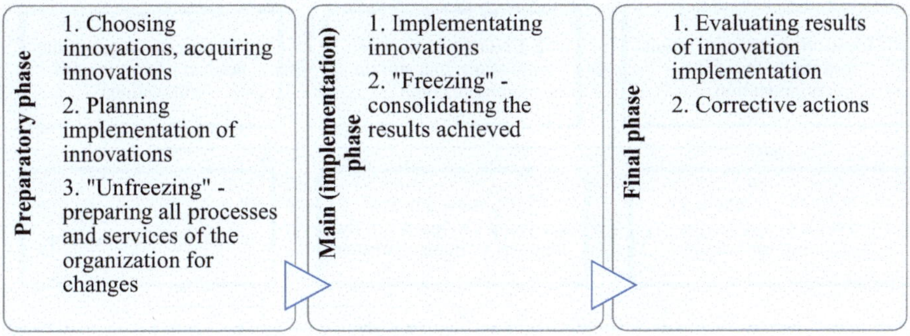

Fig. 1 Procedure for introducing innovations into organizational activities

The implementation phase involves introduction of the innovation and, in case of positive effect, fixing the results for further scaling of innovations within the organization. If the assessment of the implementation results at the final phase showed a gap between the value planned at the preparatory phase and the one actually obtained at the final phase, it is necessary to take corrective actions in relation to the processes related to the organizational innovative activities at the preparatory phase. Thus, the paramount importance of preparatory phase for success in implementing innovation is beyond doubt for construction industry organizations. One of the main issues of the preparatory phase is determining the organization's readiness for innovations and innovative development. At the same time, results of introducing innovations and costs of innovative activities depend on the internal environment and business processes of construction organizations [5].

The issue of managing the processes of implementing innovations and, in particular, assessing the organization's readiness to introduce innovations is triggered by the increased demand for innovative development in construction sector. Having studied methodological approaches to assessing the readiness of organizations for innovative development currently described in the economic and managerial literature, we offer the following their classification:

1. Approaches, considering environmental factors: external organizational environment (innovation climate) and internal organizational environment (strategic goals and objectives, organizational structure, production processes, etc.) [6–8]. The approach of Limsuwannarot et al. [9], Gusev et al., that classify external and internal factors, affecting the organizational innovative effectiveness, in terms of their controllability (uncontrollable, partially controllable and controllable) is of particular interest in the context of our study. Microeconomic factors are classified in a separate category [10].
2. Approaches, comprehensively considering all the internal organizational processes, which create opportunities for implementing innovation [5]. For example, Kandel [11], Ahmed Saleh et al. [12], Kudryavtseva offers an algorithm for decision-making on implementing innovations for industrial enterprises

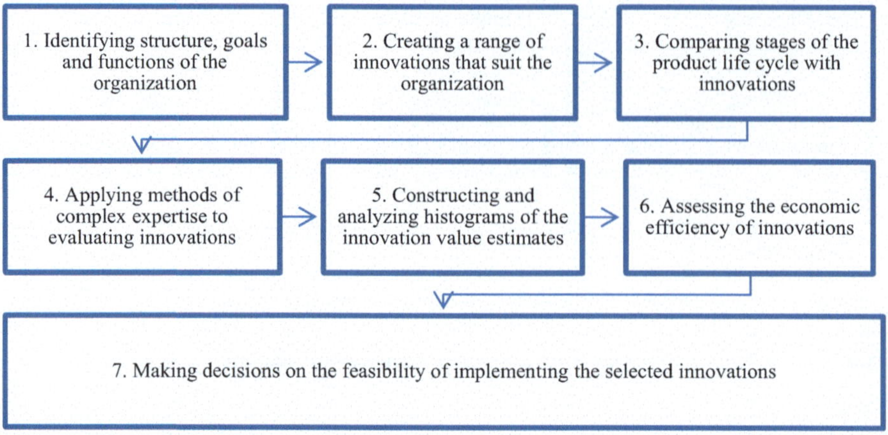

Fig. 2 Algorithm for decision-making on implementing innovations for industrial enterprises, developed by Kudryavtseva [13]

(Fig. 2), based on the analysis of internal organizational processes and the use of modeling and programming methods [13].

By summarizing approaches to assessing the readiness of organizations for implementing innovations described in the economic literature, Neverov and Dudnik [14] put forward a hypothesis that this assessment should cover three main areas: technical and technological, social and financial. The highlighting of these assessment areas is determined by the following: innovations are inevitably associated with the changes in production processes, in technological chains, and in approaches to the processes management and organization. Hence, the organizational technological component must be taken into account. As organizational changes entail the processes related to personnel management, the social component should be also considered. The financial aspect is also a significant factor when deciding on the effectiveness of implementing innovations because innovative development always involves investments.

Our research on approaches to assessing the organization's readiness for innovative development has shown that the most common assessment methods are an expert method, an optimization level assessment method, an indicator balance estimation method and a method of simulation modeling [5, 15, 16]. Thus, having analyzed methodological foundations for assessing the readiness of organizations for innovative development we have concluded that there is a sufficient methodological basis for solving the tasks set in this study. At the same time, when applying the existing approaches and methods to assessing organizational readiness for innovative development it is necessary to consider the specifics of the object of this study, that is, characteristics and peculiarities of the construction industry organizations.

3 Results

Technical and technological readiness for implementation of innovations, characterizing construction organizations involves an assessment of their technologies readiness level, analysis of their technological processes and technical condition of the equipment, determination of technological compatibility of their production system elements (old and new), as well as their digitalization level. This assessment allows management to evaluate the organization's potential to quickly reorient production capacities. Indicators and criteria used for assessing technical and technological readiness for innovation implementation in construction industry are presented in Table 1.

In accordance with the evaluation criteria presented in Table 1, the final value of the technical and technological readiness of construction industry organizations for implementing innovations takes the following form:

$$T_{readi} = \sum \left(T_{comp} + T_{procc} + T_{digit}\right)/3 \quad (1)$$

where T_{readi}—the value of technical and technological readiness for implementing innovations level, score;

T_{comp}—the value of production system compatibility level (old and new elements), score;

Table 1 Technical and technological indicators for assessing readiness of construction industry organizations to implement innovations

Indicator	Characteristics	Criteria
Level of production system elements compatibility (old and new) (T_{comp})	Level 4 (minimum)—lack of compatibility Level 3—partial compatibility Level 2—significant compatibility Level 1—full compatibility	Level 4–0 points Level 3–1 point Level 2–2 points Level 1–3 points
Level of technological and production processes organization (T_{procc})	Level 4 (minimum)—lack of mechanization and automation of processes Level 3—partial mechanization and automation of processes (total not less than 30%) Level 2—significant mechanization and automation of processes (total up to 70%) Level 1—full mechanization and automation of processes (total more than 70%)	
Level of digitalization (T_{digit})	Level 4 (minimum)—lack of digitalization Level 3—partial digitalization (disparate digital activities are carried out, but they are not formalized) Level 2—significant digitalization (digital activities are carried out systematically, but are not formalized) Level 1—full compatibility (digital activities are carried out on systematically and formalized)	

T_{procc}—the value of technological and production processes organization level, score;

T_{digit}—the value of organization's digitalization level, score.

In terms of technical and technological assessment of organizational readiness for implementing innovations the values of the aspects assessed should not be lower than average, that is, correspond to the 2nd level of readiness and higher. Therefore, the total value of T_{readi} should be not lower than 2. If the T_{readi} takes a value below 2, it is concluded that this construction organization has neither technical nor technological readiness to implement innovations.

Financial readiness of organizations for implementing of innovations is assessed by using generally accepted in financial management indicators of the organization's financial activities and its financial stability. Therefore, when assessing construction industry organizations, we use the following indicators and criteria (Table 2).

From a financial assessment point of view the final value of the organization's readiness for implementing innovations takes the following form (in accordance with the evaluation criteria presented in Table 2):

$$F_{readi} = \sum (F_{aut} + F_{stab} + F_{perf})/3, \quad (2)$$

Table 2 Financial indicators for assessing readiness of construction industry organizations to implement innovations

Indicator	Characteristics	Criteria
Financial independence ratio/ Financial autonomy ratio (F_{aut})	Level 4—the ratio is lower than the norm Level 3—the ratio takes the normative value only in one year over the past 3 years Level 2—the ratio for the last 3 years takes the normative value for two years Level 1—the ratio takes a normative value over the past 3 years	Level 4–0 points Level 3–1 point Level 2–2 points Level 1–3 points
Level of financial stability (F_{stab})	Level 4 (minimum)—the financial stability ratio takes a value lower than the norm Level 3—for the last 3 years the financial stability ratio takes the normative value only in 1 year Level 2—the financial stability ratio for the last 3 years takes the normative value for 2 years Level 1—the financial stability ratio takes a normative value over the past 3 years	
Level of financial performance (F_{perf})	Level 4 (minimum)—the dynamics of the net profit for the last 3 years is negative Level 3—for the last 3 years the net profit takes a positive value in the last year of the period Level 2—the net profit for the last 3 years takes a positive value in the last 2 years of the period Level 1—the dynamics of the net profit for the last 3 years is positive	

where F_{readi} is the financial readiness of construction industry organizations for implementing innovations, score;

F_{aut} is the value of financial independence level of the construction industry organization, score;

F_{stab} is the value of financial stability level of the construction industry organization, score;

F_{perf} is the value of financial performance level of the construction industry organization, score.

Financial assessment of the organization's readiness for innovations is focused on getting the desired or target values of the aspects analyzed that are not lower than average or correspond to the 2nd level and higher. Therefore, the total value of F_{readi} should be not lower than 2. If F_{readi} takes a value below 2, it means that this organization is not ready to implement innovations.

Organizational and social assessment of an organization's readiness for introducing innovations is aimed to identify the potential of human resources contributing to innovative development, as well as to evaluate the potential of the organizational management processes and structure for adapting to new production processes triggered by the innovative development of the organization. The following indicators and criteria are used (Table 3).

According to evaluation criteria presented in Table 3, the final value of the organization's readiness for implementing innovations, with its organizational and social assessment aspects being considered, takes the following form:

$$OS_{readi} = \sum (OS_{adapt} + OS_{cult} + OS_{comp})/3, \quad (3)$$

where OS_{readi} is the organizational and social readiness of organizations in construction industry for implementing innovations, score;

OS_{adapt} is the value of adaptability level of organizational and managerial structures of a construction industry organization, score;

OS_{cult} is the value of organizational culture level, characterizing a construction industry organization, score;

OS_{comp} is the value of the level of employee competency, characterizing a construction industry organization, score.

Organizational and social assessment of organizational readiness for implementing innovations is intended to identify the desired values, which are to be not lower than average, that is, corresponding to the 2nd level and higher. Thus, the total value of OSr_{eadi} should be not lower than 2. If OS_{readi} takes a value below 2, we conclude that this organization is not ready to implement innovations due to the poor results of its organizational and social assessment.

Considering indicators and criteria of technical, technological, financial, organizational and social assessment of construction organizations' readiness to innovations, the minimum value of their readiness for innovative development (L_{readi}, Formula 4) should be at least 6 points. If L_{readi} takes a value below 6, the construction organization is not ready to implement innovations.

Table 3 Organizational and social indicators for assessing readiness of construction industry organizations to implement innovations

Indicator	Characteristics	Criteria
Level of organizational and management structures adaptability of (OS_{adapt})	Level 4 (minimum)—the organizational and managerial structure is unable to adjust to innovative development Level 3—the organizational and managerial structure is partially able to adjust to innovative development Level 2—the organizational and managerial structure demonstrates a significant ability to adjust to innovative development Level 1—the organizational and managerial structure demonstrates ability to fully adjust to innovative development	Level 4–0 points Level 3-1 point Level 2–2 points Level 1–3 points
Level of organizational culture (OS_{cult})	Level 4 (minimum)—complete inconsistency of the organizational culture with the goals of innovative development Level 3—partial compliance of the organizational culture with the goals of innovative development Level 2—a significant compliance of the organizational culture with the goals of innovative development Level 1—full compliance of the organizational culture with the goals of innovative development	
Level of employee competency (OS_{comp})	Level 4 (minimum)—the complete mismatch of employees' competences to innovation processes Level 3—partial compliance of employees' competencies with innovation processes Level 2—a significant compliance of employees' competencies with innovation processes Level 1—full compliance of employees' competencies with innovation processes	

$$L_{readi} = \sum (T_{readi} + F_{readi} + OS_{readi}). \quad (4)$$

4 Discussion

Considering innovative potential of organizations from the position of their financial, organizational, technical and technological capabilities, we have proposed a comprehensive approach to assessing readiness of construction industry organizations for innovative development. In his studies Limsuwannarot et al. [9], Ahmed Saleh et al. [12], Akhnevsky [17] emphasizes traditionally high costs of new technologies introduction as a factor hindering innovative development of organizations in the construction industry. Limited financial resources discourage management from financing research and development, which negatively impacts effectiveness of the

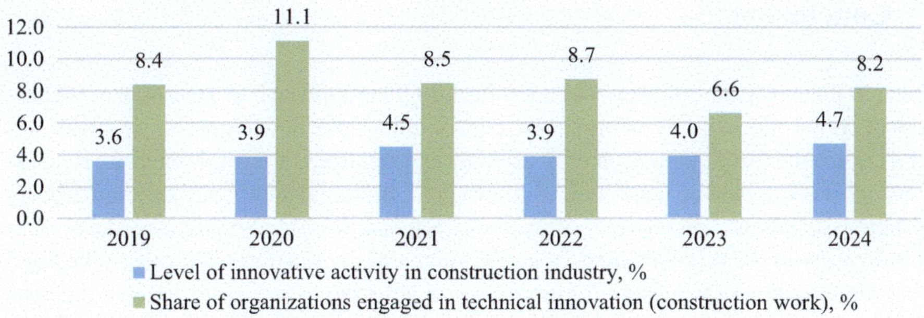

Fig. 3 Indicators of innovative activity in construction industry (according to Rosstat)

organizational innovative development (Fig. 3). This confirms the critical importance of the organization's financial readiness to implement innovations while planning its innovative development. In addition, the author identifies an excessive number of small businesses scattered throughout the construction industry as a barrier to the sector innovative development—having neither financial nor organizational readiness to introduce innovations, such organizations are often rather skeptical about them.

In fact, innovation conservatism, regarding the construction industry, characterizes studies of many researchers [18–21]. Thus, Tan et al. [22], Karapetyan and Kogotkova identify such barrier as the necessity to additionally coordinate the further use of innovations in production process, since for many innovative products there are no standards, GOSTs (state standards), and estimate norms for including innovations in the design and estimate documentation [23–25]. According to Alekseev [26] and Kisel [27], construction organizations are often characterized by chaotic processes and lack of readiness to rebuild their business processes for implementing innovations. Sometimes an obstacle to the use of new technologies is created by the poor electronic records management.

Therefore, the methodological approach proposed by the author for determining the readiness of construction industry organizations for innovative development allows for the structuring of innovative development processes and the preparation of company for implementing innovative activities. To do this, the proposed methodological approach should be supplemented with a list of activities that need to be carried out in order to move from one readiness level to another. At the same time, in the further studies researchers should consider the issues related to the organization's expansion in various directions, the readiness to which should also be taken into account in the course of its innovative development.

5 Conclusion

Thus, in the study author proposes a methodological approach to assessing the readiness of construction industry organizations for innovative development, which allows for a comprehensive evaluation of organizational readiness for implementing innovations, while considering technical, technological, financial and organizational aspects of their assessment. The novelty of the approach is determined by the use of a system of indicators in the relevant areas of assessment and their criteria that reflect the level of the organization's readiness for innovative development according to the indicators under consideration. The results of applying the author's methodological approach allow the management of construction organizations to make decisions on the expediency of introducing innovations or implementing management actions and measures in the areas of the internal environment where readiness for innovative development has not been achieved yet.

References

1. Decree of the Government of the Russian Federation dated October 31, 2022 No. 3268-r "Strategy for the Development of the Construction Industry and Housing and Utilities of the Russian Federation for the Period up to 2030 with a Forecast until 2035"
2. Decree of the President of the Russian Federation of 07.05.2024 "On the National Development Goals of the Russian Federation for the Period up to 2030 and for the Future until 2036"
3. Decree of the President of the Russian Federation of December 1, 2016 No. 642 "On the Strategy of Scientific and Technological Development of the Russian Federation"
4. Decree of the Government of the Russian Federation of May 20, 2023 No. 1315-r "Concept of Technological Development for the Period up to 2030"
5. Glazkova VV (2023) Methodical aspects of assessing the readiness of unified heat supply organizations for the introduction of innovations. Bull Altai Acad Econ Law 8–1:30–34
6. Smetanina AI (2022) Development of an innovative strategy of an organization to ensure its competitiveness. Abstract of the dissertation for the degree of Candidate of Economic Sciences in the specialty 08.00.05 Economics and Management of the National Economy (Innovation Management). Nizhny Novgorod, 22 p
7. Tursunkulov IA (2022) Improving the mechanism for managing the innovative potential of construction enterprises. World Econ Financ Bull 16:64–66
8. Xaydarova ES (2024) Development of innovative potential of construction materials industry enterprises. Int Bull Appl Sci Technol 4(4):116–119
9. Limsuwannarot O, Chanma C, Jarujittipan P (2023) Developmental guidelines to be an educational innovative organization for the basic schools in the education sandbox of Thailand. Asia Soc Iss (3):e254720
10. Gusev V, Naumkina T, Voytolovskiy N, Minaeva E, Zemskova O (2018) Methods for assessing the efficiency of innovation activity of enterprises of the electric power industry. In: MATEC Web of Conferences, vol 170, p 1–8
11. Kandel R (2022) Local level institutional self-assessment system: technological innovation in public sector. Prashasan: Nepalese J Pub Admin 52(1):171–180
12. Ahmed Saleh M, Saad M (2023) Digital transformation impact on agility and strategic risk management. J Bus Manag Sci 11(1):63–82
13. Kudryavtseva AS (2021) Algoritm otsenki i vybora innovatsii dlya predpriyatiya. Ontol Des 11, no 3 (41):364–381. https://doi.org/10.18287/2223-9537-2021-11-3-364-381

14. Viktorovich NA, Sergeevich DA (2016) Conceptual and methodological aspects of assessing the organization's readiness for the introduction of innovations. Ser: Sociol (3):623–636
15. Bencsik B, Palmié M, Parida V, Wincent J, Gassmann O (2023) Business models for digital sustainability: framework, microfoundations of value capture, and empirical evidence from 130 smart city services. J Bus Res 160:113757. https://doi.org/10.1016/j.jbusres.2023.113757
16. Vasilyeva E, Kisel T, Prohorova Y (2024) Digitalization as the factor of the improvement of quality of management in the organizations of the investment-and-construction sphere. Int J Q Res 18(3):793–806
17. Akhnevsky EV (2024) Innovative technologies as the basis of strategic development of construction companies, no. 26 (208). [Electronic resource]. Access mode: https://apni.ru/article/9724-innovacionnye-tehnologii-kak-osnova-strategicheskogo-razvitiya-stroitelnyh-kompanij
18. Iskandarov B (2025) Indicators of the innovative potential of construction industry enterprises and them improvement. Int J Artif Intell 1(5):763–767
19. Selyutina L, Pesotskaya E, Hakimov A (2023) Role of innovative development in implementation of competitive strategy of business organizations in construction. Actual problems of development economic, financial and credit systems. In: Collection of materials viii international scientific and practical conference. Belgorod, pp 310–314
20. Verstina N, Solopova N, Taskaeva N, Meshcheryakova T, Shchepkina N (2023) Evaluation of the energy efficiency class of an industrial facility: a rating system and a scale of sustainable development. Sustainability 15(22):15799
21. 宋. The innovative research on the construction of the archives of the party building of the grassroots party organizations in colleges and universities in the new era. Adv Soc Sci 12(8):4574–4579
22. Tan WL, Gangopadhyay P, Hauptman O (2021) Introduction to the special issue on "Grassroots and inclusive innovations: conceptualizing synergies and complementarities". Technol Forecast Soc Change 164:120409
23. Glazkova V (2023) Digital transformation of heat supply and unified heat supply organizations based on the introduction of digital control elements. In: Bencsik A, Kulachinskaya A (eds) Digital transformation: what is the company of today? Lecture notes in networks and systems, vol 805. Springer, Cham. https://doi.org/10.1007/978-3-031-46594-9_7
24. Gorbachevskaya EY (2020) Enhancement of the level of innovation activity of construction companies. In: IOP conference series materials science and engineering, vol 753, no 6, p 062013. https://doi.org/10.1088/1757-899X/753/6/062013
25. Karapetyan LL, Kogotkova IZ (2017) Management of the introduction of technological innovations in construction, no 3
26. Alekseev AA (2015) Mechanism of technological innovations in construction. Econ Manag Natl Econ 10(131):73–76
27. Kisel TN (2024) Level of readiness of investment and construction enterprises for digital transformation. Inform Digit Econ 5(4):567–592. https://doi.org/10.18334/ide.5.4.122429

Digital Ecosystems as New Business Models for Multi-component Environment Management

Irina Aleksandrovna Brusakova, Zumrat Talatovna Gaibnazarova, Madina Gafarovna Tuychiyeva, and Anora Norboy kizi Eshtemirova

Abstract The cybernetic aspect of managing digital transformations in digital ecosystems as new Smart business models is to describe various approaches to managing Mirabad data, processes, projects, infrastructure, architecture in a single technical and technological platform: from obtaining information about a physical object, a business process to forming a chain of output of a management decision on the effectiveness of the digital ecosystem. The purpose of the work is mechanisms for building an organizational structure model of the business model using the System of Systems methodology and ontological engineering methods. The applied methods and models allow formalizing and rebuilding the mechanisms of strategic interaction of business model participants. It is proposed to consider the processes of managing the digital transformation of digital ecosystems as a RnD project. Comparative analysis of metrics, methods and models for assessing digital maturity of digital ecosystems is presented.

Keywords Digital transformation · Smart business · Digital ecosystem architecture · Digital maturity · Artificial intelligence

I. A. Brusakova
Saint-Petersburg State Electrotechnical University "LETI", Saint-Petersburg, Russia
e-mail: brusakovai@mail.ru

Present Address:
Z. T. Gaibnazarova (✉)
University of Science and Technologies, Tashkent, Uzbekistan
e-mail: zumrat59@rambler.ru; z.gaibnazarova@usat.uz

M. G. Tuychiyeva · A. N. Eshtemirova
Tashkent State Transport University, Tashkent, Uzbekistan
e-mail: madinatuychiyeva9087@gmail.com

A. N. Eshtemirova
e-mail: eshtemirova98anora@gmail.com

© The Author(s), under exclusive license to Springer Nature Switzerland AG 2026
A. Kulachinskaya et al. (eds.), *Digital Transformation: What Kind of Smart Life do we have today?*, Lecture Notes in Networks and Systems 1895,
https://doi.org/10.1007/978-3-032-20582-7_15

1 Introduction

Modern trends in the world technological development are related to technologies of digital transformations of society and production chains. The ability to organize convenient work in a unified information environment using machine learning and predictive analytics technologies represents a set of challenges for modern development. The list of digital transformation technologies that is used in the paper covers all stages of working with data and information. Data management, data consolidation and synchronization, and the transformation of unstructured information from various sources into a management solution are associated with the introduction of data management technologies seen in the context of the transition from the technological structure of Industry 4.0 to the technological structure of Industry 5.0. Characteristic features of Industry 5.0 are big data management technologies—BIG DATA technologies, artificial intelligence technologies, ultrafast computing technologies with using quantum processors [1–4]. Digital transformations should lead to fundamental changes in existing business processes and business models. The stages of digital transformation should be preceded by the stages of preliminary audit of existing business processes, business models, and business architectures. The management of the digital ecosystem begins with the stage of assessing the digital maturity of each component. Also, a preliminary assessment of their digital maturity is required.

Acquisition of corporate data for management in new business models is complicated by the presence of strategic interactions between participants. New approaches to business organization are required. These new forms of business organization include digital ecosystems. The purpose of introducing such new organizational structures is to make simultaneous management of all components of the digital ecosystem multicomponent environment possible. Digital transformations of processes, business models, and architectures for each component of the digital ecosystem will enable organization of digital transformations of strategic interactions between various components.

The current trend in the development of methodological and scientific-practical foundations of system engineering is the use of the theory of the system of systems (SoS) to manage digital ecosystems. The ideas of the SoS make it possible to form components and interrelations between them for ecosystems of various complexity. In addition, the ecosystem approach allows us to consider the architectural features of business models of economic objects and IT architectural features of software systems as an interconnected set of components of the SoS meta-model, thus ensuring the interoperability and homeostasis of a system of systems, which are essential properties of digital ecosystems.

In the age of digital transformations, digital technologies, and requirements for the digital maturity of enterprises, industries, and companies, the tasks of forming support and decision-making systems for digital ecosystems are becoming particularly important. The term "enterprise" is already used as a complex dynamic system component of an enterprise's strategic interaction with components of the internal and external environment [1]. The term "digital ecosystem" is increasingly used, thereby emphasizing the importance of communication with external and internal environments [2–4]. The tasks of strategic planning in such ecosystems' management are directly related to the "reproduction" of various situations in digital ecosystems multicomponent environment [5]. The theory of digital ecosystem management can be identified with the theory of complex systems of systems (SoS) management [5]. Reconfiguration of the interconnections between the components rebuilds the existing business processes of resource management in such a digital ecosystem. In fact, we are dealing with the tasks of business process reengineering—a complete restructuring of business processes, business models, and business architectures. BIG DATA technologies are used to organize reengineering scenarios and decision-making algorithms for digital ecosystems, the feature and advantage of which is the ability to process structured and unstructured datasets using both SQL and NOSQL technologies.

The digital ecosystem is analyzed as a system with multiple parallel processes, which is managed by all key stakeholders, using EAI (Enterprise Application Integration) digital technologies [6]. Moreover, multi-process models also participate in the digital systems interactions. Datasets that represent the measurement information of the input–output transformation are generated for various business processes of various participants in digital ecosystems. Information about end-to-end business processes in digital ecosystems plays a special role in shaping management decisions. To support an interconnected set of structured and unstructured information, it is proposed to implement an ITSM data integration approach. The essence of this EIM (Enterprise Information Management) approach is the ability to consolidate, integrate, and synchronize the multiformat data needed to come to a management decision. The purpose of these decisions is to analyze, monitor, and predict the effectiveness of the digital ecosystem.

1.1 The Digital Ecosystem as a New Business Model for Strategic Interaction Between Business Participants

We consider the digital ecosystem to be an interconnected set of economic entities that implement all stages of the life cycle of a manufactured product. Such an ecosystem can be managed by organizing a technical and technological resource integration platform that would allow the entire IT infrastructure of the digital ecosystem to be assembled using unified integrating technologies. The components of such an IT infrastructure are the hardware and software, the operating system, database management systems (DBMS), data management technologies, and the visualization level of management decisions. In terms of platform integration, a digital ecosystem is an association of IT resources and participants working in a single information space and ensuring the integrity of the system.

The architecture of a digital ecosystem can be represented using various description techniques, for example, the Business Studio tool, the ARCHIMATE language [7]. The IT infrastructure for managing information flows in digital ecosystems consists of a standard set of technologies for acquiring measurement information, both technical (physical phenomenon-sensor-measuring chain-processes) and economic (business process, corporate information system). The components of such an IT infrastructure are software and hardware, an operating system, database management systems (DBMS), data management technologies, visualization technologies for management decisions, business process modeling technologies, data storage technologies, machine learning technologies, simulation technologies, language support technologies for management decisions, artificial intelligence technologies, etc. The purpose of modeling the architecture of a digital ecosystem is to describe the strategic interaction of participants, with the possibility of reconfiguring relationships if ineffective interaction is identified.

Here is an example of a multicomponent environment of a digital ecosystem in development. Figure 1 shows a digital ecosystem for a developer company that needs to use economic and mathematical tools for making decisions about the effectiveness of managing construction and sales processes by a commercial director.

Figure 1 shows the components of the developer's digital ecosystem and the relationships between them:

- the developer's company, which is its center and accumulates all the data within itself to make a decision;

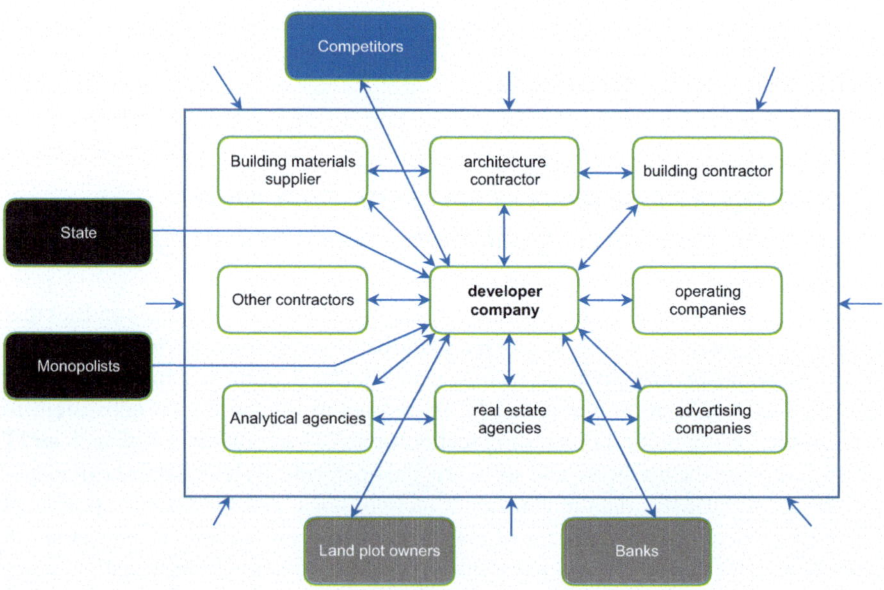

Fig. 1 The digital ecosystem of the developer's company as Smart business model

– two management circuits—the internal one, consisting of an analytical agency, a real estate agency, building materials suppliers, an operating company, an architectural contractor, a building contractor, and other contractors, and the external one, represented by land owners, banks, government agencies, and monopolists.

The arrows in Fig. 1 indicate the ways of exchanging data flows in a multicomponent environment.

To describe the architecture of the digital ecosystem, we use the approaches of ontological engineering [8–13]. The advantages of the ontological approach include the possibility of using ontological models for data mining in artificial intelligence technologies. These advantages will allow for multiple alternatives, flexibility in the functionality of support systems and decision-making on the effectiveness of managing the processes of the digital ecosystem.

Ontological engineering is a methodology for creating a semantic model of interaction, where:

- Each participant is described as an entity with roles, interests, and obligations;
- Interaction occurs according to ontologically defined scenarios (for example, "contractor ↔ contract conclusion ↔ developer");
- Logical verification of compliance of actions with standards (for example, legal, regulatory, financial) is possible;
- Chains of responsibility, dependencies, and risks can be built in the system.

Organizational and management system structure:

1. *The ontological core.*
   It is a formalized model for:
   - Roles and participants;
   - Forms of interaction;
   - Rules and restrictions;
   - Goals, interests and motivations of the participants.
2. *The developer's Management Center*
   The internal coordinating node of the ecosystem, which includes:
   - Integration functions of participants;
   - Mechanisms for evaluating the effectiveness of interaction;
   - The Center for Decision-making and conflict management.
3. *Interaction scenarios (workflow by ontology) typical patterns of interactions*:
   - The "Contractor engagement" scenario;
   - The "Joint project with another developer" scenario;
   - The scenario of "Launching a marketing campaign", etc.
4. *Relationship Management Mechanisms (ORMs) include*:
   - Role and Responsibility Matrices (RACI);
   - Targets for each interaction;

- Control of deviations and risks.

The visual model of the ecosystem (semantic graph) is represented as

[Developer] ↔ [Bank] (financing)
[Contractor] (construction)
[Architect] (design)
[Material supplier] (supply)
[Land owner] (resource acquisition)
[Real estate agency] (implementation)
[Operating company] (after-sales management)
[Other developer] (partnership)
[Analysts] (data and forecasts)
[Advertising Agency] (promotion)

All connections are ontologically described relationships, managed and tracked. The advantages of the approach are presented in Table 1.

Below is a fragment example of the developer's digital ecosystem ontology, which describes the key classes (entities), relationships, and properties of the participants. This ontology is focused on describing interactions in an ecosystem—not as a program code, but as a semantic model (conceptual structure).

A fragment of the developer's digital ecosystem ontology classes are presented in Table 2.

Object properties relationship (Object Properties) are presented in Table 3.

Datatype Properties are presented in Table 4.

An example of Ontological description (triple RDF):

:ProjectX rdf:type:Project.
:CompanyA rdf:type:DeveleperCompany.
:CompanyA:initiates:ProjectX.
:BankVTB rdf:type:Bank.
:BankVTB:finances:ProjectX.
:OOO_StroyCity rdf:type:ConstructionContractor.
:OOO_StroyCity:builds:ProjectX.

Interaction scenario is presented as:

Table 1 The advantages of the approach

Advantage	Explanation
Ecosystem formalization	A clear description of the participants, their goals, roles, and relationships
Knowledge management	Knowledge about participants, processes, projects becomes reusable
Increase compatibility	Simplify communication and reduce conflicts
Decision support	Using ontological inferences for management
Scalability and repeatability	The ability to scale the ecosystem to other projects

Table 2 A fragment of the developer's digital ecosystem ontology classes

Members	Description
Develeper company	The developer is the central participant initiating and coordinating the project
Construction contractor	Construction company—contractor of contract works
Architectural firm	Architectural organization—designs buildings
Bank	Financial institution—lends projects
Land owner	Land owners—provides land
Material supplier	Suppliers of Building Materials
Real estate agency	Real estate agencies—engaged in sales
Marketing agency	Advertising and PR agencies
Analytic agency	Provider of market and competitive analytics
Other developer	Partner/competing developers
Facility management company	The company that operates the facility
Project	Development project (management unit)
Interaction scenario	Model of interaction between subjects

Table 3 Object properties relationship

Relations	Description
Initiates (Develeper Company, Project)	The Developer initiates the project
Finances (Bank, Project)	A Bank finances the project
Designs (Architectural Firm, Project)	Architect provides the design
Builds (Construction Contractor, Project)	The Contractor carries out the design
Supplies (Material Supplier, Project)	The Supplier provides materials
Owns (Land Owner, Land Plot)	The Landowner provides plots
Sells (Real Estate Agency, Object)	The Agency sells objects
Markets (Marketing Agency, Project)	Marketing agency promotes the object
Analyzes (Analytic Agency, MarketSegment)	The Developer initiates the project
Manages (Facility Management Company, Building)	A Bank finances the project
Partners with (Develeper Company, Other Developer)	Architect provides the design

Table 4 Datatype properties

Property	Range	Example
hasINN	xsd:string	The organization's identification number
hasProjectStage	xsd:string	Design
hasRiskLevel	xsd:string	Construction
hasInteractionIntensity	xsd:integer	High
hasRegion	xsd:string	Medium

"Project launch".

Participants:

- Developer
- The owner of the plot
- Bank
- Architect

The ontological structure of the scenario is presented as:

:StartProjectScenario:includesParticipant:DeveleperCompany.
:StartProjectScenario:includesParticipant:Bank.
:StartProjectScenario:includesParticipant:ArchitecturalFirm.
:StartProjectScenario:requiresRelationship:finances.
:StartProjectScenario:requiresRelationship:designs.
:StartProjectScenario:requiresRelationship:owns.

Ontology possibilities can be represented as a set of propositions:

- Semantic search: "Which projects are being built with the participation of more than two contractors?".
- Bottleneck analysis: identifying missing links or weak ecosystem participants.
- Forecasting and risk analysis: interaction scenarios → deviation modeling.
- Formation of new coalitions: selection of potential partners based on ontological profiles.

Scenario-based game approaches to simulation modeling based on ontological engineering make it possible to plan a simulation experiment using statistical methods for a specific architecture of the digital ecosystem. The sample data for the simulation experiment are sample data representing the results of strategic interaction between the players of the digital ecosystem over a certain period of time: the forecast of demand for apartments of a certain class, expected revenue and profit, marketing ROI, and the time until the full sale of real estate.

To form complex integrated KPIs for the functioning of components of digital ecosystems and the entire system as a whole, business intelligence technologies are used to build predictive models using the following technologies, for example:

- BI (Business Intelligence)—technologies of intellectual analysis and processing of unstructured data, allowing to design analytical platforms of corporate information systems;
- OLAP (OnLine Analytical Processing, interactive analytical processing)—technologies for multidimensional data analysis in data warehouses
- Data Mining—technologies for data extraction and interpretation;
- SMART technologies for goal specification (Specific Measurable Attainable Relevant Time-Bound).
- Machine learning technologies.
- Artificial intelligence and fuzzy modeling technologies allow us to present hybrid models for managerial decision-making (AGI technologies, Advanced AIDeep

Learning (DL) technologies, GPT-40, Grock-2 technologies, Gemini Thinking technologies implement the Gemini 2.0 instant thinking model, Deep Seek-R1 DeepSeek-AI technologies, Open AI technologies, Fuzzy Tech, etc.) [14–17].

Thus, the description of the digital ecosystem as a multicomponent environment is proposed to be implemented using the mechanisms of scenario-game description of the architecture of the digital ecosystem. The use of ontological engineering mechanisms makes it possible to formalize the representation of the strategic interaction of the components of the digital ecosystem.

The stages of relational design based on a variety of specific implementations of statistical simulation experiment conditions are crucial in the formation of data sets for simulation modeling of process indicators in digital ecosystems.

1.2 Comparative Analysis of Metrics and Models for Assessing the Digital Maturity of Digital Ecosystems

Let's consider digital transformation management processes for digital ecosystems as an RnD project. This approach includes all the necessary stages of innovative design: from the pre-design stage of the successful practices market survey to the development of a digital transformation strategy and the launch of its own innovative solution on the market and the subsequent commercialization of the results. The pre-project stage allows you to determine the readiness of infrastructure, projects, human resources, and processes for digital transformation.

The application of the Capability Maturity model (CMM) [18] makes it possible to assess the digital maturity of business processes in order to achieve certain goals in terms of assessing the digital maturity of data and infrastructure. The analysis of the digital maturity of an enterprise includes the results of a comprehensive analysis of the digital maturity of data management, infrastructure, human resources, and digital culture creation processes. The set of possibilities for the CMM model is represented by five levels of achieving a specific goal: from level 1 (chaotic information resources, complete lack of digitalization) to level 5 (optimization of all business processes, full controllability of digital transformation processes).

There are models for assessing the digital maturity of data and models for assessing the digital maturity of infrastructure [18]. For example, the following models are used to assess the digital maturity of data:

Data Governance Maturity Model;

IBM—Data Governance Council, DataFlux—Data Governance Maturity Model;

Garter—Enterprise Information Management (EIM) Maturity Model;

Magnitude Software (Kalido)—Data Governance Maturity Model;

Oracle—Data Governance Maturity Levels;

EWSolutions—EIM Maturity Model;

Stanford University—The Stanford DG Maturity Model.

For example, in the Stanford model, digital data maturity is assessed at five levels [6]: the first level is initial (processes are not regulated) to level 5 (quantitative goals for improving processes are known and fully manageable).

For the DMM data management maturity model [20], 179 evaluation criteria and 28 sublevels of analysis are used.

For example, to identify the degree of customer orientation, the following indicators are highlighted

- Strategy
- Technologies (networks, security, architecture
- Management Digital Technologies (Agile Change Management, Integrated Service Man., Automated Resource Man., Real-Time Analytics& Integrated Service, Smart&Adaptive Process Man., Standard&Governance Automation).
- organization and culture (level of digitalization, leadership and management, organizational structure and leadership, management flexibility).
- Models used to assess the digital maturity of the infrastructure:
- IDC models (The IDC MaturityScape benchmark framework);
- the COBIT (Control Objectives for Information and related Technology) model [21–27].

Both models are used to evaluate the effectiveness of the stages of digital transformation implementation strategy. The movement from stage to stage is accompanied by a variety of criteria for evaluating the success of actions. Thus, the movement of the digital transformation strategy for the COBIT model from level 0 (processes have not been studied) to level 1 (activities are carried out chaotically, from case to case without a single approach, management is not organized), and further to levels 2–5 (the process is developed to practical application) is carried out according to the phased implementation of digitalization goals. Therefore, criteria should be developed to determine those points in time that make it possible to determine the completion of the previous stage and the beginning of the next stage of digitalization.

The digital competencies of human resources for digital transformations must be constantly measured. For example, the Graves model [28] makes it possible to assess the level of digital competencies of employees, the level of digital culture of the team based on the use of the color scale of the power of manifestation of competencies. For instance, the yellow color in Graves model illustrates a high degree of employees' functionality and creativity; turquoise color allows you to evaluate globalism of thinking, integrality, the ability to evolve views, and balance. The spiral view of the model demonstrates the evolutionary path of employees' development in the team.

For a preliminary cognitive analysis of the ranking of various metrics and models for assessing the digital ecosystems maturity, we will highlight the main areas of decision-making in the Industry 4.0 methodology. As you know, Industry 4.0 indexes are configured for six stages (goals) of digital transformation: 1—informatization; 2—connectivity; 3—visibility; 4—permeability; 5—predictability; 6—self-correction. You can enter scores for achieving goals for each of the stages. Moreover, it is necessary to ensure the gradual accumulation of indexes (metrics) to assess the level of digital maturity of the enterprise.

Thus, building an integrated assessment involves analyzing statements of the form:

"Enterprise Digital Maturity Assessment = Digital maturity assessments of data management processes.

Digital Maturity Assessments of infrastructure

Human Resource Digital Maturity Assessment

Organizational Structure Digital Maturity Assessment

Assessment of the digital maturity of the culture of interaction".

The cognitive image of an integrated assessment of the digital maturity of an ecosystem can be formalized using predicate algebra or neural network modeling. With this approach, the cognitive image can be described using digital twin technologies. When comparing the cognitive image with the target indicators of digital maturity, knowledge management technologies can be used. There is a gradual decrease in the vagueness of the cognitive image as we move through the levels of digital maturity.

To further formalize and structure integrated assessments of the digital maturity of a high-tech enterprise, it is proposed to use predicate algebra and frame models of knowledge representation.

The movement from one level of digital maturity to the next is associated with the accumulation of the necessary quality. Knowledge representation models should contain digital maturity indexes of previous levels.

For COBIT models, each level is described by its own model. Thus, each level is characterized by a set of indexes for different structural areas.

The structural areas of the enterprise's organizational digital model are the following: tangible and intangible resources, information systems, organizational structure, and organizational culture. Here is a description of each structural area:

- tangible and intangible resources can be characterized by the following indicators: the degree of automation of receiving data from detectors (sensors); the degree of ensuring digital competencies of personnel; the degree of convenience of data preprocessing; the degree of convenience of communications; the degree of convenience of interface functionality;
- the information system is characterized in terms of, for example, the following indicators: the availability of a specialized interface, the degree of convenience of data management.
- The degree of organization of vertical and horizontal data integration, the degree of IT security, the degree of fault tolerance, etc.;
- organizational structure: the degree of flexibility in creating project teams, the degree of convenience of managing decision-making rights, the degree of customer orientation of management decisions, the degree of convenience of cooperation within the network, the degree of flexibility of management, etc.;
- organizational culture: the degree of error recognition, the degree of openness to innovation, the degree of data-based learning, the degree of continuous professional growth, the degree of flexibility in shaping changes, the degree of openness in making managerial decisions, etc.

The integrated assessment of the digital maturity of a high-tech enterprise is formed as a set of estimates based on five COBIT models by levels:

initial—repeatable—defined—manageable and measurable—optimized.

According to the number of fulfilled criteria in [10], it is proposed to judge the digital maturity of processes.

The use of artificial intelligence mechanisms at the pre-design stage makes it possible to ensure the effectiveness of the analysis. For example, Industry 4.0 digital maturity indexes are described by data presented in ordinal measurement scales. When assessing digital maturity based on the processes of all departments of a high-tech enterprise, the following artificial intelligence mechanisms will be required: neural network modeling of real situations, a description of the situation using various models of knowledge about the digital maturity index.

The Burke-Litvin models relate to models of a comprehensive assessment of the digital maturity of an organizational structure [29]. In this model, it is assumed that all the factors of organizational structure change are interrelated. Therefore, organizational transformations are supposed to be carried out with the involvement of stakeholders. Coordination of actions and a high level of readiness of human resources for changes are necessary conditions for organizational transformations. This model fits well with the representation of the architecture of a digital enterprise as a multiprocessor model with stakeholders in the management loop.

The transformational factors in the Burke-Litvin model are:

- mission and strategy,
- leadership;
- the culture of the organization.

The transactional factors are:

- structure;
- management practice;
- information system.

Special attention is paid to motivation in human resource management.

Thus, it is proposed to consider digital transformation management processes for digital ecosystems as an RnD project. Special attention here should be paid to the pre-project stage. The pre-project stage allows you to determine the readiness of infrastructure, projects, human resources, and processes for digital transformation.

The various stages of determining the digital maturity of processes, projects, data, infrastructure, and human resources are associated with the development of criteria for assessing the levels of digital maturity. Criteria should be developed to determine those points in time that make it possible to determine the completion of the previous stage and the beginning of the next stage of digitalization.

It is proposed to implement the metrization of the decision-making space on digital maturity using Industry 4.0 indices. The structural areas of the organizational digital model of the enterprise are selected to form a system of metrics using Industry 4.0 indices. Such areas for analysis include tangible and intangible resources, information systems, organizational structure, and organizational culture. It is proposed to form an integrated assessment of the digital maturity of a high-tech enterprise as a cognitive image of digital maturity [21–29].

1.3 System Engineering Technologies in the Management of Digital Ecosystems

The digital ecosystem management paradigm in Industry 5.0 is based on new ways of forming information flows between participants in production, business, public administration, etc. New business models for managing business entities have appeared: Internet of Things platforms, a model of services in the value chain, a marketplace model, a trusted data access model. The expanding paradigm of enterprise system engineering and software systems makes it possible to extend management principles to new business models, organizational systems, production systems, and ecosystems [2].

System engineering as a science that analyses and synthesizes business entities architectures and is applicable for the analysis and synthesis of digital ecosystems. Digital ecosystems as interconnected aggregates of various economic entities have internal and external management circuits. The internal management circuit includes components of digital ecosystems united by the production of certain products (educational services, industrial products, IT services, etc.). The management of the multicomponent environment of the polystructural ecosystem model is described using the management theory of the SoS meta-model [4, 5]. The tasks of describing the architecture of a digital ecosystem should be linked to the tasks of selecting and applying the necessary technologies and tools for ecosystem management. Support and decision-making systems for the effectiveness of the digital ecosystem use the analysis of key performance indicators of all ecosystem players.

Let's define a digital ecosystem as an interconnected set of economic entities that is implemented at all stages of the life cycle of a manufactured product. A digital ecosystem can be managed by organizing a technical and technological resource integration platform that would allow the entire IT architecture of the digital ecosystem to be assembled using unified integrating technologies. The components of such an IT architecture are the software and hardware, the operating system, database management systems (DBMS), data management technologies, knowledge management technologies, visualization technologies of management decisions, etc. Ensuring the interoperability and homeostasis of a system of systems are essential properties of digital ecosystems.

The organization of information flows for digital ecosystems is associated with a polymetric system for collecting, processing, storing and transmitting data of the integrating element of the polystructure [5]. The technologies used to consolidate and synchronize formalized and non-formalized data in the context of the ITSM approach make it possible to manage data and design multiservice environments in a single integrating space.

Managing the processes of the digital ecosystem (resource process management) and managing the processes of creating the necessary IT infrastructure and software systems are two equally important tasks. The technological stacks of the tools used for these tasks must be synchronized in accordance with the requirements for managing the processes of these tasks.

System engineering policies allow you to control:

- connectivity of individual tasks;
- provide a common understanding of tasks for people of different specialties;
- ensure the complexity of the solutions obtained (taking into account the economic, environmental, etc. components of the solutions);
- eliminate "bottlenecks" and "inconsistencies" in processes.

The terminological base of the "platform approach" to process management in complex systems contains the following concepts: technical and technological platform, technological platform, software systems, technological stack, etc. In fact, all these concepts are used to form a list of necessary tools for managing the lifecycle of converting primary measurement information about the ecosystem domain into processed secondary information. Thus, a software system is considered as a system consisting of software, hardware and data, "the main value of which is created through the execution of software" [5]. A technical and technological resource integration platform is considered as an integrated mechanism for organizing relationships between participants in business processes of a specific business model, "a set of interrelated technologies, hardware and software that provides the creation, deployment and support of business applications and processes", is "the foundation for creating business applications" [6].

Nowadays, the term "technology stack" is increasingly used to refer to a set of necessary software and hardware, digital information technologies, technological platform, etc. [6]. Let's define a technology stack as a set of technologies (business modeling technologies, data storage and processing, knowledge, etc.), tools (websites, mobile applications, cloud services, etc.), software components (programming languages, frameworks, databases, etc.) for creating a software product. The main software product for managing digital ecosystems is an IT service (business application). For the technology stack, there are client and server sides, databases, infrastructure—everything that allows you to "build" a software product. For example, a software product can be an IT service for visualizing management decisions for management support and decision-making systems. In the context of process management tasks in ecosystems, we consider as a software product a set of services that implement visualization of a management decision.

A digital ecosystem can be considered, for example, as an interconnected set of components, including external players (for example, suppliers, logistics companies, marketplaces, government and municipal structures, banks, etc.) and internal players (a construction company, an architectural bureau). The requirements for the functionality of such an ecosystem are formulated by both customers (a construction company) and consumers (real estate buyers). How effective are sales at various stages of construction, how justified is the financing of the construction life cycle, and how feasible is the purchase of real estate? What are the risks of long-term construction? System engineering policies make it possible to comprehensively implement the design of such a digital ecosystem of a multicomponent environment, from requirements development to risk assessment. What kind of IT architecture will be required to serve the digital ecosystem?

When managing a digital ecosystem, the necessary IT architecture can be visually represented as an "automation pyramid" containing a set of software and hardware [8, 9] in the form (Fig. 2).

Figure 2 shows the technological stack of tools and technologies that are used in managing the digital ecosystem. Figure 2 shows: ERP (Enterprise Resource Planning)—enterprise resource management corporate information system (CIS), ERP SOA—service-oriented corporate information system; MES (Manufacturing Execution Systems)—production system; SCADA (Supervisory Control and Data Acquisition)—automation system for domain data management. PLM (Product Lifecycle Management) is an application software for managing the product lifecycle, integrated into a specific IT infrastructure, technology platform; CAD (Computer Aided Design) is a design system, production automation technology in the design of technical devices of the software and hardware of the CIS. CAE (Computer Aided Engineering)—calculation and engineering analysis technologies; CAM (Computer

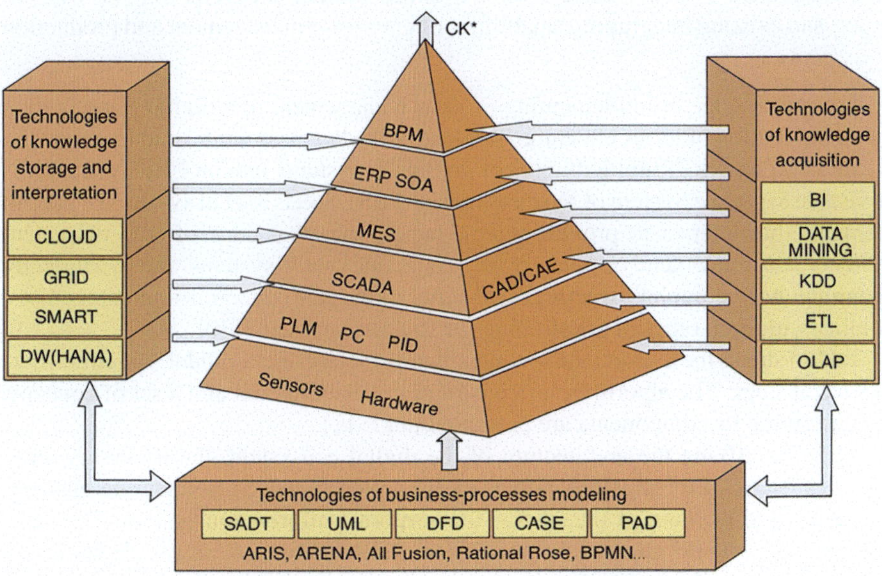

Fig. 2 Technological stack of tools and technologies of IT infrastructure of innovation activity

Aided Manufacturing)—pre–production systems; PDM (Product Data Management)—project data management systems. In Fig. 2 the management decision on the effectiveness of a digital enterprise is illustrated by the management content of CK* [8].

Business process modeling technologies in Fig. 2 are represented by SADT, UML, DFD notations, etc., modeling tools All Fusion, ARIS, ARENA, Rational Rose.

Currently, to manage digital ecosystems in the context of Industry 4.0 and 5.0, the reference model of the Industrie 4.0 (Industry 4.0) RAMI 4.0 architecture is used to describe the IT architecture and ecosystem architecture [31]. RAMI 4.0, or the industry 4.0 Reference Architectural Model, is a three-dimensional map that integrates information about the business architecture, product lifecycle, and enterprise management levels. RAMI 4.0 is a unified model for all components that ensures efficient and effective exchange of data and information between all ecosystem participants. The RAMI 4.0 architectural model displays all ecosystem players in the form of a three-dimensional space:

- the architectural layer in the form of information about the business model, organizational structure, functionality, information about communication technologies, integration mechanisms;
- information layer on the value stream of the final product life cycle in the form of information on design documentation, technological processes, auxiliary processes;
- a layer of integration of automated management systems with enterprise management systems throughout the entire system lifecycle; can be used to expand existing integration capabilities of enterprise management and production systems.

The initial stage of implementing process management of a digital ecosystem is the identification of its architecture. To identify the business model and integrate the business processes of multicomponent digital ecosystems into an ERP system, it is advisable to use the System of Systems theory (SoS). SoS is a set of systems or system elements that interact to provide unique capabilities that none of the component systems can implement on their own. "As a rule, such systems are dynamically changing, have complex structures and different targets, sometimes in antagonistic conflict with each other in the struggle for shared resources" [30].

Table 5 shows the mechanism for constructing the SoS meta-model, which consists of several steps. The algorithm for constructing a meta-model and a set of concepts for describing its components are presented in [4, 11].

After identifying the architecture of the digital ecosystem, the necessary technological stack of tools and technologies is formed (Table 5) to integrate business processes into the core of the stack—the corporate information system. The technology stack should contain, in addition to the technologies and tools shown in Table 5, technologies and tools for support and management decision-making. These are usually artificial intelligence and machine learning technologies, fuzzy selection, multidimensional data analysis technologies, etc.

Table 5 Metamodel structure algorithm

Step 1	Defining the primary SoS components – Target problem at the level (connected with SoS targets whose achievement is provided by several components' integration) – SoS infrastructure (defines SoS borders and provides net means for connection) – SoS level surrounding (extra entities that interact with SoS infrastructure and components) – SoS level organization (defining particular roles and targets of the components) – Composite systems (independent systems that perform partial service functions at the SoS level based on the assigned role)
Step 2	Defining the SoS level entity – SoS level target – Requirements (specific conditions and opportunities that must be met to achieve the goal) – Organizational role (set of services in SoS frame)
Step 3	Defining the level entity of SoS components – component level goals (independent system goals) – the ability to meet SoS requirements to create conditions for achieving SoS goals – SoS component level interface to provide the level services or exchange information – Decision making mechanisms at SoS component level – SoS component interested parties – SoS component level domain (can be presented as a domain model to define concepts and relations) – SoS component level life cycle – SoS component level environment (entities beyond SoS components, interacting with it.)
Step 4	Defining environment objects at SoS level for each external object, it is necessary to have data – SoS level resources (materials, funds, staff etc.) that can be used for the SoS targets – Condition (if the resource is available or potentially available, under what conditions) – SoS level physical environment (environmental factors in physically measurable forms)

The main feature of the SoS theory approach used to describe the architecture of a digital ecosystem is the presence of an integrating element in a polystructural system, which reflects the functions and processes occurring in its various constituent elements. In [4], it is proposed to form an integrating element of a polystructural system, the purpose of which is to manage the "history of interactions of components of the polystructure" [12–18]. The management space takes on the character of a "smart space". On the basis of the polymetric system, coordinated interaction in the space–time dimensions of such a "smart space" is conducted. The operability of such an integrating element should be ensured by selecting the appropriate technological stack.

However, building up the necessary digital technologies to manage digital ecosystems depends on the requirements for management solutions. For example, if, in addition to monitoring the performance indicators of the components of the digital ecosystem, you need to have information about advanced business analysis (text recognition, the use of artificial intelligence technologies, machine learning using vectorization and embedder models), the digital technology stack should contain artificial intelligence technologies, machine learning and knowledge management technologies [19–23]. As an example, we will give the technological stack necessary support systems and management decision-making to perform the task of language support for recognizing a set of various fragments of unstructured information from different information sources. Figure 3 shows the technological stack of tools for the task of language support for text recognition.

Thus, the application of the principles of system engineering and the theoretical basis of SoS allows us to introduce an architectural approach for designing and describing the behavior of an interconnected set of components of a digital ecosystem. The formation of technological stacks of tools for managing the components of digital ecosystems and their supporting IT architectures should be considered from the point of view of visualization and content of required management decisions [32–36].

Component	Tools / Libraries
structured data storage	Jatoba, Postgres Pro, ClickHouse
full-text search	OpenSearch, Elasticsearch, Solr, Typesense
text recognition from digital materials	Apache Tika, GROBID, Tesseract (OCR)
entity extraction	SpaCy, Flair, DeepPavlov NER
text vectorization	FRIDA (from SBER)
vector search knowledge	FAISS, Weaviate, Qdrant, Vespa
graph storage	Neo4J, ArangoDB, RDFlib, Blazegraph, Apache AGE
LLM language models for text generation	Qwen 2.5 – 7b- instruct-1M (Alibaba group)
Retrieval-Augmented Generation (RAG)	LangChain, Llamalndex, Haystack
user interface	React, Vue, Angular, Streamlit, Gradio
visualization of data graphs	D3.js, Cytoscape.js, Graphistry, NetworkX+matplotlib
orchestration, pipeline	Airflow, Prefect, Luigi, Nifi
containerization and deployment	DeckHouse, NOVA container Platform, Helm
monitoring and logging	Grafana, Zabbix, Prometheus, ELK Stack, Sentry
API integrations	FastAPI, Flask, gPRC, REST/GraphQL

Fig. 3 Technological stack of language support tools for text recognition using artificial intelligence technologies

2 Conclusions and Prospects for Development

In the paper, a scenario-based game approach to describing the architecture of a digital ecosystem as a multicomponent environment is proposed. The use of ontological engineering mechanisms will make it possible to formalize the representation of the strategic interaction of the components of the digital ecosystem and prepare corporate data for the simulation of a management decision.

Also, it is proposed to consider digital transformation management processes for digital ecosystems as an R@D project. The pre-project stage allows us to determine the readiness of infrastructure, projects, human resources, and processes for digital transformation. The various stages of determining the digital maturity of processes, projects, data, infrastructure, and human resources are associated with the development of criteria for assessing the levels of digital maturity. In order to determine those points in time, criteria should be developed that make it possible to determine the completion of the previous stage and the beginning of the next stage of digitalization.

An approach to the metrization of the decision-making space on the digital maturity of the ecosystem based on Industry 4.0 indices is proposed. To form a system of metrics using Industry 4.0 indices, structural areas of the organizational digital model of the ecosystem are selected. Such areas for analysis include tangible and intangible resources, information systems, organizational structure, and organizational culture. Also, it is proposed to form an integrated assessment of the digital maturity of a high-tech enterprise as a cognitive image of digital maturity.

The concept of system engineering and the theoretical basis of SoS for designing and describing the behavior of an interconnected set of digital ecosystem components are proposed. The formation of technological stacks of tools for managing the components of digital ecosystems and their supporting IT architectures should be considered from the point of view of visualization and content of required management decisions.

Promising research in the field of digital transformations of ecosystems is research related to the use of artificial intelligence technologies in choosing strategies for reconfiguring ecosystem components and their relationships depending on changes in external and internal factors.

References

1. Maier MW (1998) Architecting principles for systems-of-systems. Syst Eng 1(4):267–284. https://asymmetricleadership.com/wp-content/uploads/2020/04/architectingprinciplesofsystemofsystemsMAIER.pdf
2. Zinder EZ (2016) The expanding paradigm of enterprise engineering. Bus Inform 4(38):7–18. https://bijournal.hse.ru/article/view/26089/21745
3. Fang Z (2021) System-of-systems architecture selection: a survey of issues, methods, and opportunities. IEEE Syst J, T. 16, no 3, C. 4768–4779
4. Shvedenko VV (2024) Polystructural modeling of complex systems for their effective interaction. In: System analysis in design and management. St. Petersburg State Technical University

of Peter the Great, St. Petersburg. https://cyberleninka.ru/article/n/polistrukturnoe-modelirovanie-slozhnyh-sistem-dlya-ih-effektivnogo-vzaimodeystviya
5. Baek Y-M, Song J, Shin Y-J, Park S, Bae D-H (2018) A meta-model for representing system-of-systems ontologies. In: 2018 IEEE/ACM 6th international workshop on software engineering for system-of-systems (SESoS), 27 August 2018. https://ieeexplore.ieee.org/document/8445798
6. EAI (Enterprise Application Integraion). https://www.tadviser.ru/index.php/%D0%A1%D1%82%D0%B0%D1%82%D1%8C%D1%8F:Enterprise_application_integration
7. ARCHIMATE. https://www.archimatetool.com
8. Andreev DA, Voronov MV. Features of ontological engineering of technological knowledge. https://ojs.itmo.ru/index.php/IMS/article/view/74
9. Dobrov BV, Ivanov VV, Lukashevich NV, Soloviev VD (2006) Ontologies and thesauruses. Kazan
10. Resource Description Framework (RDF). http://www.w3.org/RDF
11. Web Ontology Language (OWL). http://www.w3.org/2004/OWL
12. Noy NF, Deborah L. Ontology development 101: a guide to creating your first ontology. Stanford Knowledge Systems Laboratory Technical Report KSL. https://protege.stanford.edu/publications/ontology_development/ontology101.pdf
13. Shu G, Rana OF, Avis NJ, Dingfang C (2007) Ontology based semantic matchmaking approach. Adv Eng Softw 38:59–67
14. Deep learning & Machine learning: what's the difference. https://datastart.ru/blog/read/deep-learning-machine-learning-v-chem-raznica
15. A little about Grok-2. https://habr.com/ru/companies/bothub/articles/852506
16. Google presented Gemini 2.0 Flash Thinking. https://habr.com/ru/companies/bothub/news/869370/
17. DeepSeek. https://www.deepseek.com
18. Industry Maturity Index 4.0. https://i40mc.de/wp-content/uploads/sites/22/2016/11/acatech_STUDIE_Maturity_Index_rus_WEB.pdf
19. Industry4.0. https://www.sap.com/cis/insights/what-is-industry-4-0.html
20. Levels of digital maturity of an industrial enterprise. https://cyberleninka.ru/article/n/urovni-tsifrovoy-zrelosti-promyshlennogo-predpriyatiya/viewer
21. Digital maturity Methodology for assessing the digital maturity of an organization. https://cpur.ru/wp-content/uploads/2020/10/Metodologiya-oczenki-czifrovoj-zrelosti-organizaczii.pdf
22. Technological Maturity Model CMMI. https://tenstep.com.ua/open/A1.1CMMI.htm
23. The Stanford Maturity Model. https://www.stanford.edu/
24. Data Management Maturity (DMM). https://cmmiinstitute.com/data-management-maturity.
25. IDC Investigations. https://www.channele2e.com/after-nines/state-of-cloud-computing-adoption-research-says/
26. IT Process Management Maturity Levels. https://www.itexpert.ru/rus/ITEMS/200809072359/
27. COBIT (Control Objectives for Information and related Technology). https://community.mis.temple.edu/mis5203sec001sp2019/files/2019/01/COBIT-2019-Framework-Introduction-and-Methodology_res_eng_1118.pdf
28. Graves Spiral Model. https://first-expert.ru/spiralnaya-dinamika/
29. Burke-Litwin model of organizational change. https://trainingtechnology.ru/model-organizacionnyx-izmenenij-berka-litvina/
30. Brusakova IA (2025) Theoretical innovation: textbook. Moskow, Urait
31. RAMI 4.0 (Reference Architectural Model Industry 4.0): Explained with example—KR Architecture World. https://ec.europa.eu/futurium/en/system/files/ged/a2-schweichhart-reference_architectural_model_industrie_4.0_rami_4.0.pdf
32. Gaibnazarova ZT (2023) Prospects for improving the forecasting of the gross domestic product of regions based on mathematical models. In: 9th IEEE international conference on information technology and nanotechnology (ITNT), 17–21 April 2023. Russian Federation, Samara, pp 1–4. https://doi.org/10.1109/ITNT57377.2023.10139074

33. Gaibnazarova ZT, Solieva BT, Iminova NA (2024) Forecasting the risk of bankruptcy in poorly formalized processes. Sci Tech 23(1): 80–86. https://doi.org/10.21122/2227-1031-2024-23-1-80-86
34. Gaibnazarova ZT (2024) Artificial intelligence and smart railway transport as a tool to accelerate the transformation process in Uzbekistan. In: Sari M, Kulachinskaya A (eds) Digital transformation: what are the smart cities today? Lecture notes in networks and systems, vol 846. Springer, Cham. https://link.springer.com/chapter/10.1007/978-3-031-49390-4_19, https://doi.org/10.1007/978-3-031-49390-4_19
35. Ziankova I, Gaibnazarova Z, Malei A, Sharafanova E, Gasheva O (2025) Digital technologies for comprehensive promotion of a startup. In: Skhvediani A, Kulachinskaya A, Mateeva Petrova, M (eds) Digital transformation of socio-economic and technical systems: theory and practice. Lecture notes in networks and systems, vol 1309. Springer, Cham, 39–55 p. https://doi.org/10.1007/978-3-031-85608-2_4
36. Stankevich D, Gaibnazarova Z, Naumovich O, Tuychieva M, Shadieva G (2025) The role of universities in digital transformation: challenges and prospects. In: Skhvediani A, Kulachinskaya A, Mateeva Petrova M (eds) Digital transformation of socio-economic and technical systems: theory and practice. Lecture notes in networks and systems, vol 1309. Springer, Cham, 179–193 p. https://doi.org/10.1007/978-3-031-85608-2_13

Digital Neuromarketing in Business: Assessment of the Current State and Effectiveness of Application

Nikolay Nikolaevich Molchanov, Oksana Sergeevna Muraveva, and Anait Vardanovna Grigorian

Abstract This chapter presents an analysis of the concept of neuromarketing in the context of the digital transformation of modern society. It demonstrates the differences between primary and secondary neuromarketing research, introduces a definition of digital neuromarketing, explores its application areas, and highlights trends in the use of modern technologies, enabling more accurate, personalized, and effective advertising campaigns that take into account the emotional and subconscious motivations of consumers. It also examines the current attitudes toward the use of traditional and digital neuromarketing tools by companies in the Russian Federation, as well as the correlations between traditional marketing research methods and neuromarketing and digital neuromarketing tools. The results of an analysis of the effectiveness of digital psychometrics for optimizing product promotion are presented. The authors propose a sequence of digital psychometric research stages and a wave-based approach to conducting a new product promotion campaign using digital psychometric tools.

Keywords Neuromarketing · Digital neuromarketing · Digital psychometrics · Psychotype · Digital tools · Advertising campaign · Effectiveness

1 Introduction

Neuromarketing is a modern interdisciplinary field that lies at the intersection of economics, psychology, and neuroscience. It is based on the assumption that people make most of their decisions on a subconscious level.

There are two types of neuromarketing research: primary and secondary neuromarketing research, where primary research is aimed at collecting and processing

---

N. N. Molchanov · O. S. Muraveva (✉) · A. V. Grigorian
Department of Business Economics, Entrepreneurship and Innovation, St. Petersburg State University, St. Petersburg, Russian Federation
e-mail: oksana_muraveva@mail.ru

© The Author(s), under exclusive license to Springer Nature Switzerland AG 2026
A. Kulachinskaya et al. (eds.), *Digital Transformation: What Kind of Smart Life do we have today?*, Lecture Notes in Networks and Systems 1895,
https://doi.org/10.1007/978-3-032-20582-7_16

data, and secondary research—at using and interpreting data to predict consumer perceptions [1].

Based on this, primary neuromarketing studies consumers' unconscious reactions to various marketing stimuli, such as packaging or advertising, using various triggers such as color, sound, shape, composition, etc. This subsequently helps to create an emotional connection between the company and its audience, increasing the chances of success for an advertising campaign. Secondary neuromarketing is a continuation of primary neuromarketing research and, at the present stage, in some cases relies on the use of artificial intelligence and machine learning technologies [2]. However, Byrne and co-authors note that primary neuromarketing studies supplemented by a forecasting stage using machine learning are relatively limited, as they focus mainly on product preferences [3]. Secondary neuromarketing research can be used as a standalone tool through the use of software and web applications that build predictions about consumer behavior based on collected natural digital information.

Financial companies, telecom operators, retailers, large manufacturers, and IT giants —many of them use neuromarketing and related areas in one way or another. These include emotion analysis, psychometrics, UX research, NUDGE, instrumental neuromarketing, behavioral data analysis, and so on.

For example, in the Sber client business development strategy, psychometry has become the basis for designing almost all of the bank's communications with the client. Back in 2017, Gref G., Chairman of the Board of Sberbank, noted that the introduction of the Big 5 testing system (a psychological test based on the «big five personality traits») had saved the company around $50 million in pre-screening costs when issuing loans [4].

It is important that neuromarketing allows companies to collect objective data, obtaining «true» answers from respondents and avoiding the «social desirability» effect. This plays an important role because with traditional marketing methods such as surveys, interviews, and focus groups, respondents are subject to the social desirability effect when answering questions [5], i.e., they may give «false» answers in order to appear better than they really are, but may not be aware of this.

The value of neuromarketing research lies in the fact that it helps to improve the effectiveness of marketing communications and strategies, both in the early stages of brand development and once the brand has been launched. An important factor is that the subjects of influence can be a wide variety of stimuli: sound signals, words, visual images, video materials, texts, smells, tactile sensations during product testing or consumption, as well as when performing various actions [6].

Tools that determine «brain activity, emotional states, and areas of attention» [7] play an important role in neuromarketing, helping to obtain this objective data. These tools include the following neurotechnologies: ECG, eye tracking, facial coding, fMRI, EEG, and fNIRS (functional near-infrared spectroscopy). Depending on the type of measurement, physiological and neurophysiological tools can be distinguished. Tools that measure physiological responses involve measuring the body's involuntary responses to stimuli (studying eye responses, measuring reaction speed, measuring heart rate variability, perspiration, facial expressions, etc.) [8], while neurophysiological tools measure brain activity.

However, the possibilities for studying human behavior in the digital environment are much broader than in the offline environment. Therefore, in the age of digital technology, the concept of neuromarketing is taking on new directions for development, taking the shape of digital neuromarketing.

Digital neuromarketing is the application of neuroscience principles to understand and influence consumer behavior in the digital environment.

The literature highlights the following trends in the application of technologies in digital neuromarketing [9]: chatbots that identify emotions for customer service; use of AR and VR to create immersive experiences; integration of neurodata with analytics; creation of ethical rules and recommendations; collaboration between humans and AI.

The key areas of application for digital neuromarketing are: optimization of user experience (UX), user interface (UI) of websites and applications [10], analysis of emotions and tone of statements in the digital environment, digital psychometrics based on data parsing.

The prospects for the application of neuromarketing associated with the development of [11]: the Internet of Things (for example, real-time analysis of consumer behavior in shopping malls using cameras); social networks (new forms of behavior are emerging that provide access to big data for studying mental processes); the online gaming industry (helps increase engagement and study consumer behavior through in-game metrics).

Currently, personalization and predicting future customer needs are becoming key success factors for companies. Every day, when searching for information and scrolling through social media, people leave digital traces. These traces include huge amounts of unstructured data (natural digital information) generated by people in the digital world: clicks, scrolling time, pauses, text messages on social media, internet search history, and even typing characteristics [12]. The intersection of neuromarketing and computer science has given rise to the field of digital psychometrics.

Psychometrics is the practice of using an object belonging to a person to obtain information about them, while digital psychometrics is the collection of natural digital information about consumers for the purpose of analysis and making marketing and management decisions.

It allows you to understand and analyze user behavior in the digital environment, identifying characteristics, interests, age, beliefs, and personality types based on digital data left behind [13].

Creating dynamic psychological profiles of users in real time, using AI tools based on digital traces, helps to identify the target audience more effectively, going far beyond classic user analysis based on demographic characteristics [14]. This allows for the creation of more accurate, personalized, and effective campaigns that take into account the emotional and subconscious motivations of buyers.

Thus, digital neuromarketing represents a logical development of classic marketing in the digital environment. And user research using digital psychometrics in the context of neuromarketing creates a powerful synergistic effect for understanding and predicting consumer behavior.

In this chapter the authors present:

- the results of empirical research conducted among marketing personnel at Russian companies on their awareness and use of classical and digital neuromarketing tools in their work,
- the effects of implementing digital psychometrics tools,
- as well as the authors' algorithm for using secondary neuromarketing research.

2 Materials and Methods

The theoretical and methodological basis of the work was provided by research conducted by leading authors engaged in the theory and practice of applying traditional and modern approaches in the field of neuromarketing and psychometrics. The study used informational materials and articles.

The empirical study included the following stages:

(1) A survey among business representatives was conducted to examine their attitudes toward the use of digital neuromarketing in their companies. The following tools were used for data analysis: frequency analysis and correlation analysis (Spearman's rank correlation coefficient). One hundred and ten company representatives participated in the survey. The questionnaire included single- and multiple-choice responses, as well as Likert-scale questions. The collected data was processed using Jamovi.

(2) Evaluation of the effectiveness of digital psychometrics (as a specific form of digital neuromarketing) for a company in the tourism industry. Results obtained from a standard mailing, without a focus on customer personality, were compared with those obtained through a marketing campaign personalizing emails based on each customer's psychotype. A factor analysis of the marketing campaign was conducted, and metrics such as ARPU, CAC, CLV and ROMI were calculated.

3 Results

3.1 Research on Attitudes Toward the Use of Digital Neuromarketing Tools by Business Representatives

The authors conducted a survey among business representatives to study their attitude towards the use of digital neuromarketing in their companies. Of the 110 respondents, 54% work in companies with 250 or more employees, while the rest work in companies of various sizes ranging from 15 to 250 employees. There was an approximately equal distribution of companies working in the B2B and B2C markets, respectively.

By industry, most of the respondents work in companies related to trade, education, and consulting.

The degree of agreement among respondents with statements about digital neuromarketing was assessed, and the following results were obtained.

77% of respondents are familiar with the concepts of neurotechnology and digital neuromarketing: however, 59% of respondents do not study specialized literature on this topic, and only 18% of respondents regularly read articles in magazines about neurotechnology and digital neuromarketing.

14% of respondents confirmed that the companies they work for use digital neuromarketing tools.

Among the objects targeted by neurotechnology and digital neuromarketing, the following were noted: advertising—59%, products—55%, websites—55%, interior design—9%, placement of items on shelves—14%, sensory (taste, hearing, smell) characteristics of goods (brands)—14%.

The following hypotheses were tested during the study.

Hypothesis 1: The more often a company uses digital neuromarketing, the higher the respondent rates this marketing tool.

The data (Table 1) shows a positive, significant correlation (0.658 and 0.768) between the attitude towards the use of digital neuromarketing and the opinion about the effectiveness of this tool. It is also noticeable that respondents who plan to use digital neuromarketing in the future rate its effectiveness higher. As a result, hypothesis 1 is confirmed.

Table 1 Spearman's correlation coefficient between respondents' answers about the use of digital neuromarketing (DNM) and their attitude towards this tool

Attitude towards the use of DNM	I consider DNM to be a sought-after technology capable of significantly improving a company's efficiency	I believe that the value of the DNM is greatly exaggerated; its real capabilities do not produce a significant effect	I consider the use of DNM for commercial purposes to be unethical	I believe that the use of DNM for commercial purposes should be prohibited by law
Positive, already used DNM	**0.658** ***	**−0.507** *	−0.15	0.033
Positive, we plan to use the DNM in the foreseeable future	**0.768** ***	−0.407	−0.032	0.15

Source Calculated by the authors

* - weak connection

** - moderate connection

*** - strong connection

Hypothesis 2: Companies' use of digital neuromarketing depends on their marketing budget. The larger the marketing budget, the more often digital neuromarketing tools are used in the company.

Based on the data obtained, it can be concluded that the use of digital neuromarketing in companies can vary depending on the size of the marketing budget. However, no significant correlation between the size of the marketing budget and the use of digital neuromarketing has been identified (Table 2). Thus, hypothesis 2 is rejected.

Hypothesis 3. Respondents' awareness of digital neuromarketing influences their positive assessment of its effectiveness. The more a respondent knows about digital neuromarketing, the higher they rate its effectiveness.

A moderate correlation was found between respondents' answers about their awareness and their assessment of the effectiveness of digital neuromarketing tools (Table 3). Spearman's correlation coefficient was 0.452 for respondents who indicated that their company uses digital neuromarketing tools and 0.460 for respondents who use digital neuromarketing tools in their work. Thus, respondents understand the effect that the use of digital neuromarketing tools can have. Therefore, hypothesis 3 is confirmed.

Hypothesis 4: Currently, companies use primary neuromarketing tools comprehensively, simultaneously applying different types of equipment and combining them with classic desk-based and empirical marketing research. Figure 1 shows the various types of marketing tools used in the companies where the respondents work. More than 60% of respondents noted that their companies use traditional marketing tools,

Table 2 Spearman's correlation coefficient between respondents' answers about the size of the company's marketing budget and the use of digital neuromarketing

The marketing budget at the company	Distribution of companies using DNM by budget level	Number of companies planning to use DNM
	%	%
Less than 100,000 rubles ($1244)	25%	11%
100,000–500,000 rubles ($1244–6219)	0%	11%
500,000–1,000,000 rubles ($6219-$12,438)	25%	11%
1000–5000 thousand rubles ($12438–62,189)	0%	11%
Over 5,000,000 rubles ($62,189)	25%	33%
I don't have any information	25%	22%
Spearman's correlation coefficient	**0.038**	**–0.09**

Source Calculated by the authors

Table 3 Spearman's correlation coefficient between respondents' answers about awareness and assessment of the effectiveness of digital neuromarketing tools

Respondents' assessment of the potential increase in effectiveness through the use of DNM	I had not encountered the terms neurotechnology and DNM before	I know about the existence of the DNM, but I don't specifically collect information about it	I regularly review articles in magazines about DNM	Our company practices the use of DNM	I personally use DNM in my work
Less than 10%	29%	57%	14%	0%	0%
10–30%	25%	75%	0%	0%	0%
30–50%	13%	38%	13%	13%	25%
50–70%	0%	50%	50%	0%	0%
70–90%	0%	50%	0%	25%	25%
Over 90%	0%	0%	0%	0%	0%
Will not change	25%	0%	25%	25%	25%
Spearman's correlation coefficient	−0.035	0	0.23	**0.452** *	**0.460** *

Source Calculated by the authors
* - weak connection
** - moderate connection
*** - strong connection

such as desk research (68.2%) and field research (63.6%). These are used much more frequently than others (once a week or once every few months).

To a lesser extent, tools such as eye tracking (27.3%), fMRI (18.2%), and EEG (18.2%) are used. This suggests that classical neuromarketing tools are less widely used, which may be due to the high cost of the equipment.

Table 4 presents the results of a correlation analysis on the use of various neuromarketing tools by companies. It can be seen that there is a strong correlation between fMRI and other neuromarketing tools: EEG (1.000), eye tracking (0.855), facial analysis (0.751), digital psychometrics (0.691), and UX analysis (0.809). This indicates that companies using fMRI tend to use other neuromarketing methods in a comprehensive manner.

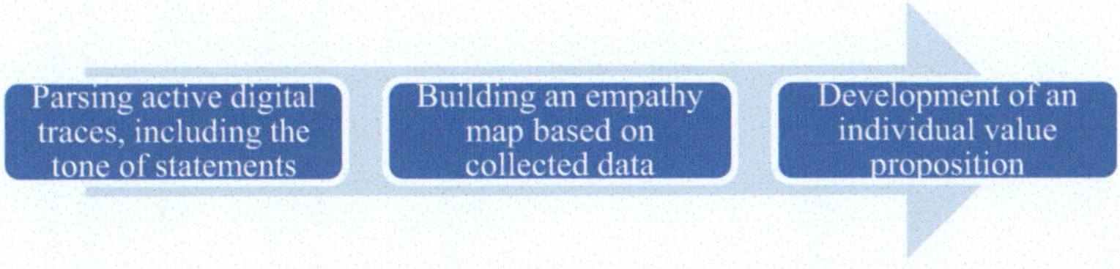

Fig. 1 Stages of digital psychometry of potential customers. *Source* Compiled by the authors

Table 4 Spearman's correlation coefficient between respondents' answers regarding the use of various marketing tools in companies

	Field research	Desk research	Digital psychometrics	fMRI	EEG	Eye-tracker	Facial analysis, galvanic skin response	User experience (UX) analysis
Field research	1	**0.599** **	**0.765** ***	**0.552** **	**0.552** **	**0.557** **	0.27	0.503 *
Desk research	**0.599** **	1	**0.593** **	0.372	0.372	0.345	0.193	0.24
Digital psychometrics	**0.765** ***	**0.593** **	1	**0.691** ***	**0.691** ***	**0.736** ***	**0.431** *	**0.619** **
fMRI	**0.552** **	0.372	**0.691** ***	1	**1.000** ***	**0.855** ***	**0.751** ***	**0.809** ***
EEG	**0.552** **	0.372	**0.691** ***	**1.000** ***	1	**0.855** ***	**0.751** ***	**0.809** ***
Eye-tracker	**0.557** **	0.345	**0.736** ***	**0.855** ***	**0.855*****	1	**0.721** ***	**0.792** ***
Facial analysis, galvanic skin response (biometrics), lie detector	0.27	0.193	**0.431** *	**0.751** ***	**0.751** ***	**0.721** ***	1	**0.657** ***
User experience (UX) analysis	**0.503** *	0.24	**0.619** **	**0.809** ***	**0.809** ***	**0.792** ***	**0.657** ***	1

Source Calculated by the authors
* - weak connection
** - moderate connection
*** - strong connection

Traditional marketing methods—field and desk methods—have a moderate correlation with neuromarketing tools, which indicates that NEUROMARKETING is not always used alongside traditional methods.

As a result, neuromarketing tools are used in a comprehensive manner, complementing each other. Meanwhile, traditional marketing methods remain more widespread and are used more often.

3.2 Analysis of the Effectiveness of Using Digital Psychometrics Technology to Optimize Product Promotion (Using the Example of a Travel Agency)

The authors evaluated the effectiveness of one of the tools of digital neuromarketing—psychometrics. The study was conducted using the example of a travel agency. The results obtained from a regular mailing, without any bias toward psychotype, were compared with the results obtained from a marketing campaign that personalized letters according to each customer's psychotype. The psychotypes of customers were determined by parsing open digital information from customers' social networks.

When classifying customers by psychotype and developing personalized emails based on the specific psychotype obtained through data parsing, the open rate of emails increased (from 34 to 45%), which is 11 percentage points in absolute terms and 32.35% in relative terms.

The company's CRM system contains 3750 customer email addresses. Below are the results of an assessment of the effectiveness of a marketing campaign using personalized mailings based on personality type and general mailings that do not take temperament type into account (Table 5).

After implementing the project using digital psychometrics technology, the percentage of recipients who clicked on the link in the email increased (from 28 to 33%, which is 5 percentage points in absolute terms and 17.86% in relative terms).

After the implementation of the project using digital psychometrics technology, the conversion rate (the percentage of people who performed the target action) increased from 2 to 5% (which is 3 percentage points in absolute terms and 150% in relative terms, which is undoubtedly a positive trend). The increase in click-through rates is also a positive result, although less significant than the increase in open rates. This may indicate that the content has become more appealing to recipients. In addition, the percentage of unsubscribes from the mailing list decreased from 10 to 6% (which indicates a decrease in the number of unsubscribes by 4 percentage points in absolute terms and 40% in relative terms). There has also been an increase in customer feedback in terms of the number of reviews left (the number of reviews after the mailing increased from 340 to 576, which is 69.41%) (Table 6).

Next, a factor analysis of the marketing campaign was conducted (Table 7), which shows the key metrics of the mailing with their relative and normalized change, as well

Table 5 Assessment of the effectiveness of a marketing campaign using digital psychometrics technology in emails

Criterion	Unit of measurement	Before implementation	After implementation	Absolute change	Relative change
1	2	3	4	5 = 4–3	6 = 5/3
Increasing email open rates	% of recipients who opened the email	340	450	110	32.35%
Click-through rates	% of recipients who clicked on links in emails	280	330	50	17.86%
Conversion	% of recipients who completed the target action	20	50	30	150.00%
Unsubscribe rate	% of recipients who unsubscribed from the mailing list after receiving emails	100	60	−40	−40.00%
Feedback	Number of reviews received	340	576	236	69.41%

Source Calculated by the authors

Table 6 Evaluation of the effectiveness of a marketing event

Criterion	Unit of measurement	Before implementation	After implementation	Absolute change	Relative change
Email open rates	%	340	450	+110	+32.35%
Click-through rates	%	280	330	+50	+17.86%
Conversion rates	%	20	40	+20	+100.00%
Unsubscribe rates	%	100	60	–40	+40,00% (inversion)
Feedback	Quantity	340	550	+210	+61.76%
Expenses	₽($)	1,772,173 ₽ ($22,152)	3,653,846 ₽ ($45,673)	+1,881,673 ₽ ($23,521)	+106.20%
Revenue	₽($)	2,040,000 ₽ ($25,500)	4,760,000 ₽ ($59,500)	+2,720,000 ₽($34,000)	+133.33%
ROI	%	15%	30%	+15 p.p	+100%

Source Calculated by the authors

as their weight and contribution to the final effect. ROMI is given the highest weight (0.30), conversion is second in importance (0.25), and clickability and unsubscribes are given the lowest weight (0.05).

Consider the indicators in more detail:

- **Openability.** The open rate shows the percentage of subscribers who opened the email. In our example, it increased by 32.35% (normalized to 3.23%). With a weight of 0.05, this contributed only 0.16% to the final effect. This small contribution indicates that although interest in the newsletter topic increased, it was not a decisive factor in the campaign's success.
- **Click-through rates (CTR)**. Click-through rate (CTR) is the percentage of recipients who clicked on links in the email. An increase of 17.86% (normal to 1.78%) with a weight of 0.05 contributed 0.09%. The small effect indicates that increasing the number of clicks within the email had a negligible impact on the final result compared to more important metrics.
- **Conversion**. Conversion reflects the percentage of subscribers who performed a target action (purchase, registration, etc.). Our conversion increased by 100% (normal 10.00%). With a weight of 0.25, this contributed 2.50% (≈28.1% of the total). The significant increase in conversion means that the advertising materials effectively attracted the audience, and this was one of the main drivers of the result.
- **Unsubscribes**. The unsubscribe metric shows the number of subscribers who have opted out of the mailing list. Unsubscribes decreased by 40% (normal: 4.00%). With a weight of 0.05, the contribution was only 0.20%. This means that the decrease in unsubscribes is positive, but in itself had little effect on the overall effect (perhaps the initial unsubscribe rate was low).

Table 7 Factor analysis of the marketing campaign

Criterion	Relative change %	Normalized change (%)	Weight	Weighted change (%)	Share in total %
Openability	32.35	3.23	0.05	0.16	0.16
Click-through rates (CTR)	17.86	1.78	0.05	0.09	0.09
Conversion	100.00	10.00	0.25	2.50	2.50
Unsubscribes (improved version)	40.00	4.00	0.05	0.20	0.20
Feedback	61.76	6.18	0.15	0.93	0.93
Revenue	133.33	13.33	0.15	2.00	2.00
ROMI	100.00	10.00	0.30	3.00	3.00
Total				**8.88**	**8.88**

Source Calculated by the authors

- **Feedback**. The amount of feedback received in one form or another increased by 61.76% (normal 6.18%). Weight 0.15 contributed 0.93%. The increase in feedback indicates greater audience engagement. This is a positive sign (more comments, questions, etc. indicate interest), although the direct effect on commercial results was moderate.
- **Revenue**. Total revenue from the mailing (income) increased by 133.33% (normal 13.33%). With a weight of 0.15, the contribution was 2.00% (≈22.5% of the total). The significant increase in revenue directly enhanced the overall effect of the campaign.
- **ROMI**. This indicator grew by 100% (normal 10.00%). With a weight of 0.30, the contribution was 3.00% (≈33.8%). ROMI shows how effectively marketing costs are recouped. The largest contribution among all metrics confirms that the main efforts brought significant profitable results. A high ROMI indicates that every ruble invested brought a tangible profit.

The analysis of indicators clearly shows that **ROMI, conversion, and revenue** accounted for a significant share of the total effect (≈84.4% of 8.88%):

- ROMI—contribution 3.00 (33.8% of the amount)
- Conversion—contribution 2.50 (28.1%)
- Revenue—contribution 2.00 (22.5%).

The remaining metrics contributed <16% (feedback 0.93; unsubscribes 0.20; open rate 0.16; CTR 0.09). Thus, the main winning factors are financial results (profitability and sales), rather than engagement metrics. This is consistent with the general understanding that conversion and ROMI are among the key marketing KPIs.

Further, ARPU, CAC, and CLV indicators were calculated (Table 8).

Customer Acquisition Cost decreased by 15,556 ₽($194), which is a positive indicator. This means that the company has become more effective in attracting new customers, possibly due to improved marketing strategies or process optimization.

CLV increased by 4,835,556 ₽($60,444), which indicates that customers are generating more revenue over the entire period of their interaction with the company. This factor can be attributed to increased customer loyalty.

Thus, based on the data obtained, the following conclusions can be drawn:

Table 8 Calculation of ARPU, CAC, and CLV indicators

	Units	Before implementation	After implementation
ARPU	₽($)	408,000 ₽($5,100)	528,889 ₽($6,611)
CAC	₽($)	60,000 ₽($750)	44,444 ₽($5,555)
CLV	₽($)	16,320,000 ₽($204,000)	21,155,556 ₽($264,444)
Revenue	₽($)	2,040,000 ₽($25,500)	4,760,000 ₽($59,500)
Marketing expenses	₽($)	300,000 ₽($3,750)	400,000 ₽($5,000)

Source Calculated by the authors

- All three key indicators (ARPU, CAC, CLV) show positive dynamics after the implementation of changes.
- The increase in ARPU and CLV, combined with a decrease in CAC, indicates that the company has become more profitable and efficient in customer management.
- These changes may indicate a successful strategy aimed at increasing revenue and optimizing customer acquisition costs.

The analysis clearly shows the high effectiveness of using digital psychometrics tools.

4 Discussion

4.1 *The Wave Approach to Digital Promotion of a New Product on the Market*

For the most effective campaign to promote a new product on the market, the authors of this article proposed an innovative approach based on digital psychometrics technology. The innovative approach lies in the fact that when launching a new product on the market, for more effective promotion, it is necessary to influence certain psychotypes in stages: first of all, choleric types, then sanguine types and only then melancholic and phlegmatic types.

First, you need to identify your potential buyers by psychological type (temperament type), then create empathy maps based on the data collected, and finally develop individual value propositions with targeted mailings for each specific psychological type (Fig. 1).

1. Parsing active digital traces, including the tone of statements.

Various types of digital traces can be used. First, there is text information from various sources (consumer profiles, blogs, etc.). Second, there are images (profile photos, as well as images posted by users on the network). Third, the results of internet surfing analysis and metadata. Fourth, other types of digital traces: «likes», financial transactions, changes in a person's geolocation.

Sources of digital traces include: social networks, messengers, review sites, etc. Based on digital traces, it is possible to determine what a person usually writes and does in the online space: for example, searches for reviews of a product or leaves comments about a product, watches videos with expert advice, etc. It is also necessary to analyze the tone of their statements using special services, such as sentiment analysis, content analysis, and emotional component analysis. When using an open approach to analyzing consumer texts, it is possible to analyze «affective processes» and discover correlations with consumer personality traits.

«Likes» can be used to determine personality traits, intelligence, and life satisfaction [15]. Digital traces can also be used to explore a person's emotional state (anxiety, depression, pain). A wide variety of digital traces are used to identify these:

texts, visual material including photos and images, mobile phone call patterns, location changes, etc. However, the accuracy of predicting emotional states is somewhat lower than that of personality traits.

2. Building an empathy map based on the collected data. This part involves systematizing what the respondent (potential customer) thinks, feels, how they make purchases, what they say/write about the purchased goods/services. What are their values, pain points, and barriers in life in general and in relation to the selected product.
3. Developing an individual value proposition based on a specific psychotype and the empathy map. An individual value proposition is built on the basis of one of the following strategies: joint value creation, professionalism, tradition, a store without a salesperson, expertise, innovation, bias, imitation, depending on the buyer's psychotype identified during psychometry (Table 9).

After identifying the temperament types, the marketing campaign is divided into three stages (promotion waves). Let's take a look at these three stages (waves) (Fig. 2).

Table 9 Consumer psychotypes depending on the level of independence from the opinions of others about the product

Level of dependence on the opinions of others about a product	Temperament (personality type)
Independence from the opinions of others	Choleric (individual)
Selective dependence	Sanguine (realist)
Demonstration of trends	Phlegmatic (conservative)
Reliance on the opinion of the majority	Melancholic (imitator)

Source Compiled by the authors

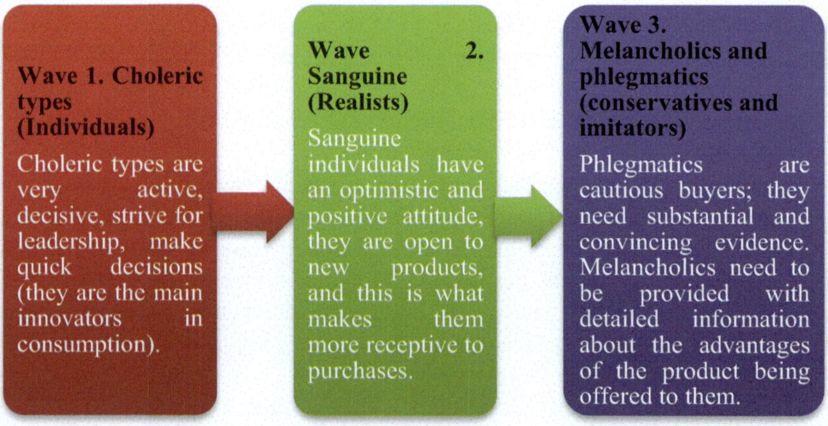

Fig. 2 Wave effect in an advertising campaign. *Source* Compiled by the authors

Stage No. 1. Impact on choleric individuals (first wave of promotion). Key characteristics of choleric individuals include being very active, decisive, striving for leadership, and making quick decisions (they are the main innovators in consumption). Thus, if this psychotype is initially interested, this group of people can become the first buyers and disseminators of information about this new product. Choleric people are characterized by leadership traits. They take active positions in social groups and can influence the opinions of others. If choleric people speak positively about a new product, this can influence the opinions of other consumers, including sanguine people, who may follow their example. For example, sanguine people may be inspired by the energy and enthusiasm of choleric people, which may lead to their interest and purchase. In addition, because choleric people tend to make quick decisions, this can help the manufacturing company increase sales. In the first stage, it is necessary to use bright, dynamic advertising elements and tools that emphasize the advantages of the product. It is also necessary to highlight the unique and innovative aspects of your product that can help it stand out. You should also offer limited-time promotions and discounts to speed up the decision-making process for sanguine people. Use bright, fast videos that choleric people respond to.

Stage No. 2. Impact on sanguine people (second wave of promotion). The second stage in the wave approach concept can be described as the impact on sanguine people. This is because this group of people has an optimistic and positive attitude, they are open to new products, and this is what makes them more receptive to purchases. Thus, the choleric types, who were the target of the first wave, will act as a catalyst for sanguine types, who are also receptive to new products and will want to buy new products they like. It is also worth noting that sanguine people are often the center of attention in various social groups and have good communication skills. Sanguine people will begin to share their experiences with friends and family, resulting in the so-called «word of mouth» effect. If sanguine people start actively using a product and sharing their experiences, this can help create a positive image for the product. Their enthusiasm and energy can attract the attention of other consumers, including phlegmatic and melancholic people, who are more cautious in their decisions regarding innovations.

Next, we will briefly consider possible tools for influencing people in the first two stages. Influencing choleric people (strategies for influence):

- **Promotions and testing**. It is necessary to offer choleric people the opportunity to test the product for free or at a reduced price. This can be done through special events or online campaigns.
- **Emphasizing effectiveness**. It is necessary to use aggressive marketing messages, focusing on the fast and effective action of the product.
- **Creating a community**. In order for choleric people to feel comfortable sharing their experience of using the product, it is necessary to create a platform for exchange. Create a platform for sharing experiences where choleric people can share their results and impressions of using the product.

Impact on sanguine people (impact strategies):

- **Create vivid content**. Develop attractive advertising materials that emphasize the emotional aspects of using the product.
- **Social media**. Use platforms to create viral content. Sanguine people love to share interesting finds, so create challenges or contests related to using the new product.
- **Promotional events**. Organize events where sanguine people can try the product, socialize with each other, and share their impressions. These can be both offline and online events.

Stage No. 3. Impact on phlegmatic and melancholic people (third wave of promotion). At the final stage, phlegmatic and melancholic people (the most numerous market segments) become the targets of influence. Phlegmatic people may be more cautious when making purchases; they need substantial and convincing evidence. That is why they need to be convinced that they are being offered the best value for money. Melancholic people need to be provided with detailed information about the advantages of the product being offered to them. They need to be convinced that this brand is currently in high demand. The campaign to promote the new product will be multifaceted and consistent, starting with the active involvement of choleric people and moving on to sanguine people, and then to other groups. The use of emotional content, social networks, and community building will create a positive image of the product and increase its popularity in the market. It is also important to take into account feedback from users in order to continuously improve both the product and marketing strategies.

5 Conclusion

As a result of the conducted research on awareness and readiness to use digital neuromarketing tools, the following conclusions were reached. On the one hand, three hypotheses (1) the more frequently a company uses digital neuromarketing, the more highly the respondent values this marketing tool; (2) the more the respondent knows about digital neuromarketing, the more highly they rate its effectiveness; (3) neuromarketing tools are used in a comprehensive manner, complementing each other, were confirmed. On the other hand, the hypothesis the larger the marketing budget, the more frequently a company uses digital neuromarketing tools was rejected. Next, a rationale was given for the positive effects of using digital psychometric tools.

For discussion purposes, the authors proposed a wave-based approach to digitally promoting a new product on the market. The developed algorithm involves clustering potential customers based on digital psychometrics by parsing natural digital information, followed by targeted, time-lag-based targeting of choleric, sanguine, phlegmatic, and melancholic individuals. This contributes to increased advertising campaign effectiveness.

The authors propose the following future research areas in digital neuromarketing:

- Using eye-tracking technology to analyze advertising layouts tailored to consumer personality types in the fashion industry. A wave-based approach will be tested in the digital promotion of innovations in the fashion market.
- Creation of a method for developing and evaluating a neurobrand in the fashion industry.
- Deeper monitoring of the use of digital neuromarketing methods (primarily in the Russian market).

References

1. Berezka SM, Kovalenok AY, Shamalova AY, Starchenko VM, Shirokova PI, Kosonogov VV, Vetrova TV, Shestakova AN (2023) The neuromarketing research market: companies, technologies, application practices, and development prospects. In: Berezka SM, Kovalenok AY (eds) National research university higher school of economics. Publishing House of the Higher School of Economics, Moscow, 108 p. ISBN 978–5–7598–2799–3 (in cover). https://doi.org/10.17323/978-5-7598-2799-3
2. Hakim A, Klorfeld S, Sela T, Shabat-Simon M, Levy D (2020) Machines learn neuromarketing: improving preference prediction from self-reports using multiple EEG Measures and Machine learning. Int J Res Mark. https://doi.org/10.1016/j.ijresmar.2020.10.005
3. Byrne A, Bonfiglio E, Rigby C, Edelstyn N (2022) A systematic review of the prediction of consumer preference using EEG measures and machine-learning in neuromarketing research. Brain Inform. https://doi.org/10.1186/s40708-022-00175-3
4. Aleinik N (2020) Who conducts neuromarketing research in Russia? https://rb.ru/longread/neuromarketing-companies/?ysclid=l8uh0qgzaz871213352
5. King Mю, Bruner G (2000) Social desirability bias: a neglected aspect of validity testing. Psychol Mark. https://doi.org/10.1002/(SICI)1520-6793(200002)
6. Kalenskaya NV (2022) Application of neuromarketing technologies in the study of consumer behavior in the digital environment. In: Minnikhanov RN (eds) International Forum Kazan Digital Week-2022: Collection of materials of the International Forum, Kazan, September 21–24, 2022. Scientific Center for Life Safety, Kazan, pp 342–347
7. Mikhailova DD (2025) Neuromarketing: efficiency and ethical challenges in the digital age. In: Effective support of scientific and technological progress: analysis of problems and search for solutions: collection of articles from the International scientific and practical conference, Yekaterinburg, August 1, 2025. Aeterna, UFA, p 67–71
8. Rawnaque FS, Mahmudur Rahman K, Anwar S, Vaidyanathan R, Chau T, Sarker F, Mamun K (2020) Technological advancements and opportunities in neuromarketing: a systematic review. Brain Inform. https://doi.org/10.1186/s40708-020-00109-x
9. Neuromarketing in the digital age: can algorithms understand emotion? One Data Software. Electronic resource. https://www.onedatasoftware.com/blog/neuromarketing-digital-age-algorithms-emotion
10. Millagala K (2023) Neuromarketing as a digital marketing strategy to unravel the evolution of marketing communication. In: 5th Nordic International Business, Export Marketing and Tourism Conference 2023. https://www.researchgate.net/publication/377489276_Neuromarketing_as_a_Digital_Marketing_Strategy_to_Unravel_the_Evolution_of_Marketing_Communication/stats
11. Voziyanova NY, Kameneva NV (2025) Neuromarketing in digital epoch: situation and propsects of development. Vestnik Plekhanov Russ Univ Econ (3):231–243 (In Russian). https://doi.org/10.21686/2413-2829-2025-3-231-243

12. D'Alfonso S, Coghlan S, Schmidt S et al (2025) Ethical dimensions of digital phenotyping within the context of mental healthcare. J Technol Behav Sci. 10:132–147. https://doi.org/10.1007/s41347-024-00423-9
13. Syropyatov VV, Eliseeva VS, Makhar DK (2023) Brand personification: applying psychometric methodology to social networks' big data (the Russian market data). Creat Econ 17(5):1705–1730. https://doi.org/10.18334/ce.17.5.117809
14. Nozari H (2025) Cognitive targeting and neuromarketing applications in AI-driven digital advertising. Appl Innov Ind Manag 5:52–58. https://doi.org/10.63630/aiim.52.52
15. Latynov VV, Ovsyannikova VV (2020) Forecasting psychological characteristics of a human based on their digital traces. Psychology. HSE J (1). https://cyberleninka.ru/article/n/prognozirovanie-psihologicheskih-harakteristik-cheloveka-na-osnovanii-ego-tsifrovyh-sledov

Features of Management Decision-Making Processes Digitalization in the Environmental Sphere

M. A. Plakhotnikova, O. V. Zakharova, T. Y. Chernysheva, and K. E. Chelyshev

Abstract The sustainable development of the regions largely depends on the effective management of environmental risks, which is carried out on the basis of the use of modern digital technologies. Decision-making at the regional and municipal levels is key to the successful functioning and development of the region. Today, the level of information technology makes it possible to bring the decision-making process to a new level, allowing regional and municipal leaders to choose the most relevant alternatives that are optimal in the current environment. In many regions of Russia municipal authorities are striving to digitalize decision-making processes, while at the moment they have not developed a unified approach to choosing digital platforms to create management decision-making support systems. Given the nature of management decisions in the field of environmental risks and the nature of information resources used to make them, in general, the most reasonable methods that can be used as a basis for management decision support systems are qualitative decision-making methods based on expert assessments. Thus, the formation of a reasonable approach to the development of systems for supporting management decision-making in the field of environmental risks for regional and municipal governments at the moment requires the identification of the features of this process. The purpose of this study is to identify the features of digitalization of the process of making management decisions in the environmental sphere and develop on their basis a prototype system for supporting management decisions. The result of the study is a prototype of a system for supporting management decisions taking into account environmental risks, intended for use by regional and municipal governments.

Keywords Decision support system · Management decisions · Environmental risks · Digitalization of management processes · Sustainable development · Information systems development

---

M. A. Plakhotnikova (✉) · O. V. Zakharova · T. Y. Chernysheva · K. E. Chelyshev
University of Tyumen, Tyumen, Russia
e-mail: erelda@mail.ru

© The Author(s), under exclusive license to Springer Nature Switzerland AG 2026

A. Kulachinskaya et al. (eds.), *Digital Transformation: What Kind of Smart Life do we have today?*, Lecture Notes in Networks and Systems 1895,
https://doi.org/10.1007/978-3-032-20582-7_17

1 Introduction

The introduction of digital technologies in the processes of environmental risk management in the regions has been leading to noticeable improvements in their activities for quite a long time, reducing the cost and accelerating activities to prevent and eliminate the consequences of environmental risks [1]. However, over time, the very processes of choosing and implementing digital technologies have become more complicated and absorbed new factors, for example, such as the use of artificial intelligence methods, therefore, it became necessary to combine different technologies in one management task. At the moment, the digital transformation of management processes is an objective reality and a key factor that has a significant impact on the sustainable development of regions, both in Russia and around the world [2].

The use of artificial intelligence to manage environmental risks is currently a popular area of scientific and practical activity [3]. In addition, trends in the development of socio-economic processes show the increasing importance of the concept of sustainable development, which, along with the rest, includes an environmental component [4]. The effectiveness of the use of digital technologies in general to manage environmental risks, including threats of floods, is currently a key issue that determines the prospects for sustainable development of regions with increased risks in the environmental sphere. Highlighting the features of such an assessment allows us to take into account the trends in the development of digital technologies, which, ultimately, contributes to the development of successful strategies for digitalizing environmental risk management processes at the level of regional and municipal management.

The development of artificial intelligence creates certain effects of digital transformation, which to some extent affected many aspects of risk management at the regional and municipal levels. For the sustainable development of regions it is necessary to take into account both the positive and negative effects of the digital transformation of management processes and formulate strategies to increase the effectiveness of their digitalization [5]. Solving situations arising in the process of expanding digitalization of environmental risk management, among other things, can be carried out by regulating the effects of digital transformation [6], which leads to the need to develop tools to smooth (or reduce) the impact of negative effects. The development of such tools cannot be carried out without analyzing the current state of digitalization of environmental risk management in Russia, identifying the effects of digital transformation and assessing their impact on the effectiveness of regional management.

The current situation is complicated by the fact that at the moment the IT market offers a wide range of digitalization tools that can give various effects in different conditions. Accordingly, it is necessary to develop a strategy for digitalizing business processes for environmental risk management, which could provide the greatest effect and reduce the potential risks of introducing digital technologies.

First of all, there was a need to form adapted methods that make it possible to assess the effectiveness of the introduction of digital technologies in the management processes of regional and municipal authorities, which makes the topic of research relevant and demanded.

2 Materials and Methods

The research methodology consists in the consistent use of general scientific and specific methods. First of all, comparative analysis of sources was applied, which determined the effective directions for the development of the information product. Based on the results of comparative analysis, performance indicators were identified, on the basis of which the need to increase the level of cyrovization of management business processes of municipal and regional governments in the field of environmental risks was determined.

Statistical research methods made it possible to identify the main trends in the development of digitalization of management business processes. The dynamics and structure of indicators that determine the main trends in the field of digitalization of management business processes were revealed.

The developed information product is based on the Saati hierarchy analysis method, which involves the selection of a strategic alternative according to the criteria using the formula [7]:

$$P(x) = \sum_{i=1}^{n} vi * bi(x) \tag{1}$$

where $P(x)$ is the overall utility of the evaluated strategy x;

vi is criterion weight i;

$bi(x)$ is utility of the evaluated strategy x by criterion i;

i is number of criteria, varies from 1 to n.

Modeling as a research method was used to create models of the developed information system, as a result its inputs and outputs, and its immediate environment were determined.

3 Results

Assessing the effectiveness of the use of digital technologies in municipal and regional government bodies, first of all, should take into account those indicators that generally characterize the processes of digitalization of the economy. One of the most significant indicators in this area is the number of Internet users, as well as the degree of accessibility of the Internet and its resources [8]. Speaking about the digital economy, it is necessary to note its relationship with such a concept as the Fourth Industrial Revolution (Industry 4.0), which can be defined as "a change in formation towards absolute digitalization of business processes of production with the participation of artificial intelligence" [9, 10]. In addition, the fourth industrial revolution is at the very beginning of its implementation and not all of its consequences have yet manifested themselves in full. At the beginning of 2025, the number of Internet users in Russia amounted to 127.6 million people, while Internet penetration is 88% (for comparison, in the USA 91.8%, and in China 73.7%) [8]. Compared to the 2010s, the annual growth of Internet users has noticeably decreased, this is due to the high percentage of already connected users. In addition, the time spent by the average users on the network also decreased by almost 20 min compared to 2022 [11].

Another important indicator of the digitalization of the economy is the number of users of social networks. At the same time, the number of social networks is constantly growing, which attracts new unique users who were previously not satisfied with something in existing social networks [12]. The share of users who prefer to communicate through social networks is constantly growing, therefore, the digital business processes of companies must take into account the ability to find and maintain a target audience on social networks.

A significant parameter is also the structure of preferences of Internet users. Studies show that in 2024, 47% of the Russian population are users of social networks, while 90% of them actively use them. The next indicator is the number of mobile Internet users and its ratio with the Internet on personal computers (PCs) and laptops.

In addition, the dynamics of traffic from all the above-listed devices is very indicative. According to analysts in 2024 91.4 million people use mobile Internet in Russia. If we compare the shares of 2024 with the indicators of 2023, then the share of the smartphone increased by more than 33 percentage points, the share of traffic from tablets decreased by 9 percentage points, and the share of traffic from PCs and laptops decreased by 5 percentage points [13].

All of the above trends influenced the development of digitalization of municipal and regional management, mainly in terms of automation of management business processes using various information systems, including management decision support systems (MDSS) and other software products for management automation. The effectiveness of the use of digital technologies at the level of regional and municipal authorities can be assessed in different ways, depending on the approach used.

The essence of calculating the effectiveness of digital technologies at the level of regional and municipal authorities is expressed in the fact that they should receive a positive effect from the introduction of digital technologies with a minimum of costs for them [14]. The above essence of the effectiveness of digital technologies is highly generalized, in practice, the methodology for calculating the effectiveness of digital technologies largely depends on their type and nature of application.

At the same time, it is necessary to take into account the external environment in which the regional and municipal government operates, for example, a very important factor is the degree of digitalization of business processes of other similar structures and the pace of introduction of digital technologies. This area is very dynamic, so the methods for assessing the effectiveness of digital technologies must be constantly reviewed and improved.

Based on the comparative analysis [8, 11, 13], the author's definition of the effectiveness of digital technologies was formulated. The effectiveness of digital technologies is a positive result of the introduction of digital technologies into the activities of the organization attributed to the costs incurred during the implementation. The effectiveness of digital technologies for business processes of regional and municipal authorities can be assessed within three groups of indicators [15].

(1) The efficiency of the product obtained as a result of the digital business process;
(2) Efficiency of the business process;
(3) The achieved degree of consumer satisfaction with the digital business process product.

The Russian information and communication market is constantly developing, this is a prerequisite for maintaining its competitiveness in comparison with the world market. The development of regional and municipal authorities should be carried out in the current directions set by the transformation of the external environment, i.e. in this case, the information and communication market. As practice shows, technological factors are, in some cases, as significant for the functioning of digital business processes in municipal and regional government bodies as economic and political ones.

The functioning of digital business processes in regional and municipal authorities is influenced by many different factors that can form entire influence trends. However, at the moment, the development of the digital economy is a determining factor that largely determines the state of digital business processes.

When assessing the effectiveness of the introduction of digital technologies in regional and municipal authorities, it is necessary to remember about the technical means with which the functioning of digital business processes is supported. Equipment for digitalization of business processes also has its own indicators and characteristics, which may become irrelevant. Therefore, it is necessary to monitor devices that support digital business processes and adapt the physical environment of the digital business model to modern conditions.

The first trend is to shift the focus to the use of smart backgrounds for e-business purposes, it has been traced for several years and at the moment it is a very promising direction. Already, most consumers of e-business goods and services prefer mobile options for the services provided, so the presence of a convenient mobile application is one of the key success factors in this area.

The second important trend that needs to be paid attention to when assessing the impact of the information and communication market on e-business is the high degree of user activity in social networks. Thus, e-business should use social networks with maximum effect, i.e. the digital technologies used should allow including social networks in the business model.

Given the trends considered, we can conclude that the introduction of innovative decision-making support systems for the digitalization of business management processes in the activities of municipal and regional governing bodies is one of the directions of their sustainable development.

The development and implementation of the Management Decision Support System (MDSS) in the field of environmental risks was carried out on the example of the Department of Informatization of the Tyumen Region. Decision support systems integrated with geo-information tools have proven effective in environmental risk management. In particular, they help to assess the quality of air, water and land, simulate scenarios of changes, and also take into account multi-criteria goals (environmental, social, economic) [16].

In particular, models based on ecosystem indicators and GIS are used to assess the vulnerability of watersheds and make decisions on the sustainable development of territories [17]. At the same time, the development of digital solutions for municipal and regional governments shows that the automation of business processes for environmental protection is a key tool for improving management efficiency, especially in the face of rapidly changing legislation [18]. The specifics of environmental risks taken into account in the sustainable development system require strict management principles [19], therefore, the management decision-making support system should be adapted to them.

The proposed management decision support system allows for:

- Building a pipeline for collecting data from various sources (monitoring, reports, GIS) and normalizing them, which will improve the quality of data;
- Real-time data reception (for example, through IoT sensors), which will reduce the delay in recognizing excesses and quickly increase the response;
- GIS visualization: interactive maps combining spatial characteristics of emission sources, pollution density, meteorological data and administrative boundaries;
- Scenario analysis module (multi-criteria), which allows assessing the consequences of various measures (emissions reduction, filtration, policy change) taking into account environmental, social and economic criteria;
- Anomaly detection algorithms to detect unexpected outliers or critical events;
- User reports adapted for various roles (analyst, administrator, official), as well as a notification system when standards are exceeded;
- A prototype with testing on real data from the region (monitoring stations, enterprise data, open GIS layers), which will demonstrate the effectiveness of the system and its practical usefulness.

The authors have defined functional requirements for the RMS based on the established SNW analysis (Strengths, Neutral, Weaknesses) of existing analogues:

1. Collection of data from various sources.
2. Creation of a database of environmental indicators.
3. Automatic processing and aggregation of data according to the specified time and territorial parameters.
4. Determination of "problem areas" based on threshold values of indicators.
5. Displaying environmental indicators on an interactive map of the region.
6. Generation of graphs, diagrams and reports for users.
7. Generation of recommendations for improving the environmental situation (for example, reducing emissions, introducing filtration).

8. Modeling of scenarios for the development of the situation when individual parameters change.
9. The presence of a system of notifications about exceeding the permissible standards.
10. Friendly web interface with differentiation of access rights (administrator, analyst, user).
11. Support for authorization and data protection.

The authors developed the MDSS model using the "Black Box" method presented in Fig. 1.

The authors have developed a model of the external environment showing data sources for the work of the MDSS and consumers of its information products (Fig. 2).

The proposed MDSS currently has a developed user interface.

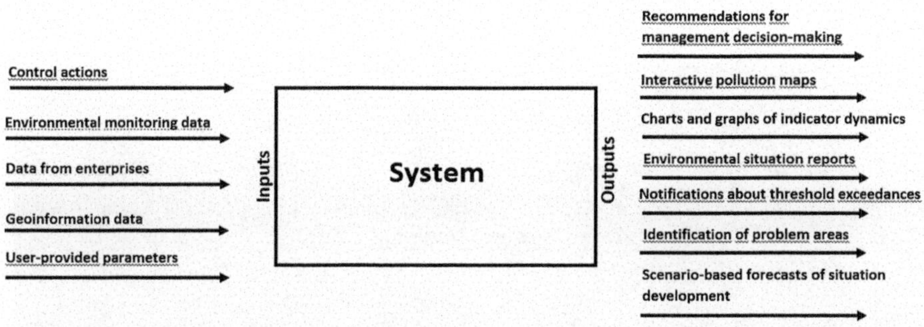

Fig. 1 Black box model of the developed MDSS

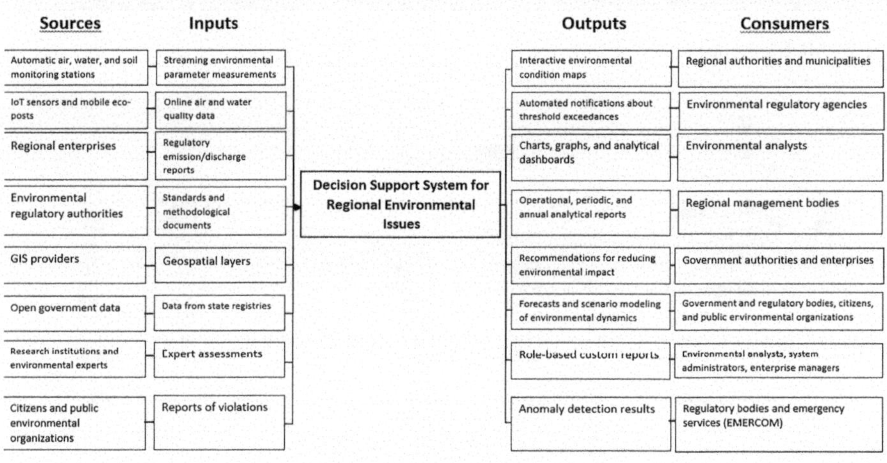

Fig. 2 Environmental model for the proposed MDSS

The interface presented in Figs. 3, 4, 5 and 6 demonstrates high values for all criteria for its assessment, namely: clarity and unambiguity of perception, an indication of the amount of information necessary for the operation of the system, visual simplicity and attractiveness of the image, functionality corresponding to functional requirements highlighted on the basis of SNW analysis, etc. Figure 3 demonstrates the initial page that appears when starting the MDSS.

Figure 4 shows the sections of the MDSS menu, showing the methods that can be used to make management decisions using this information product.

Figure 5 illustrates the possibility of introducing the necessary criteria and alternatives, while the number of criteria and alternatives can be any, it is set at the previous stage and can be adjusted by the user himself. The capabilities of this information product allow you to describe the necessary criteria and alternatives in sufficient detail to facilitate their visual perception in the process of further.

Figure 6 shows the evaluation form of each alternative for criterion 1.

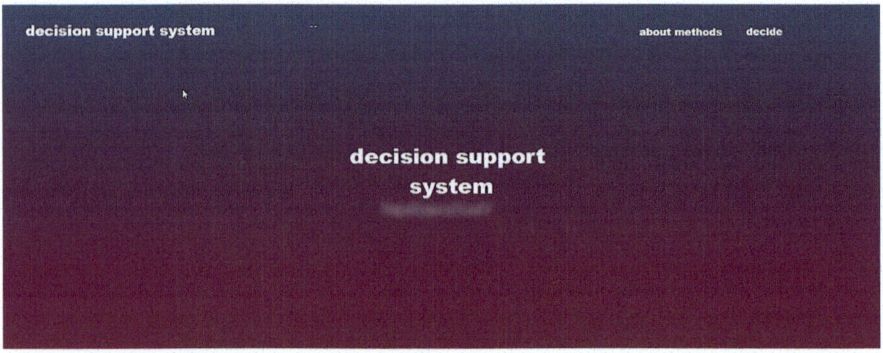

Fig. 3 MDSS home page

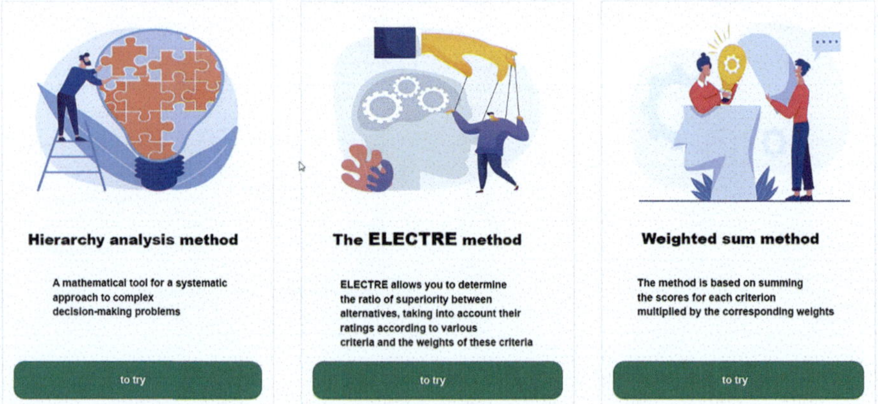

Fig. 4 MDSS menu sections

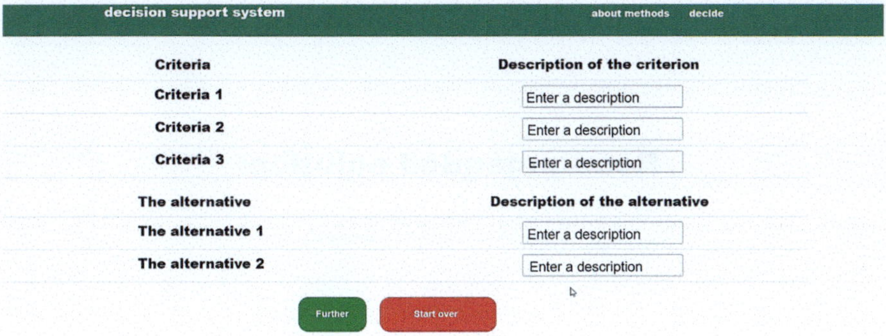

Fig. 5 Criteria and alternatives form

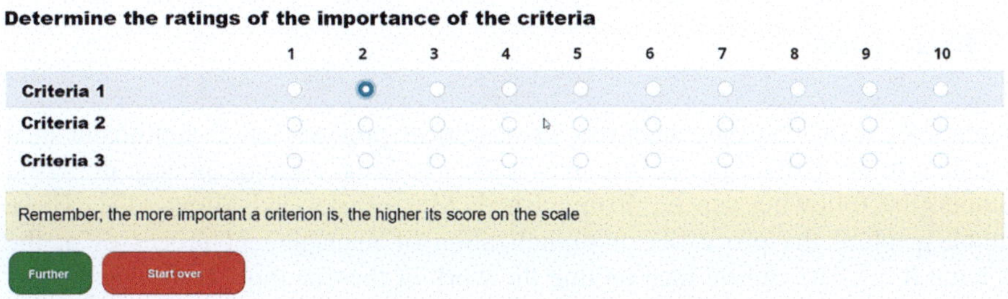

Fig. 6 Form for alternative evaluation by criterion 1

After evaluating all alternatives for all selected criteria, one of the methods presented above selects the best alternative and provides a justification for its selection. Figure 7 shows the recommended solution form. The authors developed a prototype of the MDSS for environmental risk management at the level of municipal authorities, which is an information product with limited functionality. At the same time, this prototype is ready for testing in real conditions and represents the minimum functionality of the future information system. In general, a plan has been developed for the further development of this information product based on a spiral life cycle model. At the moment, several development iterations have already been completed and the expected improvements have been identified based on the specified users requirements.

A preliminary assessment of the effectiveness of the implementation of MDSS in the activities of the Tyumen Department of Informatization using an expert assessment based on the trends highlighted above showed that the implementation of this information system is appropriate. Thus, the peculiarities of digitalization of management decision-making in the environmental sphere are the need to apply support for management decisions based on the retrieval augmented generation technology and the use of strategic reactions to the most likely environmental risks as alternatives, which in turn can be updated on the basis of strategic environmental risk management plans.

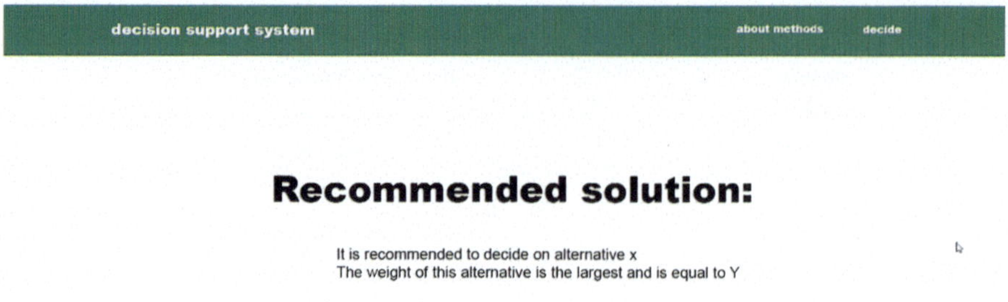

Fig. 7 Recommended solution form

4 Discussion

Digitalization of environmental risk management processes is a significant area of research, both among domestic and foreign authors. Among modern domestic authors, the following can be distinguished: Makovetsky and Sitova [17]. These authors explore the processes of digitalization of management processes at the regional level. The central idea uniting the work of these authors is to identify and describe the role of digital and hybrid technologies in transforming environmental risk management models.

In addition, some authors pay considerable attention to issues of assessing the effectiveness of digital transformation of management processes in general. The following authors can be distinguished in this direction: Seliverstov and Lazarev [11]. These authors analyze existing methods for assessing the effectiveness of digital transformation of management processes characteristic of regional and municipal governments.

Among modern foreign authors who study the issues of digitalization of business processes of various organizations and systems of different levels, as well as their effectiveness, the following can be distinguished: Agyei et al. [20]. These authors are united by attempts to classify digitalization trends characteristic of regional management processes.

As the analysis of sources shows, the search, summarization and generation of responses to requests must be carried out in the system without leaving the text of regulatory documents. The use of retrieval augmented generation (RAG) technology provides solutions to key issues related to the relevance, personalization and availability of information in emergency situations. The main advantages of the proposed architecture are highlighted below.

(1) Increased accuracy of answers: generation is carried out on the basis of indexed official documents and instructions, which minimizes the likelihood of errors and inaccurate information [21];

(2) Relevance of data: when updating sources (for example, official sites of the Ministry of Emergency Situations and regional portals), the system can automatically decode new materials, maintaining the relevance of the knowledge base [22];
(3) Personalization of recommendations: the responses take into account the nature of the request (evacuation, flood situation, need for urgent assistance) [23];
(4) Scalability: RAG architecture allows you to process large volumes of regulatory documents and messages, covering several scenarios at once [24];
(5) Versatility and flexibility: the system can adapt to different types of emergencies by replacing data sources [25].

The implementation of the above advantages is achieved due to the architecture built on the principles of RAG. It is based on the interaction of vector search modules, language models and sources of reliable information [26].

Despite the wide interest in the problem of implementing decision support systems and assessing the effectiveness of their application at the regional management level, the main trends that need to be taken into account when introducing and assessing the effectiveness of the use of digital technologies in environmental risk management processes are not explicitly highlighted, which makes it necessary to initiate such studies.

Thus, the existing technology allows us to support the required functionality of the MDSS, while an important area of further research is the introduction of agent-based artificial intelligence (AAI) to form relevant strategic reactions, taking into account the change in the importance of certain environmental risks.

5 Conclusions

The study confirms that digitalization of management decision-making processes in the environmental sphere is a key factual mechanism for ensuring sustainable development of Russian regions. Analysis of current trends in the digital economy shows significant potential for transforming environmental management: a high degree of Internet penetration, the dominance of mobile traffic and the active use of social networks create the prerequisites for the introduction of innovative solutions. The developed management decision support system (MDSS), implemented in the activities of the Department of Informatization of the Tyumen Region, demonstrates the practical value of implementing digital technologies for environmental risk management. The system integrates data from various sources, provides GIS visualization, scenario analysis and operational anomaly detection, which can significantly improve the quality and speed of decision-making in crisis situations.

Of particular value is the system architecture based on RAG (Retrieval Augmented Generation) technology, which ensures high accuracy of responses by working with official documents, data relevance and personalization of recommendations.

This approach is especially effective in managing risks such as floods, when responsiveness and reliability of information are critical.

An expert assessment of the effectiveness of the implementation of the DMSS shows its high potential for regional and municipal governments. The system allows not only to automate routine processes, but also to form prognostic scenarios taking into account multi-criteria factors: environmental, social, and economic.

In the context of geopolitical instability and ever-increasing environmental risks, the development of domestic digital solutions, such as the presented MDSS, is of strategic importance. Further development of decision support systems should be aimed at increasing their adaptability, integration with other digital control platforms and expanding the functionality of artificial intelligence to predict complex environmental situations.

Acknowledgements This study was supported by the Ministry of Science and Higher Education of the Russian Federation within the framework of the Carbon Measurement Test Area in Tyumen' Region (FEWZ-2024-0016).

References

1. Calabrese R, Dombrowski T, Mandel A, Pace RK, Zanin L (2024) Impacts of extreme weather events on mortgage risks and their evolution under climate change: A case study on Florida. Eur J Oper Res 314(1):377–392
2. Chabot M, Bertrand J-L (2025) Adaptive flood risk management: A decision support system integrating deep learning, digital twins, and economic risk assessment. Glob Environ Chang 95. https://www.sciencedirect.com/science/article/pii/S0959378025001062. Last accessed 08 Dec 2025
3. Ghaffarian S, Taghikhah FR, Maier HR (2023) Explainable artificial intelligence in disaster risk management: Achievements and prospective futures. Int J Disaster Risk Reduct 98. https://www.sciencedirect.com/science/article/pii/S2212420923006039. Last accessed 08 Dec 2025
4. Anisimov AY, Plakhotnikova MA, Grabsky AA (2023) Digitalization of business processes of a metallurgical enterprise as a direction for ensuring sustainable development. Bull Surgut State Univ 11(4):6–15
5. Wanamaker C (2018) The environmental, economic, and social components of sustainability: The three spheres of sustainability: Adapted from the U.S. Army Corps of Engineers. https://soapboxie.com/social-issues/The-Environmental-Economic-and-Social-Components-of-Sustainability (last accessed 8 Dec 2025)
6. Strunk J, Nissen A, Smolnik S (2025) All risks ain't the same—a risk facets perspective on AI-based decision support systems. Decis Support Syst 199. https://www.sciencedirect.com/science/article/pii/S0167923625001587. Last accessed 08 Dec 2025
7. Saaty TL (2008) Decision making with the analytic hierarchy process. Int J Services Sciences 1(1)
8. Shahzad U, Muhammad S (2025) Digital pathways to sustainability: The role of Europe-Central Asia region's digitalization, trade, and geography maximizing sustainable development in China. J Environ Manag 394. https://www.sciencedirect.com/science/article/pii/S0301479725033171. Last accessed 08 Dec 2025
9. Stovba EV, Sharafutdinov AG, Shvetsov MS, Stovba AV, Safiullina NA, Prakhova EV (2025) Digital transformation of business processes of enterprises in modern conditions for the development of an innovative economy. Bull Altai Acad Econ Law 10–1:124–131

10. Ustinova LN, Mukharramova ER (2025) Digital transformation as a tool for optimizing business management systems Issues of an innovative economy. 15(2):505–518
11. Seliverstov DY, Lazarev MP (2025) Digital business transformation: the process of creating a digital transformation strategy, taking into account the methodology of I. Adizes, Progive Econ 1:104–112
12. Pavlov AA (2025) Strategic aspects of the digital transformation of modern organizations: a managerial approach. Bull Eurasian Sci 17(4). https://esj.today/PDF/30FAVN425.pdf. Last accessed 8 Dec 2025
13. Digital business transformation: models and solutions, a collection of scientific papers based on the materials of the All-Russian scientific and practical conference with international participation November 28–29, 2024. Yukonf Publishing House, Tambov, p 170
14. Agyei A, Baah G, Anowuo I, Yeboah E, Peprah WK (2025) Policy strategies for balancing digital financial inclusion with environmental sustainability and economic growth: lessons for sustainable development in Sub-Saharan African economies. Sustain Futur 10. https://www.sciencedirect.com/science/article/pii/S2666188825010640. Last accessed 08 Dec 2025
15. Zhang W-H, Qin J, Lu D-G, Pan Y, Faber MH (2025) Towards effective decision support for structural design and risk management: An information-dependent probabilistic system representation enhanced with support vector machine and unfair sampling. Reliab Eng & Syst Saf 266, Part A. https://www.sciencedirect.com/science/article/pii/S0951832025008002. Last accessed 8 Dec 2025
16. Si F (2022) Development of the concept of sustainable development. Innov Invest 1:4–9
17. Makovetsky MY, Sitova SV (2022) Development of approaches to the interpretation of the concept of sustainable development. Bulletin of Moscow University named after S. Yu. Witte. Ser 1: Econ Manag 2 (41). https://cyberleninka.ru/article/n/razvitie-podhodov-k-interpretatsii-kontseptsii-ustoychivogo-razvitiya (last accessed 8 Dec 2025)
18. Litvinenko AN, Sutyagin MS (2023) Conceptual model of sustainable development of the organization. Criminol J 2:195–202
19. Margolin AM, Vyakina IV (2022) Risks, challenges and mechanisms of ESG transformation of management systems. MIR (Modernization. Innovation. Development) (3). https://cyberleninka.ru/article/n/riski-vyzovy-i-mehanizmy-esg-transformatsii-sistem-upravleniya (last accessed 8 Dec 2025)
20. Agyei A, Baah G, Anowuo I, Yeboah E, Peprah WK (2025) Policy strategies for balancing digital financial inclusion with environmental sustainability and economic growth: lessons for sustainable development in Sub-Saharan African economies. Sustain Futur 10. https://www.sciencedirect.com/science/article/pii/S2666188825010640. Last accessed 2025/12/08
21. Glukhikh IN, Chernysheva TY, Shentsov YA (2024) Decision support in a smart greenhouse using large language model with retrieval augmented generation. In: Proceedings of SPIE—the international society for optical engineering, vol 13217, p 132170S
22. Ozaki T, Jin L, Wu S, Li X (2024) Understanding the impact of confidence in retrieval augmented generation, arXiv: preprint. https://arxiv.org/abs/2412.20309
23. Wei J, Huang Y, Chen M, Zhang R (2025) Talk before you retrieve: agent-led discussions for better RAG in medical QA, arXiv: preprint. https://doi.org/10.48550/arXiv.2504.21252
24. Xie Y, Jiang B, Mallick T, Lee C, Kuo P (2022) A RAG-based multi-agent LLM system for natural hazard resilience and adaptation, arXiv: preprint
25. Lee M, Wang Z, Sharma R, Patel N (2025) ER-RAG: Enhance RAG with ER-based unified modeling of heterogeneous data sources. arXiv: preprint
26. Gupta S, Ranjan R, Singh SN (2024) A comprehensive survey of Retrieval-Augmented Generation (RAG): Evolution, current landscape and future directions, arXiv: preprint

The Genesis of the Economic Security Parameters of Intelligent Systems

Victoria Telyatnikova, Elena Zudina, Oleg Kolyshev, Ivan Chusov, and Flor Zakharchenko

Abstract Intelligent systems are complex software and hardware complexes capable of self-learning, adaptation, and decision-making based on the analysis of large amounts of data. Currently, intelligent systems are used in various sectors of the economy, including finance, manufacturing, logistics, and management. The study reveals the main directions of formation of parameters of economic security of intelligent systems, which allow determining the sustainable state and economic development. It is proposed to use eight parameters, the consideration of which will make it possible to form an information resource to ensure the economic security of intelligent systems. In particular, the criteria for the effectiveness of university complexes in terms of the economic security of intelligent systems are identified; the parameters of the economic security of intelligent systems in the context of the effectiveness of university complexes are analyzed. A matrix of expert assessments of the economic security of intelligent systems in the context of the effectiveness of university complexes in terms of the parameters of innovation and investment role interaction is being developed.

Keywords Genesi · Intelligent systems · Economic security · Parameters · Innovative systems · University complex · Small business

V. Telyatnikova (✉) · E. Zudina · O. Kolyshev
Volgograd State Socio-Pedagogical University, Volgograd, Russia
e-mail: shokhnekh@yandex.ru

I. Chusov
Volgograd Cooperative Institute (Branch), Russian University of Cooperation, Volgograd, Russia

F. Zakharchenko
Volgograd State University, Volgograd, Russia

© The Author(s), under exclusive license to Springer Nature Switzerland AG 2026
A. Kulachinskaya et al. (eds.), *Digital Transformation: What Kind of Smart Life do we have today?*, Lecture Notes in Networks and Systems 1895,
https://doi.org/10.1007/978-3-032-20582-7_18

1 Introduction

An intelligent system is currently considered as a system that has the ability to reason, learn, adapt, and solve problems that require intelligence, similar to a human one. It can include various methods and technologies such as machine learning, neural networks, natural language processing, expert systems, and others. It is obvious that the intelligent system seeks to mimic human cognitive functions in order to automate complex tasks and make informed decisions. The genesis of the economic security parameters of intelligent systems is defined as a continuous process driven by the evolution of technology, threats, and socio-economic factors. Understanding this process is essential for effective risk management, protecting economic interests, and ensuring the sustainable development of the digital economy. It is important to constantly monitor new trends, analyze emerging threats, and adapt security measures to changing conditions.

It is in this direction that the conceptual aspects of achieving a state of economic security are relevant. The economic security of an intellectual system is determined by the level of protection of economic interests, resources, and assets associated with the development, implementation, operation, and protection of an intellectual system from economic threats.

The essence and concept of economic security determine the directions of the formation of a complete cycle for its achievement and control indicators.

Economic security as a stable state ensures the operation of the system, focused on internal indicators, which take into account the external economic demands of the environment.

In research, economic security begins to be considered as a sphere of scientific knowledge, including concepts about the state of the economy, which guarantees sustainable economic growth and quality of life in conditions of meeting the needs of society [1, 2].

Many economists consider economic security in their research as ensuring social needs in conditions of economic independence and stability, allowing for self-development and progress, while at the same time highlighting criteria including: resource potential as an opportunity and the ability to use it effectively; the presence of a minimum level of threats, both external and internal; the presence of concreteness mechanisms in the economy; high level of social stability and quality of life; availability of flexible tools that identify threats and mitigate risks [3].

A significant and qualitative characteristic of the economic system is the level of its integration with the intellectual system (IP). Ensuring innovation and economic security in the Internet space of the cyber economy on the platform of cognitive artificial intelligence assistants is the result of the evolution of threats and countermeasures, as well as the development of intelligent systems themselves. The following stages of genesis can be distinguished:

(1) The threats are mainly related to technical failures, programming errors, and basic cyber attacks;

(2) With the expansion of the use of IP and their integration with other systems, more complex threats related to data leaks, fraud, and attacks on intellectual property have emerged [4];
(3) With the advent of self-learning IP, new risks have emerged related to the inexplicability of decisions (black box), algorithm bias (bias), and the possibility of using IP to cause damage (deepfakes, automated cyber attacks);
(4) With the establishment of interaction between multiple IP systems belonging to different organizations and dependence on critical infrastructure (energy, transport, communications), there are risks of system failures, cascading effects, and large cyber attacks capable of causing significant damage to the economy [5].

In conditions of economic security and security, it is possible to stimulate innovative activities aimed at the digital development of the macro, micro and meso levels. But the presence of innovation in economic systems also allows for independence and other types of security, including environmental and social.

Consequently, at present, economic security at the national, regional, local government and company levels is ensured by intellectual and innovative approaches to sustainable development, as the tasks of technological growth and environmental conservation, including biodiversity, are being solved. In turn, innovative development always requires resource security, both financial and industrial, as well as human and intellectual.

2 Methods

To study the genesis of the economic security parameters of intelligent systems, integrated and systematic approaches were used, including various methods of collecting, analyzing and synthesizing information. Source materials: articles in scientific and professional economic journals devoted to the development of intelligent systems and economic security; materials of international conferences. Research methods: system analysis and modeling; content analysis to identify basic concepts and relationships; synthesis and interpretation of results.

1. The first parameter of the economic security of intelligent systems may be an assessment of the sufficiency of resource availability. It is possible to create, develop and maintain an intelligent system in conditions of investment support, the processes of which require monitoring of all stages of development and implementation. That is why the formation of a system of parameters for the economic security of intelligent systems is key in the research of innovation and investment role interaction between various forms of business.
2. The second parameter of the economic security of intelligent systems is the analysis of compliance with the requirements of sustainable development. Sustainable development, announced in 1987, increased interest in intelligent systems, and also stimulated innovation and investment activity at the global level, where new

approaches are required to ensure balanced resource management [6, 7]. In this concept, sustainability routes are split off in the direction of development: (1) ecological concepts; (2) social concepts; (3) economic concepts [7].

3. The third parameter of the economic security of intelligent systems is the assessment of the risk of readiness for innovative entrepreneurship. Research confirms the insufficient formation of the readiness and ability of human resources to conduct innovative entrepreneurial activities. Solving problems in the process of professional and entrepreneurial competence formation requires a new approach to higher education at the university. The need to form a platform for university complexes suggests increasing the personal competencies of human resources, which are both subjects that determine consumer needs and objects involved in the economic chain of providing society's needs in the form of small businesses [6–8].
4. The fourth parameter of the economic security of intelligent systems is the analysis of the concept of effective interaction of university complexes in the development of technological entrepreneurship. Currently, it is becoming advisable to form the parameters of economic security for the platform of university complexes from the perspective of studying the innovation and investment role interaction of intelligent systems in which small business occupies a key place. Small business as a form of self-employment and an alternative to wage labor is a platform for improving future personal competencies in the field of innovative entrepreneurship. The relevance of the formation of future personal competencies is manifested in the conditions of transition to a new technological order, taking into account current trends in sustainable development. The lack of favorable conditions for the identification and implementation of new professions in education is becoming a fundamental problem [8, 9].
5. The fifth parameter of the economic security of intelligent systems is the assessment of the risks to the development of the younger generation in the direction of innovation and investment activities from the perspective of lean economics, sustainable development and the digital economy. Such an approach to the genesis of the parameters of economic security of intelligent systems will take into account the requirements of the world community for sustainable development, involving the study of the paradigm of personal growth, and, consequently, the formation of a new concept in the field of education.

It is obvious that threats to economic security are manifested in the staffing of future public demands for new professions, which implies answering a number of important questions.: "Who should I teach? What to teach? Who should I teach for? The scope of implementation of competencies?". All issues are being reinforced in the context of the global community's transition to the digital economy by innovative facilities involved in the economic chain of meeting society's needs in all spheres of life.

That is why minimizing threats to intelligent systems in professional and entrepreneurial activities requires a new approach to higher education at the university. It should be noted that today the problems in the formation of competencies in professional and entrepreneurial activities are not solved at the site of

the "flagship university". The study shows that there is only one "flagship university" in the region. Consequently, only one "boiling point" is being created in the region in the process of scientific and practical research, which will not allow fulfilling all the parameters necessary for the formation of new concepts of effective innovation and investment role interaction of small businesses [10–12].
6. The sixth parameter suggests an assessment of the risks of the formation and implementation of the university complexes platform. Thus, in order to minimize threats to economic security, a parameter is needed that indicates the formation of a platform for effective interaction between universities as a target conglomerate in the training of future personnel is determined by an institutional approach. The institutional approach is currently being effectively applied in all sectors of the economy, including science and education [5, 8, 12].
7. The seventh parameter is an assessment of the quality of foresight development of intelligent systems [5, 13]. In such a system, global and national education needs foresight development or, as suggested in the study, "Stepgorizont forecast". Stepgorizont-forecast is aimed at dividing the goal into micro-steps, the achievement of which will make it possible to achieve the goal.
8. The eighth parameter suggests an assessment of risks for the development of small businesses in the direction of flexible development of innovation processes in the context of government support. The turbulent environment of the business environment aims to strengthen the innovation and investment role interaction of small businesses in the context of sustainable development and digitalization. In a disordered reality, it is difficult to ensure innovation and investment role-based interaction of small businesses outside the platform of university complexes, relying solely on a rational basis [14, 15].

3 Effective Part

The expediency of switching to a platform of effective interaction between university complexes is determined by the peculiarities of the educational process synthesizing social and economic directions [5, 6]. Innovation and investment design in small businesses requires competent personnel who will be able to identify the future demands of the digital and sustainably developing global and national community. Education in the higher education system as a mechanism of diverse interaction between universities from the perspective of perception of role interaction.

The criteria for the effectiveness of university complexes in the economic security of intelligent systems are shown in Fig. 1.

In this study, we can identify ten parameters of the innovation and investment role interaction of small businesses in the parametric risk analysis of the effectiveness of university complexes.

X1—request for a risk map of intelligent systems (macro-, meso-, micro-level);
X2—request for parameters of economic security of investment processes (macro-, meso-, micro-level);

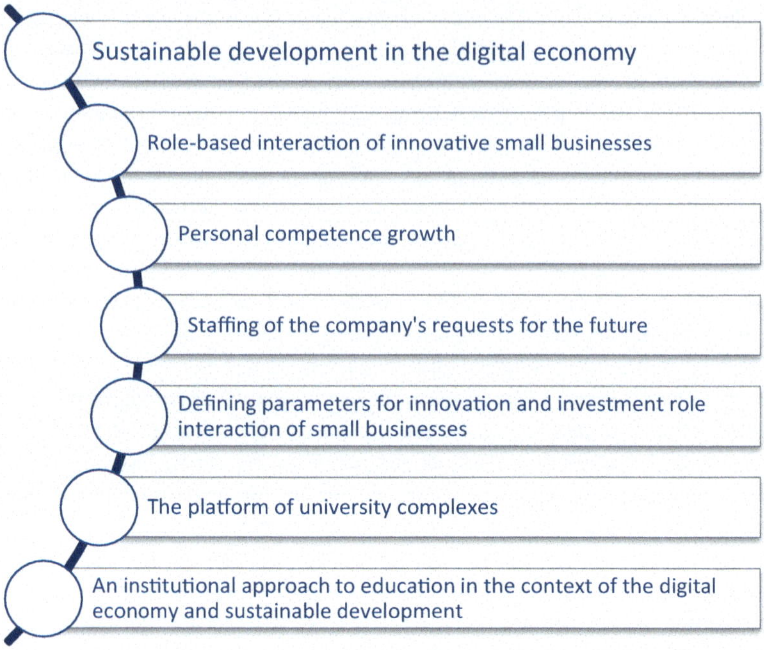

Fig. 1 Criteria for the effectiveness of university complexes in the economic security of intelligent systems

X3—society's demand for sustainable development (macro-, meso-, micro-level);
X4—society's request for the integration of business entities into the digital economy;
X5—request for the formation of human capital as an intellectual resource (macro-, meso-, micro-level);
X6—the company's request for Stepgorizont Forecast in the preparation and implementation of the future, projected for 30–40 years (macro-, meso-, micro-level);
X7—request for project-based innovative thinking and project-based learning (macro-, meso-, micro-level);
X8—request for a distance form of innovative learning (macro, meso, micro level);
X9—request for team projects (macro, meso, micro level);
X10—request for the construction of Stepgorizont Forecast scenarios in the direction of the topics of the educational process of territorial development (macro-, meso-, micro-level);
X11—request for the development of innovative small businesses (macro, meso, and micro levels).

The parameters of the economic security of intelligent systems in the context of the effectiveness of university complexes are shown in Fig. 2.

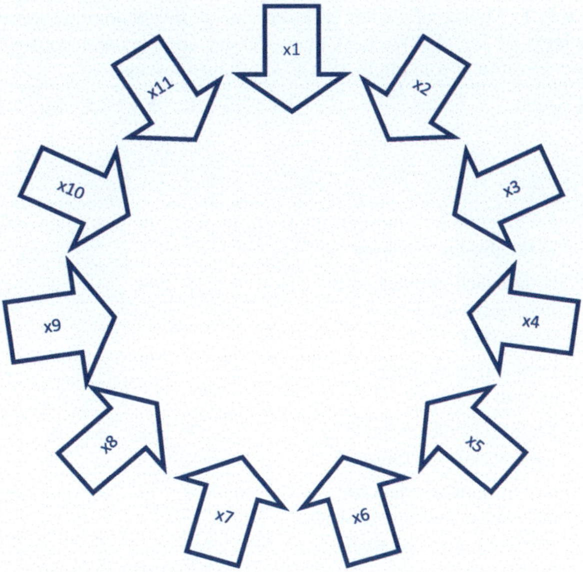

Fig. 2 Parameters of economic security of intelligent systems in conditions of efficiency of university complexes

The development of economic security parameters for intelligent systems involves the inclusion of an analysis of the effectiveness of university complexes, which is functionally necessary for:

(1) Identifying risks in the defining role of small business in the innovation infrastructure of the region;
(2) Highlighting the risks in human resources in Stepgorizont-Forecast;
(3) Substantiation of improvement, formation and adjustment of small Business support and development Programs;
(4) Selection of target teams for innovative projects, research and practical staff, students, postgraduates, doctoral students with the participation of innovative small businesses.

The matrix of expert assessments of the economic security of intelligent systems in the context of the effectiveness of university complexes in terms of the parameters of innovation and investment role interaction of small businesses is presented in Table 1.

The matrix of expert assessments will allow to identify the risks of economic security of the effectiveness of the platform of university complexes according to the parameters of innovation and investment role interaction of small businesses for subsequent adjustment.

Small business plays a key role as a driver of innovative development in the context of sustainability. It is obvious that Stepgorizont-forecast of personnel policy is aimed at identifying new professions and new types of activities for small businesses, which will allow for a smooth transition to sustainable development in the context of striving for the sixth technological order.

Table 1 Matrix of expert assessments of the economic security of intelligent systems in the context of the effectiveness of university complexes according to the parameters of innovation and investment role interaction of small businesses

Criteria	X1	X2	X3	X4	X5	X6	X7	X8	X9	X10	X11	The amount of points
1. Sustainable development in the digital economy	+	+	+	+	+	+	+	+	+	+	+	11
2. Role-based interaction of innovative small businesses	+	+	+	+	+	+	+	+	+	+	+	11
3. Personal competence growth	+	+	+	+	+	+	+	+	+	+	+	11
4. Staffing of the company's requests for the future	+	+	+	+	+	+	+	+	+	+	+	11
5. Defining parameters for innovation and investment role interaction of small businesses	+	+	+	+	+	+	+	+	+	+	+	11
6. The platform of university complexes	+	+	+	+	+	+	+	+	+	+	+	11
7. Institutional approach to education in the context of the digital economy and sustainable development	+	+	+	+	+	+	+	+	+	+	+	11
Total	7	7	7	7	7	7	7	7	7	7	7	77

Source Author's development

Consequently, the economic security of intelligent systems can be significantly influenced by the effective operation of university complexes, which involves the management of the competence development of innovative human resources with the potential of a leader. Also, a criterion analysis of the effectiveness of innovation and investment role interaction of small businesses in the industry/sector can be formed on the intellectual platform of university complexes.

4 Discussion

The development and justification of the security of intelligent systems is currently the most urgent task. Intelligent systems are designed to be used in all spheres of life, including economic sectors. However, many literary sources provide plenty of evidence that any threats to intelligent systems that serve people's needs will have economic consequences. That is why, ensuring economic security is classified by many researchers as: (1) simple internal threats; (2) simple external threats

[4]; (3) complex internal threats from an intelligent system that damages itself; (4) external complex threats of system failures and cascading effects [5, 15]. However, in order to establish the parameters of economic security, it is important to pay special attention to resource security (financial, industrial, personnel, intellectual). Special attention is paid to the long-term foresight forecast when developing an intelligent system from a position of further support and maintenance. Intelligent systems that help simplify routine work in professional and entrepreneurial activities require appropriate training of all human resources, which can be carried out at the level of creating university complexes, without singling out one core university in the region, which often manipulates its position, especially in the direction of grant support [8, 12–14]. The role of small businesses, which can be involved in highly specialized but fragmented tasks, is also obvious. Stimulating and using the partner's narrow specialization of small business competencies in relation to the maintenance of intelligent systems will not allow to concentrate complete information in one system as a future threat of manipulative influences [16, 17]. Many parameters are individually or partially synthesized in the works of scientists and researchers [18].

However, the present study has made it possible to identify the main parameter of the economic security of intelligent systems—this is the work of university complexes, which involves managing the development of competencies of innovative human resources with the potential of a leader, as well as setting criteria for monitoring the effectiveness of innovation and investment role interaction of small businesses in the industry./the sector. The study also pays special attention to the structure of information support for the economic security of intelligent systems, namely: (1) information on the sufficiency and availability of resources in the development and maintenance of intelligent systems; (2) information on the targeted fulfillment of the requirements of digital and sustainable development; (3) information on readiness for innovative entrepreneurship; (4) information on understanding the role of partnership with universities as complexes; (5) information about the long-term value and foresight of the development of intelligent systems in terms of the ability to receive government and business support.

5 Conclusion

In drawing conclusions, it is important to note that there is an obvious need for a new approach to improving the fundamental theoretical and methodological provisions of the economic security of intelligent systems, including an intelligent platform for the foresight development of university complexes for the strategic activation of innovative small businesses. The economic security of an intellectual system is aimed at developing economic interests, preserving resources and assets related to the development, implementation, operation and protection of an intellectual system.

Of course, setting the parameters of the economic security of intelligent systems is a multidimensional institutional approach that takes into account many areas of untenable and destructive activities. However, there are eight main parameters that can be identified, taking into account which will make it possible to form an information resource to ensure the economic security of intelligent systems, namely: (1) assessment of the sufficiency of resource availability; (2) analysis of compliance with the requirements of digital and sustainable development; (3) assessment of the risk of readiness for innovative entrepreneurship; (4) analysis of the concept of effective interaction of university complexes in the development of technological intellectual entrepreneurship; (5) assessment of risks to the development of the younger generation towards the formation of intellectual innovation and investment activities; (6) assessment of risks in the formation and implementation of the intellectual platform of university complexes; (7) assessment of the quality of foresight development of intelligent systems; (8) assessment of risks to small business development as a potential for flexible application of innovative intellectual processes in the context of government support.

References

1. Shokhnekh A, Melnikova Y, Dugina T, Nemchenko A, Mironov A (2020) The mechanism for identifying and responding to real and potential threats in the system of ensuring economic security in innovation and investment policy. In: E3S web of conferences 2020, vol 161. EDP Sciences, Prague, p 02002. https://doi.org/10.1051/e3sconf/202016102002. EDN NWNQCU
2. Vasilyeva L (2020) Economic security: definitions and essence. J Appl Res 3:6–13. https://doi.org/10.475776/2712-7516_2020_3_6
3. Shokhnekh A, Telyatnikova V, Pozdeev V, Nosov A, Mushketova N (2020) Concept strategy of innovative small business in the system of economic security of the digital economy. Growth poles of the global economy: Emergence, changes and future perspectives. Springer, Cham, pp 1397–1406. https://doi.org/10.1007/978-3-030-15160-7_143. EDN QRKKBJ
4. Shokhnekh A, Sidunova G, Glinskaya O, Orlova N, Vasilyev E (2017) Approaches to parametric modeling of the socio-economic institution of the efficient self-organization and control of small business in the system of financial security. Espacios 38(33):23. EDN XNWGRP
5. Shokhnekh A, Mironova O, Moiseeva L, Yakovleva L, Evstafieva A (2019) Provision of innovational and economic security of small business in the internet space of cyber-economy on the platform of cognitive assistants of artificial intelligence. Stud Comput Intelligence 826: 1023–1029. https://doi.org/10.1007/978-3-030-13397-9_106. EDN WZJWJZ
6. Melnikova Y, Shokhnekh A, Telyatnikova V, Naumova E, Glinskaya O (2017) Managing transformations of economic security of educational systems in conditions of leveling the risks of foresight development. Audit Financ Anal 3–4:665–667. EDN: YOWIRW
7. Naumova E, Shokhnekh A, Telyatnikova V, Melnikova Y, Gomayunova T, Glinskaya O (2017) A cognitive approach to strategic management of small business development in the digital economy, taking into account the risks of drift. Audit Financ Anal 3–4:82–385. EDN: YOWICH
8. Telyatnikova V, Shokhnekh A, Kolyshev O, Bespyatykh V, Azarskaya M (2020) Approaches to formation of an innovative platform of university complexes in the strategy of human resources development as small business entrepreneurs.: Growth Poles of the global economy: Emergence, changes and future perspectives. Springer, Cham, pp 1363–1372. https://doi.org/10.1007/978-3-030-15160-7_139. EDN JHUHTO
9. Shokhnekh A, Melnikova Y, Gomayunova T (2022) Concepts and determinants of cyclical nature innovation and investment policy in strategic economic security in the agricultural sector. Book: Smart innovation in agriculture: Part of the smart innovation, systems and technologies book series, vol 264, pp 89–101. Springer Nature Singapore Pte Ltd., Singapore. https://doi.org/10.1007/978-981-16-7633-8_10. EDN KEFUUQ

10. Oudah AlM, Shokhnekh A, Glinskaya O, Shokhnekh M, Chusov I (2019) Regional mechanisms of modernization of infrastructure of regions and the country in the system of innovational development. Book: Specifics of decision making in modern business systems: Regularities and tendencies. Emerald Publishing Limited, pp 209–227. EDN BENXZB
11. Lomakin N, Sazonov S, Polianskaia A, Shokhnekh A, Lukyanov G, Gorbunova A (2020) Hadoop and deductor based digital ai system for predicting cost of innovative products in conditions of digitalization of economy. In: Conference, 2020, Saint-Petersburg, vol 1, p 3373810. https://doi.org/10.1145/3372177.3373810. EDN WIVAYK
12. Lomakin N, Petrukhin A, Shokhnekh A, Maximova O, Samorodova I (2020) Dynamics and structure of investment portfolios of Russian banks studied by the artificial intelligence system. In: Conference 2020. Springer Nature Switzerland AG, vol 1100, pp 365–373. https://doi.org/10.1007/978-3-030-39319-9_42. EDN VDOQBW
13. Lomakin N, Petrukhin A, Shokhnekh A, Gavrilova O, Golodova O (2020) Artificial intelligence system for predicting the price of the SIU8 basic asset of the SI-9.18 option in the “long straddle” strategy. In: Conference 2020. Springer Nature Switzerland AG, vol 1100, pp 739–746. https://doi.org/10.1007/978-3-030-39319-9_82. EDN HIIUJR
14. Lomakin N, Sazonov S, Shokhnekh A, Maramygin M, Tkachenko D, Angel O (2019) Digital Ai “decision tree” for predicting Russian Gdp value based on big data mining to ensure balanced and sustainable economic growth. In: conference 2019, Saint Petersburg, SPBPU, vol 1, p 3373351. https://doi.org/10.1145/3372177.3373351. EDN CIXDKM
15. Krokhin A (2025) Security risks when using external LLMs in a corporate environment. Sci Alm France-Kazakhstan Assoc 4:81–91. EDN JXAWHG
16. Shokhnekh A, Mironova O, Hanafeev F, Kuzmenko O, Shilova L (2019) Indicators of artificial intelligence of financial evaluation of small business investment attractiveness. Stud Comput Intelligence 826:1031–1041. https://doi.org/10.1007/978-3-030-13397-9_107. EDN EFXSYM
17. Garbuk S (2021) The tasks of regulatory and technical regulation of intelligent information security systems. Cybersecur Issues 3(43):68–83. https://doi.org/10.21681/2311-3456-2021-3-68-83. EDN UVMXAZ
18. Jalilov A, Toktorbaev A (2024) Combining sustainability and explainability in the development of secure artificial intelligence systems. Bull Sci Pract 10(12):167–171. https://doi.org/10.33619/2414-2948/109/23. EDN BFKTOO

Smart Education

Intellectual Education as the Basis of Technological Sovereignty: Lessons from Chinese Model in the Era of Digital Transformation

S. E. Demidova

Abstract In the era of global digital transformation, technological sovereignty is becoming a key element of national security and economic sustainability. This study argues that its achievement is contingent upon the development of intellectual education, a system that purposefully forms human capital for the needs of strategic industries. Using the case of China, this paper analyzes an integrated approach that combines leadership in international education quality ratings (PISA), large-scale STEM training, programs to attract global talent, and direct fiscal support for research and development. As a result of the analysis, the pivotal role of integrating strategic planning, the educational ecosystem and business innovations provided by models of public–private partnership and the "triple helix" is highlighted. The article reveals how the integration of educational programs with the needs of the labor market, implemented through the national initiatives "Made in China 2025" and "Belt and Road", creates an ecosystem for technological breakthroughs in the field of AI, 5G and microelectronics. It is concluded that technological sovereignty in the Chinese model is derived from intellectual sovereignty, and strategic investments in education and science represent the most effective way to build a competitive knowledge economy. China's experience offers a valuable framework for countries seeking to build a competitive knowledge economy and secure their technological sovereignty.

Keywords Intellectual education · Technological sovereignty · Human capital · China · STEM education · Digital transformation · Innovation · PISA · Public–private partnership

S. E. Demidova (✉)
Financial University Under the Government of the Russian Federation, Moscow, Russia
e-mail: demidovapsk@gmail.com

© The Author(s), under exclusive license to Springer Nature Switzerland AG 2026

A. Kulachinskaya et al. (eds.), *Digital Transformation: What Kind of Smart Life do we have today?*, Lecture Notes in Networks and Systems 1895,
https://doi.org/10.1007/978-3-032-20582-7_19

1 Introduction

Technological sovereignty is a country's ability to independently develop, implement, and control critical technologies, ensuring economic security and competitive advantages amidst geopolitical challenges. In the modern world, the achievement of this sovereignty is inextricably linked to the development of human capital, which is the basis for generating innovation. Digital transformation has further intensified this relationship, highlighting the need to form an intellectual education system that not only transmits knowledge, but also purposefully trains personnel for the needs of high-tech industries. The purpose of this study is to analyze, using the case of the Chinese model of the formation of a national innovation system, how the development of intellectual education becomes a key factor in ensuring technological sovereignty. China was selected for this case study because it represents one of the most successful cases of achievements in the field of international educational ratings (PISA) and large-scale personnel training for high-tech industries. The country is one of the five world leaders in terms of R&D spending and patent applications. The rise of globally competitive high-tech corporations (Huawei, BYD, Tencent) is not coincidental successes, but direct results of a consistent policy implemented over decades.

2 Materials and Methods

The methodological basis of the research is a comprehensive approach combining qualitative analysis of documents and quantitative analysis of statistical data. The empirical basis was provided by official strategic documents of the PRC ("Made in China 2025", "Belt and Road"), reports of international organizations (OECD, World Bank, WIPO) and data from China's national statistics. Methods of comparative and structural–functional analysis were used to process the information, which made it possible to identify systemic relationships between educational policy, investment in R&D and technological results. The validity of the conclusions was strengthened through data triangulation: the indicators of the effectiveness of the educational system (PISA results) were correlated with macroeconomic indicators (dynamics of patent activity, the share of STEM graduates, the amount of R&D funding). The central analytical structure for data interpretation was the "triple spiral" model, which made it possible to analyze the institutional interaction of the state, universities and business within the framework of the Chinese innovation system.

3 Literature Review

Human capital, understood as a set of knowledge, skills, health, and competencies that form an individual's economic value, is a fundamental driver of the modern knowledge-based economy. The theory of human capital was developed by Gary Becker and Theodor Schultz. The theory's core premise lies in the relationship between investments in education and public health and the growth of labor productivity and economic growth. Gary Becker in his work "Human Capital: A Theoretical and Empirical Analysis" (1964) proved that investments in education and training increase individual incomes and promote economic growth. It was Gary Becker who introduced the concept of "human capital" and considered it as a factor of production [1]. Theodore Schultz in his article "Investment in Human Capital" (American Economic Review, 1961) emphasized that education and health are key components of human capital, and investments in human capital explain a significant part of economic growth in developed countries [2]. These studies constitute classic works on the theory of human capital.

Empirical research has shifted the focus from the formal length of study to the quality of education. In their article "The Role of Cognitive Skills in Economic Development" (Journal of Economic Literature, 2008), Hanushek and Vossmann proved that the quality of education (measured through cognitive skills) is more important for economic growth than the number of years of study. Countries with a higher level of human capital demonstrate higher GDP growth rates [3]. In the study "International Comparisons of Educational Attachment" (Journal of Monetary Economics, 1993), Barro and Lee revealed a direct positive correlation between the level of education of the population and the growth rate of GDP: an increase in the average number of years of study by one year leads to an increase in GDP by 0.3–0.5% per year [4].

In the context of technological sovereignty, the connection of human capital with innovation is of key importance. Paul Romer's model of endogenous growth, reflected in the study "Endogenous Technological Change" (Journal of Political Economy, 1990), integrates human capital as the main factor in the generation of technological change and innovation [5]. In the contemporary context, highly specialized skills are particularly valuable. As Acemoglu showed in his work "Skills, Tasks and Technologies: Implications for Employment and Earnings" (Handbook of Labor Economics, 2011), the development of competencies in the field of STEM (science, technology, engineering, mathematics) is a critical condition for the growth of innovation activity and productivity in high-tech sectors of the economy [6].

In addition to education, the health of the population is an integral component of human capital. Research by Bloom and Canning [7], as well as World Bank reports [8]. These works demonstrate that improving health and increasing life expectancy directly contribute to economic development through increased labor productivity. The development of human capital is also closely linked to the problem of inequality. Goldin and Katz in the book "The Race between Education and Technology" (Harvard University Press, 2008) showed that differences in access to education reinforce socio-economic stratification in society. At the same time, the results obtained by Heckman in the article "Policies to Foster Human Capital" (Research in Economics, 2000) prove that targeted investments, especially in the early stages of a child's life, not only bring the greatest economic return, but also help to reduce this gap [9, 10]. International organizations (UNDP, OECD) recognize human capital as a central element in achieving the Sustainable Development Goals, emphasizing the need for investments in education, health, and skills development of the twenty-first century [11, 12].

Collectively, the literature confirms that the transition to a knowledge economy and ensuring technological sovereignty requires targeted investments in high-quality education, the formation of specialized STEM competencies [13–16] and public health. These investments are not social costs, but strategic investments in the main productive asset of our time.

4 Results

China's success in ensuring technological sovereignty is underpinned by a deliberate policy of human capital development, the starting point of which is the creation of a powerful educational foundation. China's results in the PISA international study serve as an objective confirmation of the effectiveness of this policy. Chinese provinces and megacities (Beijing, Shanghai, Jiangsu, Zhejiang) consistently demonstrate systemic superiority in the quality of school education, especially in mathematics and natural sciences (Tables 1 and 2).

Table 1 China's PISA results in 2015, 2018, 2022

Year	Participating regions	Mathematics	Reading	Science	Key positions in world ranking
2022	Beijing, Shanghai, Jiangsu, Zhejiang	570	550	560	Math leadership (1st place) High positions in reading and science (3rd place)
2018	Beijing, Shanghai, Jiangsu, Zhejiang	591	555	590	Absolute leadership (1st place in all domains)
2015	Beijing, Shanghai, Jiangsu, Guangdong	531	494	518	Math leadership (1st place) Consistently high level in Reading and Science (10th place)

Source Compiled by the author based on data from OECD PISA 2015, 2018, 2022
Note For comparison, the average score for OECD countries in 2022 was: mathematics—493, reading—487, natural sciences—490

Table 2 Positions of the leading countries in the PISA-2022 ranking by training areas

Country/region	Mathematics	Reading	Science	Educational profile
Singapore	600 (1)	600 (1)	590 (1)	Absolute leader—top positions in all areas
China (Macao)	552 (2)	525 (4)	543 (3)	Mathematical emphasis—leadership in exact sciences
China (Hong Kong)	540 (4)	537 (2)	520 (9)	Humanities focus—strong performance in reading
China (Taiwan)	547 (3)	515 (7)	537 (4)	Technical orientation—excellence in science
Japan	536 (5)	516 (6)	547 (2)	Balanced development—high results across all areas
South Korea	527 (8)	515 (7)	528 (6)	Stable quality—traditionally strong positions
Estonia	510 (11)	511 (9)	526 (7)	European leader—best performance in Europe
Canada	497 (15)	507 (10)	515 (11)	North American leader—balanced results
Finland	484 (20)	490 (14)	511 (12)	Innovative approach—strong pedagogical traditions
Russia	453 (39)	432 (42)	467 (37)	Growth potential—requires modernization of the educational system

Source Compiled by the author based on data from OECD PISA 2022
Note the country's position in the world ranking in the relevant field is indicated in parentheses

The high PISA results are a consequence of the well-established "educational ecosystem", which includes the high social status of teachers, the deep integration of the traditional approach to knowledge consolidation (drill and practice) with the development of critical thinking, as well as a culture of high educational expectations in families. This system cultivates not only high test scores but also forms a disciplined, hardworking and mathematically literate pool of future engineers and scientists capable of mastering the most complex technological disciplines.

This rigorous and efficient system creates a solid foundation for future technological breakthroughs. China's performance in mathematics and natural sciences is particularly successful, i.e. those disciplines that form the fundamental basis for innovation. Relying on strong school preparation, China is systematically and significantly increasing the number of graduates in the STEM field, with an ambitious plan to reach 10 million people per year by 2025. This is a direct consequence of the orientation of the entire educational system towards the needs of technological development, as defined in national strategies.

China is pursuing a purposeful policy to create a world-class education system by 2035, which is seen as the basis for technological sovereignty and national revival.

The Education Strategy until 2035 sets goals to create a high—quality education system by 2027, and a world-class system accessible to the entire population by 2035. The renewal of higher education programs is linked to the opening of new specialties (1,673 bachelor's degree programs were added in 2024), corresponding to national strategies, such as "intelligent marine equipment" and "smart materials technologies". The Double First Class program acts as a key initiative to support university science, aimed at Chinese universities entering the top ranks of the world rankings. The national platform "Smart Education of China", officially launched in 2022, has become the central element of the digital transformation of education in China. This project is a large-scale centralized digital ecosystem that provides free access to educational resources for millions of students and teachers across the country. The platform provides educational materials, virtual simulations, advanced training courses for teachers and covers all levels of education—from elementary school to universities and vocational training [17]. In 2025, version 2.0 of the platform was introduced, which made navigation more convenient and placed a special emphasis on personalized learning based on artificial intelligence. For example, in Hunan Province, the Smart Fun Classroom program was implemented as part of the platform, which uses animated personalities and interactive stories to support the individual needs of students. This project has gained over 50 million views in just a short time and serves over 520,000 students [18]. China's strategic priority is the total integration of artificial intelligence into the educational process. The authorities have launched the AI-empowered Education Initiative, aimed at transforming every element of learning through AI. Under this initiative, AI-based educational centers were established in 184 schools and 50 best cases for using AI+ higher education technologies were selected. [18]. In addition, China is implementing universal compulsory education programs in the field of AI, aimed at creating a multidisciplinary and structured curriculum that covers all levels of the educational system. Support measures are part of a coordinated government policy aimed at solving key tasks: ensuring technological sovereignty (support for science and education is directly linked to the objectives of technological dominance and stimulating innovative economic growth) and developing human capital (through exchange programs and attracting foreign scientists, China strengthens its scientific potential). Since 2021, China has been outpacing the growth in the number of STEM graduate students compared to the United States [19].

The talent attraction policy, such as the "Thousand Talents Plan" program, functions on a principle of "dual exposure". It not only makes it possible to compensate for the current shortage of highly specialized competencies and accelerate the transfer of advanced knowledge but also creates the effect of an "intellectual catalyst". Visiting leading scientists and engineers raise the level of research at local universities, become mentors for Chinese graduate students and set new standards for scientific activity, thereby improving the quality of the entire national human capital. This domestic "educational pipeline" is supplemented by a proactive and determined global talent acquisition strategy of attracting the best minds from all over the world, which makes it possible to fill any gaps in competencies in the short term.

Despite its impressive achievements, the Chinese model faces systemic challenges. The key problem is the fundamental tension exists between comprehensive state control over educational and research activities and the need for creative and entrepreneurial freedom, which is a source of breakthrough rather than catch-up innovations. This dilemma was vividly reflected in the implementation of the reforms proposed in the World Bank's report "China 2030: Building a Modern, Harmonious, and Creative Society." The document called for greater openness, flexibility and the role of the market, but in practice in recent years there has been a reverse trend—strengthening state control, centralizing management and focusing on technological sovereignty at any cost, which may limit the potential for fundamental discoveries and the creation of truly new markets.

Nevertheless, the Chinese experience clearly demonstrates that quality education alone is not enough. The key role is played by the creation of an innovation ecosystem through public–private partnership (PPP) mechanisms, which provide a direct link between universities, researchers and the real sector of the economy.

The "Made in China 2025" program places a strategic emphasis on integrating educational programs with specific industry needs. It directly stimulates the creation of technology parks, staff retraining centers, and industry laboratories at universities, ensuring that graduates have exactly the skills required by high-tech companies.

The Belt and Road Initiative (BRI) take this model to a global level. By creating joint universities and research centers in partner countries, China not only strengthens its own human capital through international exchange, but also forms a loyal global network of specialists, standardizing technological chains and educational approaches to suit its needs.

To complete the "education-science-production" cycle, the state actively uses a set of budgetary and tax instruments, creating a highly attractive environment for business investment in R&D and human capital (Table 3).

Government support measures in China aimed at stimulating domestic demand, developing high technologies and creating one of the world's strongest education and science systems are based on the principle of an "innovation funnel": at the initial stage, the government assumes the main financial risks through grants and direct financing, "funneling" fundamental research and start-ups. With the expansion of the commercialization of work, tax incentives come into effect, which encourage companies to independently increase investments in R&D and personnel development. This approach establishes a virtuous cycle of growth, where public investments "trigger" the process, while private investments maintain its sustainability. For the implementation of large technological projects that require coordination of resources and competencies, the mechanism of public–private partnership (PPP) is actively being used in China. Key drivers of this process include the Belt and Road Initiative and Made in China 2025 programs, where the government sets strategic directions and creates infrastructure, while business provides flexibility, market expertise and human resources.

Successful industry PPP projects are associated with the creation of research centers in the field of AI and robotics, such as the Artificial Intelligence Center in Shenzhen. Such centers bring together government research institutes, leading

Table 3 State system and fiscal instruments for stimulating technological sovereignty used in China

Strategic direction	Specific tools	Effect and impact on human capital
Direct government funding	• R&D spending: 2.5% of GDP → 3% by 2025 • Grants covering up to 70% of R&D expenses (startups) • Development funds (National Integrated Circuit Fund)	Creates demand for highly qualified specialists, builds research infrastructure, supports young scientists
Tax incentives for business	• 50% tax credit on R&D expenses (SMEs) • Reduction of profit tax from 25 to 15% (high-tech) • VAT exemption on imported research equipment	Stimulates corporate investment in staff retraining, creates high-tech jobs, facilitates technology transfer
Innovation infrastructure	• Technology parks (Zhongguancun, Zhangjiang) • Innovation clusters • AI and Robotics centers	Concentrates talent, enables cross-industry collaboration, commercializes research, creates a knowledge ecosystem
International cooperation	• Belt and road initiative (educational programs) • Digital silk road • Joint research centers	Attracts foreign talent, exports educational standards, creates global competency networks
Talent development	• STEM education (10 million graduates by 2025) • Thousand talents plan • Retraining programs for traditional industries	Builds a talent pipeline, enables adaptation to technological changes, creates a personnel reserve

Source compiled based on the National Bureau of Statistics of China, 2023; OECD, 2021; State Council of China, 2015; official data from development programs

universities and private companies, which allows not only to conduct breakthrough developments, but also to train specialists directly on the advanced technological front. This represents a practical implementation of the "triple helix" model [15, 16], which reflects the transition from isolated sources of innovation to their conjugation with institutional elements in the context of the development of the knowledge economy.

Figure 1 shows the author's visualization of the methodology for the development of China's technological sovereignty through the prism of the "triple helix" model [20, 21]. The Chinese model is developing in an environment of intense international competition, but at the same time it forms its own standards and rules of the game, which constitutes the essence of technological sovereignty. The systemic integration of three key elements—government strategic planning, the educational ecosystem, and business innovation—creates a synergistic effect that puts China at the forefront of the global technology race. Unlike Western models, where one of the elements often dominates (for example, the US market), the Chinese approach is characterized

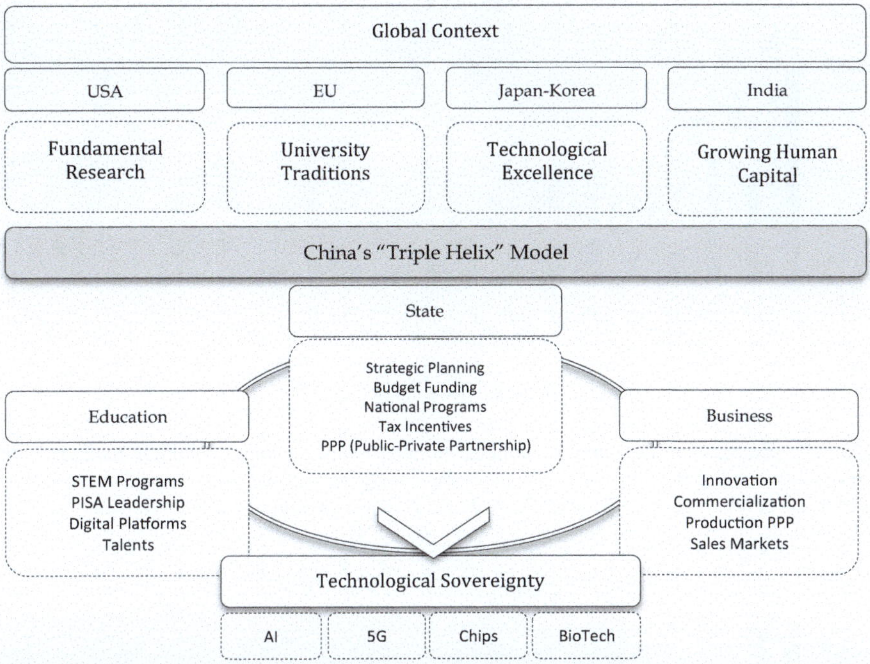

Fig. 1 The "triple helix" model of China's technological sovereignty

by a dynamic equilibrium of all three components. This allows you to simultaneously achieve short-term practical results and long-term strategic goals. The implemented comprehensive strategy for the development of human capital and the innovation ecosystem has brought tangible results to China in the form of technological breakthroughs and a sharp increase in innovation activity, which is clearly confirmed by patent statistics.

Patent activity is an important indicator not only of the current technological level, but also of long-term potential. China's high performance indicates that the country has created an effective "innovation factory"—a system that constantly generates new technical solutions and converts knowledge into patentable assets (Fig. 2). This is a direct result of the massive training of engineers and scientists, as well as powerful fiscal incentives for businesses.

The data from the World Intellectual Property Organization shows impressive dynamics. Thus, the number of resident patent applications per 100 billion USD of GDP increased from 4,030 in 2013 to 4,875 in 2023. The increase in the number of patent applications from residents per million of the population is even more significant—from 517 to 1,080 over the same period (Fig. 3). This significant increase indicates that innovation activity in China has become widespread and has ceased to be the domain of a narrow group of elite research centers.

One of the most revealing cases is the launch of the DeepSeek platform. Its creation became possible due to the confluence of several elements of the national strategy:

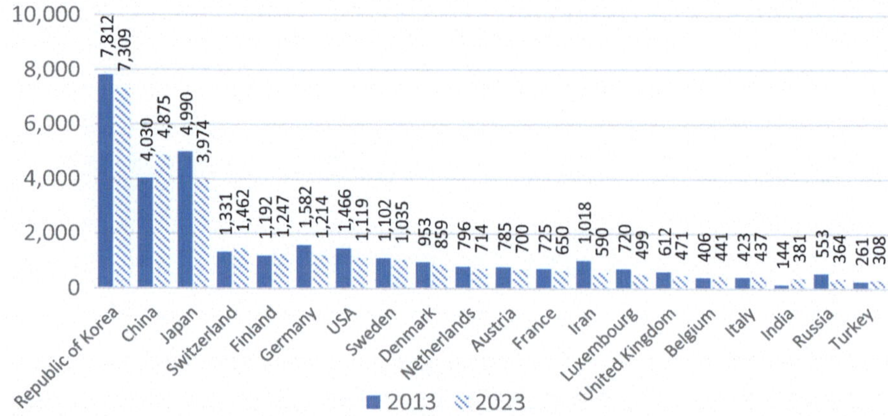

Fig. 2 The number of resident patent applications per 100 billion dollars. US GDP for the top 20 countries of origin, 2013 and 2023. *Source* Compiled by the author based on data from WIPO. World Intellectual Property Indicators. 2024. https://www.wipo.int/publications/en/details.jsp?id=4759

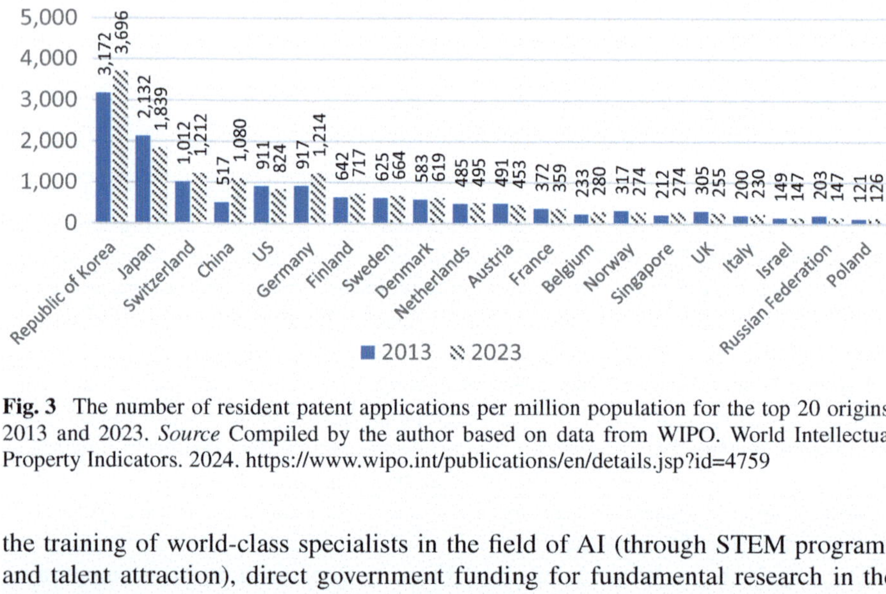

Fig. 3 The number of resident patent applications per million population for the top 20 origins, 2013 and 2023. *Source* Compiled by the author based on data from WIPO. World Intellectual Property Indicators. 2024. https://www.wipo.int/publications/en/details.jsp?id=4759

the training of world-class specialists in the field of AI (through STEM programs and talent attraction), direct government funding for fundamental research in the field of big data and machine learning, as well as the creation of a favorable environment for testing and implementing such solutions in as part of digital transformation programs. These efforts have allowed China to take a leading position in key areas for technological sovereignty: artificial intelligence—the country ranks first in the world in terms of patents in the field of AI; telecommunications—China is the undisputed leader in the development and deployment of 5G networks, and Huawei and ZTE are actively promoting these technologies on global markets. Microelectronics—despite its continued dependence on imports, the country is showing impressive growth—the volume of semiconductor production has increased by 30% over the past five years, which was facilitated by the creation of specialized state development funds. Thus, patent statistics and specific technological achievements serve as verification of the effectiveness of the Chinese model, in which human capital development is not supplementary, but central of a strategy to ensure technological sovereignty.

5 Discussion

The uniqueness of the Chinese model, therefore, lies not in its individual elements, but in their systemic integration and mutual reinforcement (Fig. 1). It is precisely the synergy between the targeted formation of human capital, strategic state planning and the creation of a favorable business environment that allowed China to create a self-sustaining innovation ecosystem capable of achieving technological self-sufficiency.

The discussion of the results reveals several fundamental features that can serve as a guide for other countries.

1. Synergy as a system-forming principle. The uniqueness of the Chinese approach lies in creating a self-reinforcing development contour, where achievements in one area immediately feed others. PISA leadership creates a pool of talent for STEM education, which, in turn, provides personnel for the implementation of national strategies ("Made in China 2025"), which stimulates patent activity and the growth of high-tech business. This model is a practical implementation and development of the theory of the "triple helix" [20, 21], elevated in China to the status of a national strategy with clear KPIs.
2. Resolution of the "planning versus innovation" dilemma. Chinese experience shows that the state can channel innovation activity rather than suppress it, creating powerful financial and infrastructural "channels" for it (technology parks, grants, tax incentives). However, the fundamental contradiction between control and freedom of creativity revealed during the study remains the main challenge to the long-term sustainability of the model and its ability to generate breakthrough rather than incremental, innovations.
3. Human capital as the dominant asset. The results of the research demonstrate a paradigm shift: in the knowledge economy, it is intellectual sovereignty that precedes technological sovereignty. The thesis that human capital is becoming the dominant asset receives further confirmation when analyzing alternative development scenarios. The study by Russian scholars serves as a stark warning: countries that fail to transition from a model dependent on demographic dividends or commodity revenues to a knowledge economy model will face growing fiscal pressure [27]. Their forecast of pension expenditure growth, which, while below inflation, still significantly outpaces the capabilities of a traditional economy, points to a "window of opportunity" for reforms. In this context, the Chinese model appears as a proactive strategy to transform a demographic challenge

into a competitive advantage by creating a high-productivity technology sector. Thus, investments in STEM education and R&D represent not merely social expenditures, but a strategic diversification of the economy and the only way to ensure long-term fiscal balance in an aging society. Massive investments in education and R&D, confirming the conclusions of the classics of the theory of human capital [1, 2], in the Chinese interpretation turn from social expenditures into strategic investments in the main productive asset, directly convertible into technological leadership and geo-economic influence.

The results also allow us to interpret the Chinese model as a strategic response to the phenomenon of "technological democratization" described by T. Friedman because of the convergence of innovations in the 1980s in the field of telecommunications, computerization and miniaturization [22]. If initially these technologies objectively led to the erasure of borders and the formation of a "flat world", then modern states, and China first, purposefully use them to create "sovereign digital spaces", thereby transforming the initial globalization effects of these technologies into an instrument for strengthening national technological sovereignty.

Limitations of research and directions for future research. It should be noted that this study focuses on the macro-level and institutional aspects of the model. Promising areas for future work are micro-level analysis: the study of the motivation of participants in programs such as "Thousand Talents", the effectiveness of specific PPP tools, as well as a comparative analysis of the adaptation of Chinese practices in countries with other political and economic contexts, for example, in Russia.

6 Conclusion

China's experience in building technological sovereignty is a comprehensive and multi-level strategy, where the development of human capital is not just one of the elements, but a system-forming foundation. This strategy, initiated as part of the World Bank's "China 2030: Building a Modern, Harmonious, and Creative Society" program [23], has proven effective in practice, transforming the country into a global leader in innovation. The achievements demonstrated by China in international education quality ratings such as PISA, scientific research [24], and expert assessments [25] became the starting point, creating a pool of talented youth with outstanding knowledge in mathematics and natural sciences. This potential has been purposefully developed through a large-scale government policy focused on training millions of graduates in STEM fields critical to technological breakthrough, as well as attracting the world's leading minds through programs such as the Thousand Talents Plan.

A critical success factor, however, has been the deep integration of the educational system with the real sector of the economy, which is ensured by the active use of public–private partnership tools within the framework of such ambitious initiatives as the Belt and Road Initiative and Made in China 2025. It is the synergy between powerful fiscal incentives for R&D, the creation of innovation clusters, and an adaptive education system that continuously supplies personnel for strategic industries that has allowed China to make a qualitative leap.

The result of this systematic approach is not only an absolute leadership in the number of patent applications in the field of artificial intelligence and advanced telecommunications technologies, but also the creation of a sustainable innovation ecosystem capable of self-development in the face of global competition. Thus, the Chinese model demonstrates that technological sovereignty is predicated on intellectual sovereignty, and investments in education, science, and research commercialization are not costs, but the most effective strategic investments in the future of the national economy, which is a valuable lesson for many countries seeking to strengthen their positions in the global technological arena.

China is actively broadcasting its model of digital education in the international arena [26], using it as a tool of "soft power". As a leader in the World Digital Education Alliance, China set a global framework of standards for digital education and AI in learning in 2024. Certain attractors promote the development of interoperable, ethical and safe educational systems around the world, which allows China to directly influence the formation of future educational paradigms.

Acknowledgements The article was prepared based on the results of research carried out at the expense of budgetary funds on the state assignment of the Financial University.

References

1. Becker G (1964) Human capital: a theoretical and empirical analysis. https://doi.org/10.7208/chicago/9780226041223.001.0001
2. Schultz TW (1961) Investment in human capital. Am Econ Rev 51(1):1–17. https://doi.org/10.1257/aer.51.1.1
3. Hanushek EA, Woessmann L (2008) The role of cognitive skills in economic development. J Econ Lit 46(3):607–668. https://doi.org/10.1257/jel.46.3.607
4. Barro RJ, Lee JW (1993) International comparisons of educational attainment. J Monet Econ 32(3):363–394. https://doi.org/10.1016/0304-3932(93)90023-9
5. Romer PM (1990) Endogenous technological change. J Polit Econ 98(5):S71–S102. https://doi.org/10.1086/261725
6. Acemoglu D, Autor D (2011) Skills, tasks and technologies: implications for employment and earnings. In: Handbook of labor economics, vol 4, part B, pp 1043–1171
7. Bloom DE, Canning D (2000) The health and wealth of nations. Science 287(5456):1207–1209
8. World Bank (1993) World development report 1993: investing in health
9. Goldin C, Katz LF (2008) The race between education and technology. Harvard University Press
10. Heckman JJ (2000) Policies to Foster human capital. Res Econ 54(1):3–56
11. UNDP (2020) Human development report 2020: the next frontier
12. OECD (2018) The future of education and skills: education 2030
13. Hu H (2023) Development of STEM education in China. In: Proceedings of the international conference on education, humanities, and management (ICEHUM 2022), pp 75–88. https://doi.org/10.2991/978-2-38476-040-4_8

14. Yu X, Yang Z (2022) STEM education in China: a five years review report. Asian J Educ Soc Stud 32(1):1–11. https://doi.org/10.9734/ajess/2022/v32i130758
15. Meng N, Yang Y, Zhou X (2022) STEM education in Mainland China. In: Concepts and practices of STEM education in Asia, pp 43–62. https://doi.org/10.1007/978-981-19-2596-2_3
16. Long J (2023) The localization of STEAM education in China: problems and improvement. In: Lecture notes in education psychology and public media, vol 28, no 1, pp 265–270. https://doi.org/10.54254/2753-7048/28/20231360
17. Bell YM (2025) China's digital education drive more promise than peril. https://asiatimes.com/2025/06/chinas-digital-education-drive-more-promise-than-peril/
18. Tech revolution powers new leap in China's digital education. May 13, 2025. http://en.moe.gov.cn/features/2025WorldDigitalEducationConference/News/202505/t20250518_1191049.html
19. Zwetsloot R, Corrigan J, Weinstein E, Peterson D, Gehlhaus D, Fedasiuk R (2021) China is fast outpacing U.S. STEM PhD growth. CSET Data Brief 28 P. https://cset.georgetown.edu/wp-content/uploads/China-is-Fast-Outpacing-U.S.-STEM-PhD-Growth.pdf
20. Leydesdorff L (1996) Emergence of a triple helix of university-industry-government relations. Sci Public Policy 23(5):279–286. https://doi.org/10.1093/spp/23.5.279
21. Etzkowitz H (2008) The triple helix: university–industry–government innovation in action. Routledge, New York and London, 176 pp
22. Friedman TL (1999) The Lexus and the olive tree: understanding globalization. Farrar, Straus and Giroux, N.Y., 394 p
23. World Bank (2013) China 2030: building a modern, harmonious, and creative society
24. Loyalka P, Liu OL, Li G, Kardanova E (2021) Skill levels and gains in university STEM education in China, India, Russia and the United States. Nat Hum Behav 5(7):1–13. https://doi.org/10.1038/s41562-021-01062-3
25. STEM Education Reform Needed To Compete With China (2024) ByAriel Cohen. Forbes. https://www.forbes.com/sites/arielcohen/2024/04/11/stem-education-reform-needed-to-compete-with-china/. Accessed 08 Nov 2025
26. Liu J (2024) Research on empowering high-quality development of international Chinese education through digital transformation. Educ Reform Dev 6(7):39–44. https://doi.org/10.26689/erd.v6i7.7723
27. Sharov VF, Balynin IV, Sedova ML (2024) Forecast of the volume of expenses for the payment of old-age insurance pensions in the Russian Federation until 2035. Financ Theor Pract 28(1):122–132. https://doi.org/10.26794/2587-5671-2024-28-1-122-132

Sociological Analysis of Smart Education in the Context of Digital Transformation

Lyudmila A. Stepnova, Elena A. Litvintseva, and Tatiana A. Savich

Abstract This chapter presents a comprehensive sociological analysis of educational system transformation processes in the era of digitalization through the lens of management sociology. The study examines smart education as a complex social institution undergoing profound structural changes under the influence of digital technologies. Based on an interdisciplinary approach integrating theories of institutional change, organizational sociology, and innovation management, the authors analyze key mechanisms of educational organization adaptation to the requirements of the digital era. Particular attention is paid to the transformation of management models, changes in stakeholder roles, evolution of organizational culture, and formation of new forms of social interaction in the educational environment. The study is based on the analysis of global Smart Education trends, Russian and international practices of digital transformation, as well as theoretical concepts of organizational change management. The findings demonstrate that successful implementation of smart education requires not only technological modernization, but also fundamental rethinking of social relations, management practices, and institutional mechanisms in the educational sphere.

Keywords Smart education · Digital transformation · Management sociology · Institutional change · Organizational culture · Management models · Educational stakeholders · Digital learning environment

---

L. A. Stepnova
Peoples' Friendship University of Russia (RUDN), Moscow, Russia

L. A. Stepnova · E. A. Litvintseva · T. A. Savich (✉)
Russian Presidential Academy of National Economy and Public Administration, Institute of Public Administration and Civil Service, Moscow, Russia
e-mail: i@tsavich.ru

L. A. Stepnova
Moscow Institute of Psychoanalysis, Moscow, Russia

© The Author(s), under exclusive license to Springer Nature Switzerland AG 2026
A. Kulachinskaya et al. (eds.), *Digital Transformation: What Kind of Smart Life do we have today?*, Lecture Notes in Networks and Systems 1895,
https://doi.org/10.1007/978-3-032-20582-7_20

1 Introduction

The global digital transformation of society determines radical changes in all social institutions, and education is no exception. The concept of Smart Education (intelligent education) emerged as a response to the challenges of the information society and represents not merely a technological innovation, but a fundamental restructuring of the educational system as a social institution [1–3].

The term "smart education" is interpreted multidimensionally in scientific discourse. Etymologically, the word "smart" means "intelligent," "clever," which indicates the necessity of achieving a certain level of digital society development for the application of intellectualized technologies. The contemporary understanding of smart education includes four complementary dimensions: the functioning of a multi-environment networked educational environment that unites teachers, specialists, and students from various organizations for collaborative activities based on common standards and intelligent technologies; an adaptive educational process utilizing artificial intelligence technologies; the formation of new smart competencies for self-realization in conditions of continuous technological innovation; the concept of S.M.A.R.T. education as a Self-Directed, Motivated, Adaptive, Resource-enriched, Technology-embedded process. These definitions do not contradict but rather complement each other, reflecting different stages of digital society development [1, 4].

According to the definition formulated within the framework of global consensus, smart education represents a common strategic vision of various countries for addressing key challenges of the artificial intelligence era and achieving high-quality education goals [5].

From the perspective of management sociology, the transformation of education toward Smart Education represents a complex process of institutional change affecting management structures, organizational culture, decision-making mechanisms, and forms of social interaction. As V.N. Minina notes, the digitalization of higher education affects not only individual social groups of teachers and students, but also the most important social institutions as a whole [6].

The relevance of sociological analysis of smart education is determined by several factors. First, the digital transformation of education represents a global trend encompassing various countries and cultural contexts [7, 8]. Second, the implementation of Smart Education technologies generates new forms of social stratification, changes patterns of access to educational resources, and transforms traditional institutional mechanisms. Third, managing digital transformation processes requires the development of new management models that account for the specificity of social interactions in the digital environment.

The analysis of Russian experience acquires particular significance, as the Russian Federation is implementing large-scale programs of educational digital transformation, which creates a unique opportunity for studying the sociological aspects of this process [3, 9–11]. The COVID-19 pandemic significantly accelerated digitalization processes, demonstrating both the capabilities and limitations of existing management models [12, 13].

2 Materials and Methods

2.1 Theoretical Framework

This study employs an interdisciplinary sociological approach integrating three major theoretical frameworks to analyze the transformation of education in the digital era: neo-institutional sociology, organizational change theory, and management sociology.

Neo-Institutional Sociology Approach. From the perspective of neo-institutional sociology, smart education can be considered as an emerging institution characterized by its own norms, rules, practices, and mechanisms of their reproduction. The institutional approach allows analyzing how digital technologies not merely integrate into the existing educational system, but transform its very institutional structure.

The key question is whether smart education represents an evolutionary development of traditional education or constitutes a qualitatively new institution. Analysis shows that a process of institutional isomorphism is occurring—educational organizations, facing the demands of the digital environment, are compelled to adopt similar organizational forms and practices. This includes coercive isomorphism (through state requirements and standards), as well as mimetic (copying successful models) and normative isomorphism (through professional communities) [14].

A crucial aspect is the legitimation of new institutional practices. As researchers note, smart education must gain recognition not only from the state and management structures, but also from the main actors—teachers, students, and employers [13, 15]. The legitimation process involves forming a common understanding of the goals, values, and principles of smart education, which represents a complex social task.

Organizational Change Theory. Organizational change theory provides a conceptual apparatus for analyzing the transformation of educational organizations. Digital transformation represents a systemic organizational change affecting all levels and subsystems of an educational organization [16].

According to K. Lewin's model, the change process includes three stages: "unfreezing" existing practices, implementing changes (changing), and "refreezing" new practices [17]. With regard to smart education, the unfreezing stage is associated with recognizing the inadequacy of traditional learning forms to the requirements of digital society. The changing stage includes the implementation of digital technologies, curriculum reform, and organizational structure restructuring. The refreezing stage involves the institutionalization of new practices and their integration into organizational culture.

However, Lewin's model is criticized for assuming the possibility of achieving a stable state. In conditions of constant technological evolution, the concept of continuous change appears more adequate, whereby the organization exists in a state of permanent transformation [9, 18]. This requires the formation of a special organizational culture oriented toward adaptability and innovativeness.

Management Sociology Framework. Management sociology examines management processes as a form of social relations realized within a specific institutional and cultural context [10]. With regard to smart education, sociological analysis of management must account for the specificity of the educational sphere as a social institution characterized by a high degree of professionalization, academic autonomy, and special value orientations [1, 19–23].

Key categories of management sociology as applied to smart education are:

- social structure of management—a system of positions, roles, and relationships in the process of managing digital transformation;
- management culture—a set of values, norms, and attitudes determining management practices;
- social technologies of management—methods and tools for influencing social processes in the educational environment;
- legitimacy of management decisions—the degree of their recognition and support by various social groups;
- management effectiveness—the relationship between achieved results and social costs.

The concept of reflexive management acquires particular significance, whereby management interventions take into account the reactions of managed subjects and are adjusted accordingly [24]. In the context of smart education, this means the necessity of constant monitoring of how changes are perceived by different stakeholder groups and flexible adaptation of management strategies.

2.2 Research Design and Data Sources

This study employs a multi-method qualitative research design combining:

- Systematic literature review and theoretical analysis of smart education concepts, institutional change theories, and organizational transformation frameworks;
- Secondary analysis of global empirical data from the 2024 UNESCO IITE & Global Smart Education Network study involving experts from 41 countries [UNESCO IITE & Global Smart Education Network: Global Understanding of Smart Education: Research Report. Beijing Normal University, Beijing (2024)];
- Comparative analysis of national models and regional variations in smart education implementation across five global regions (Asia–Pacific, Western Europe and North America, Eastern Europe, Africa, Latin America and the Caribbean);

- Document analysis of Russian federal programs, strategic documents, and policy initiatives on educational digitalization [9, 11, 18];
- Critical examination of international practices and case studies of digital transformation in educational organizations.

Data Collection and Analysis. The primary data source is the comprehensive global study conducted by UNESCO IITE and the Global Smart Education Network in 2024, which synthesized input from educational experts, policymakers, and practitioners across 41 countries representing diverse socio-economic and cultural contexts. This study employed a consensus-building methodology to identify universal characteristics of smart education while acknowledging regional variations.

Secondary data sources include:

- Academic literature on institutional change, organizational sociology, and educational innovation (n = 49 references);
- Policy documents and strategic plans from Russian federal educational authorities;
- International comparative reports on educational digitalization;
- Empirical studies on educational ecosystems, stakeholder dynamics, and organizational culture in digital learning environments.

Analytical Approach. The analysis follows a three-stage process:

Stage 1: Theoretical synthesis—Integration of neo-institutional, organizational change, and management sociology frameworks to develop a comprehensive analytical lens for examining smart education transformation.

*Stage 2: Empirical pattern identificati*on—Systematic coding and thematic analysis of global and national data to identify: universal characteristics of smart education (performative and constructive features); regional variations in implementation priorities and challenges; transformation patterns in management models, stakeholder systems, and organizational structures.

Stage 3: Critical interpretation—Sociological analysis of findings through the lens of institutional legitimacy, power relations, social inequality, and ethical implications of educational digitalization.

Limitations. This study acknowledges several methodological limitations: reliance on secondary analysis of existing global data rather than primary empirical research; potential language and cultural bias in international literature predominantly published in English; rapidly evolving nature of digital technologies requiring continuous updating of findings; challenges in establishing causal relationships in complex sociotechnical systems.

3 Results

3.1 Global Consensus on Smart Education Characteristic

A large-scale study conducted in 2024 by the Global Smart Education Network with the participation of experts from 41 countries revealed the formation of a global consensus regarding the key characteristics of smart education [5]. This represents a unique example of transnational institutional isomorphism—a process whereby institutional practices spread across the boundaries of nation-states.

The study identifies two groups of smart education characteristics (Table 1).

Performative features focus on student-centered learning, encompassing comprehensive assessment, ubiquitous smart environments, continuous improvement culture, and commitment to inclusion and equity. From a sociological perspective, this classification reflects the dialectic of goals and means, structure and agency in the process of institutional transformation. Performative characteristics establish a normative ideal that performs the function of social legitimation of changes. Constructive characteristics describe specific mechanisms for institutionalizing new practices.

Analysis reveals significant regional differences in emphases and priorities for smart education development, which testifies to the influence of sociocultural context on institutional changes [25].

The Asia–Pacific region emphasizes building technological infrastructure and developing partnerships between educational institutions and technology companies. However, the region faces challenges in the area of teacher professional development and formation of competencies for working with technologies. This reflects the specificity of the Asian modernization model, characterized by rapid technological development while preserving traditional pedagogical practices.

Western Europe and North America prioritize gender and cultural equality, and global professional development practices. The region focuses on creating a dynamic, personalized educational environment. Key challenges are associated with technological dependency, digital inequality, resource constraints, and ethical issues. This

Table 1 Key characteristics of smart education

Performative features	Constructive features
Describing the ideal model of future education	*Defining pathways to achieving smart education*
Comprehensive learning assessment	Social learning communities
Ubiquitous and smart learning environment	Ethical adoption of ICT
Culture of continuous improvement	Teacher professional development priority
Commitment to inclusion and equity	Sustainable planning
	Effective multi-sectoral collaboration

corresponds to Western society's value orientations toward individualization and social justice.

Eastern Europe emphasizes the importance of aligning smart education policy and practice through meaningful technology integration and continuous professional learning. Countries in the region face problems of insufficient infrastructure, deeply rooted traditional teaching methods, and high costs of educational technologies.

African region countries face the most serious challenges: limited funding, resource shortages, need for teacher training, digital inequality, high infrastructure costs, data privacy issues, and resistance to systemic change.

Latin America and the Caribbean are reconceptualizing education through aligning curricula with social needs. Challenges include high technology costs, insufficient infrastructure, traditional teaching methods, and resistance to digital literacy.

These differences demonstrate that smart education cannot be implemented through a universal model—contextualization of global principles is necessary, taking into account the specificity of national educational systems, sociocultural features, and level of economic development.

The Russian Federation occupies a special place in the global smart education landscape. On one hand, the country possesses a high level of digitalization and a developed IT industry; on the other, it maintains strong academic traditions and a specific institutional structure of education [18].

Sociological analysis of Russian experience reveals several key features:

- Centralized Management Model. Unlike the decentralized models of Western countries, in Russia the digital transformation of education is implemented predominantly through state programs and initiatives [26]. This ensures the scale and systematic nature of changes, but may limit flexibility and adaptability to local conditions.
- Dualism of Traditions and Innovations. Russian education is characterized by strong academic traditions and orientation toward fundamental knowledge. Simultaneously, active implementation of digital technologies is observed. This dualism creates both opportunities for synthesis and risks of conflict between "old" and "new" [27].
- Sociocultural Barriers. Studies reveal significant sociocultural barriers in the development of smart education in Russia: insufficient digital literacy among some teachers, resistance to change, entrenchment of traditional pedagogical practices, deficit of trust in digital technologies [7, 20, 28].
- Regional Differentiation. Russia is characterized by significant regional differences in the level of educational digitalization. This creates a problem of inequality in access to quality education [18].

3.2 Transformation of Management Models and Organizational Structures

The traditional model of educational organization management was based on the principles of bureaucratic rationality according to M. Weber: hierarchy, formalization, functional specialization, impersonality of rules [29]. However, digital transformation requires a transition to more flexible, adaptive forms of organization and management.

The network management model is becoming dominant in the context of smart education. An educational organization is viewed not as a closed hierarchical system, but as a node in a network of interactions with multiple partners: technology companies, other educational organizations, employers, government structures [23]. This corresponds to M. Castells' concept of "network society," where the key resource becomes the ability to coordinate and integrate into network structures [30].

Project-based organizational form complements the traditional functional structure. The development and implementation of smart education elements is carried out through temporary cross-functional teams, which ensures flexibility and results orientation [31].

Distributed leadership replaces one-person management. In conditions of high technological complexity and environmental dynamism, it is impossible to concentrate all competencies in the hands of one leader. Leadership teams are formed possessing different competencies: technological, pedagogical, managerial [32].

Digitalization radically changes management decision-making processes in educational organizations. The key trend is the transition to data-driven management. Digital platforms generate enormous data arrays about the educational process: learning outcomes, student activity patterns, effectiveness of various pedagogical methods. This creates opportunities for predictive analytics—forecasting educational trajectories, early identification of at-risk students, optimization of curricula [16, 33, 34].

Digital transformation requires radical change in the competencies of educational organization management personnel. A demand is forming for hybrid competencies combining understanding of educational processes, technological literacy, and management skills.

Studies reveal a deficit of such competencies in Russian educational organizations. Management personnel often possess either pedagogical or administrative experience but do not have sufficient competencies in the field of digital technologies [35]. This creates a gap between the strategic goals of digital transformation and the ability to implement them.

New management roles are emerging:

- Chief Digital Officer—a leader responsible for digital transformation strategy;
- Data engineer in the education system—a specialist in educational data analysis;
- Educational process architect—a designer of digital educational trajectories;
- Network interactions coordinator.

Professional development of management personnel becomes a critical task, including both formal training and informal practices of experience exchange, participation in professional communities [13, 27].

3.3 Evolution of Stakeholder Systems in Smart Education

Stakeholder theory views an organization as a system of interactions with multiple stakeholders having different, often conflicting interests [36]. This theoretical framework provides valuable insights for understanding the complexity of educational transformation in the digital age. With regard to smart education, the circle of stakeholders expands significantly beyond traditional actors, creating a more complex and dynamic environment for decision-making and policy implementation.

Traditional stakeholders include students and their families, who are the primary beneficiaries of educational services; teachers and administrative staff, who deliver and manage educational programs; state educational management bodies, which regulate and oversee the educational system; and employers, who have historically shaped educational policy and practice through their demand for skilled workforce. However, the digital transformation introduces new stakeholders that fundamentally alter the educational landscape and power dynamics within it. These new actors include technology companies developing educational platforms, learning management systems, and digital content, who bring both innovation and commercial interests to education; educational data operators carrying out data collection and analysis, who possess significant influence through their control of information flows; digital educational platforms such as MOOC providers and educational content aggregators, which challenge traditional institutional boundaries; and professional communities in the digital environment including media and influencers on educational topics, who shape public discourse and user expectations regarding educational technologies.

Each stakeholder has specific interests and expectations from smart education, which often diverge and sometimes conflict [37]. Students expect personalized learning experiences tailored to their individual needs, preferences, and learning styles, moving away from the one-size-fits-all model of traditional education. They seek flexibility of educational trajectories that allows them to navigate diverse pathways and customize their learning journeys according to personal and professional goals. Integration of formal and informal learning is increasingly important to students who recognize that knowledge acquisition occurs across multiple contexts and platforms. Furthermore, they expect development of in-demand competencies that prepare them for rapidly changing labor markets and enable career success in the digital economy. Teachers, meanwhile, are interested in preserving their professional status and authority in the educational process, which may be threatened by automated systems and platform-based instruction. They seek to obtain tools that genuinely facilitate pedagogical work rather than creating additional burdens or undermining their professional judgment. A critical concern for teachers is avoiding excessive administrative workload that detracts from their core teaching mission and reduces time available for meaningful interaction with students.

The state is oriented toward multiple objectives that reflect broader social and economic priorities. These include increasing accessibility and quality of education across diverse populations and geographic regions, ensuring that digital transformation promotes rather than hinders educational equity. Governments are focused on training personnel equipped for the digital economy with the technical, cognitive, and social skills required for twenty-first century employment. Additionally, states seek to ensure international competitiveness of the educational system in the global knowledge economy, viewing educational excellence as a strategic national asset. Technology companies, by contrast, are interested in expanding the educational technology market and establishing their products as industry standards, which creates potential for vendor lock-in and dependency. They seek to gain access to valuable educational data that can inform product development, enable predictive analytics, and potentially be monetized through various business models. Companies aim to form long-term partnerships with educational organizations that provide stable revenue streams, opportunities for product development and refinement, and favorable conditions for market expansion. These divergent interests create inherent tensions that must be managed through appropriate governance mechanisms and policy frameworks.

3.4 Strategic Models of Digital Transformation Management

Analysis of international and Russian experience allows identifying several strategic models of digital transformation management in educational organizations, each with distinct characteristics, advantages, and limitations (Table 2). The choice of model depends on multiple contextual factors that must be carefully considered by educational leaders and policymakers.

Model selection must consider multiple contextual factors that shape the feasibility and effectiveness of each approach. The scale and complexity of the organization influence the capacity to implement comprehensive change versus incremental adjustments. Large, multi-campus universities with diverse academic units typically require different strategies than smaller, more homogeneous institutions. Organizational culture determines the readiness of institutional actors to embrace different types of transformation, with some cultures more amenable to radical change and others requiring gradual evolution. Institutions with strong traditions of innovation and experimentation may successfully adopt "Big Bang" approaches, whereas those characterized by conservatism and risk aversion benefit from incremental strategies that build confidence through early successes.

Available resources—financial, technological, and human—constrain the range of viable options and affect implementation timelines. The "Big Bang Approach," while rapid, demands substantial upfront investment in infrastructure, training, and change management support, which may exceed the capacity of resource-constrained

Table 2 Strategic models of digital transformation management

Model	Description	Advantages	Risks/limitations
"Big Bang Approach"	Radical, rapid transformation of all aspects of the educational process	Speed of changes; avoiding prolonged transitional periods	High resistance; possibility of systemic failures; participants' unpreparedness
"Incremental Approach"	Step-by-step implementation of digital technologies, starting with pilot projects and gradually scaling successful practices	Less resistance; opportunity to learn from mistakes; adaptation to context	Slow pace; risk of getting stuck in a "transitional state"
"Parallel Development"	Creating "digital branches" or programs alongside traditional ones	Minimizing risks for core activities; opportunity for experimentation	Risk of dualism; competition between "old" and "new"; inefficient resource use
"Ecosystem Transformation"	Embedding the educational organization into a broader interaction ecosystem	Access to external resources and competencies; flexibility; innovativeness	Coordination complexity; partner dependence; risks to autonomy

institutions. In contrast, the "Incremental Approach" distributes costs over extended periods, allowing organizations to learn from pilot projects and adjust resource allocation based on demonstrated effectiveness. External requirements and constraints, including regulatory frameworks, stakeholder expectations, and competitive pressures, shape the strategic imperatives facing educational institutions. Government mandates for digital transformation may necessitate accelerated implementation schedules, while accreditation standards may require demonstration of systematic quality assurance processes that favor more deliberate approaches.

The "Parallel Development" model offers particular advantages in contexts where risk mitigation is paramount. By maintaining traditional programs alongside digital initiatives, institutions create safe spaces for experimentation without jeopardizing core educational functions. This approach proves especially valuable in situations where stakeholder skepticism is high or where the evidence base for digital transformation remains contested. However, as sociological analysis reveals, parallel structures generate inherent tensions: resource competition between "old" and "new" programs, status hierarchies that may privilege one mode over the other, and challenges in knowledge transfer across organizational boundaries. Without careful management, parallel development risks creating isolated "islands of innovation" that fail to transform the broader institutional culture.

The "Ecosystem Transformation" model represents the most radical departure from traditional organizational forms, reconceptualizing the educational institution as a node within broader networks of collaboration. This model aligns with theories of open innovation and networked governance, emphasizing permeability of organizational boundaries and co-creation of educational value with external partners. Successful ecosystem strategies require sophisticated coordination mechanisms, trust-building across organizational boundaries, and tolerance for ambiguity regarding control and intellectual property. While offering access to external expertise, technologies, and resources that individual institutions cannot independently develop, ecosystem approaches also introduce dependencies on partner organizations whose priorities may shift over time.

Participants' readiness for changes, encompassing both willingness and capability to adapt, represents a critical factor that can enable or impede transformation regardless of the chosen model. Sociological research demonstrates that technical readiness alone is insufficient; cultural readiness—including beliefs about the desirability of change, trust in leadership, and confidence in collective efficacy—proves equally important. The psychological contract between institutions and their faculty members significantly influences receptivity to transformation initiatives. When faculty perceive digital transformation as imposed from above without adequate consultation or support, resistance intensifies. Conversely, participatory change processes that engage faculty as co-designers rather than passive recipients of change generate higher levels of commitment and more sustainable outcomes.

Educational leaders must carefully assess these factors to select and adapt the most appropriate strategic approach for their specific context. Hybrid strategies that combine elements of multiple models often prove most effective, as they allow customization to address the unique configuration of opportunities and constraints facing each institution. For example, an institution might adopt an incremental approach for curriculum transformation while simultaneously pursuing ecosystem partnerships for technology infrastructure. The key insight from sociological analysis is that no universal "best practice" exists; rather, effective digital transformation requires contextually intelligent strategy that aligns with institutional capabilities, cultural values, and environmental demands.

3.5 Key Areas of Management Interventions

Sociological analysis of successful digital transformation practices reveals six key areas of management interventions corresponding to the constructive characteristics of smart education [5, 38]:

1. Formation of National Vision and Plan. The presence of clear strategic planning shared by various stakeholders is a critical success factor. Statistical analysis showed strong correlation ($r > 0.7$) between the presence of sustainable national digital transformation policy and education quality.
2. Infrastructure Capacity Development. This area includes not only technical infrastructure (equipment, networks, platforms), but also organizational infrastructure (procedures, regulations, support mechanisms). Studies show moderate

positive correlation ($r \approx 0.4$) between the development of a reliable digital educational environment and education quality.
3. Investment in Human Potential. Professional development of educators is critically important. However, many countries underestimate this aspect, focusing predominantly on technological infrastructure. A systematic approach to developing digital competencies is necessary, including both formal training and peer support, experience exchange, and creation of professional learning communities.
4. Ensuring Inclusivity and Equity. Statistical analysis revealed significant correlation ($r > 0.4$) between commitment to inclusivity and equity and education quality. This requires targeted policy to overcome digital inequality, adaptation of technologies for students with special needs, and gender balance.
5. Developing a Culture of Continuous Learning. This presupposes creating mechanisms for regular monitoring and assessment, feedback, practice reflection, and openness to change. Studies show that sustainability of school development using digital technologies has not yet received due attention in many countries.
6. Developing Multi-Sectoral Collaboration. Effective collaboration between educational organizations, government, business, and civil society demonstrates strong correlation ($r > 0.7$) with education quality. This requires creating institutional partnership mechanisms, sharing responsibility, and coordinating efforts.

4 Discussion

While data-driven management creates opportunities for evidence-based decision-making, sociological analysis reveals significant risks:

- Reductionism. Reducing complex social and pedagogical processes to quantitative indicators leads to loss of qualitative aspects of education [25]. The multidimensional nature of educational outcomes—including critical thinking, creativity, social skills, and ethical development—cannot be adequately captured by metrics alone.
- Gaming the Metrics. When participants in the educational process focus not on real educational outcomes but on manipulating metrics, the system becomes dysfunctional [39]. This phenomenon, known as "Campbell's Law," occurs when quantitative indicators become targets rather than measures of success.
- Intensification of Control. Digital technologies create opportunities for total monitoring, which can transform the educational environment into a "panopticon" according to M. Foucault [40]. Constant surveillance may undermine trust, autonomy, and the intrinsic motivation essential for deep learning.
- Ethical Dilemmas. Issues of data confidentiality, algorithm use, potential discrimination based on automated decisions raise fundamental questions about privacy, fairness, and human agency [26]. The "black box" nature of many AI systems makes it difficult to ensure accountability and transparency.

A sociologically grounded model of data-driven management must combine quantitative analytics with qualitative assessment methods, ensure transparency of decision-making algorithms, and include mechanisms of ethical control.

The multiplicity of stakeholders generates conflicts of interest that must be resolved through management mechanisms:

1. Conflict Between Academic Autonomy and External Control. Digital technologies create opportunities for detailed monitoring of the educational process, which may be perceived by teachers as an encroachment on academic freedom [41]. The tension between accountability demands and professional autonomy requires careful negotiation.
2. Conflict Between Commercialization and Educational Mission. Involvement of technology companies may lead to commercialization of education, subordination of educational goals to business interests [42]. When profit motives drive platform design and content selection, educational values may be compromised.
3. Conflict Between Standardization and Personalization. State standards require unification, whereas smart education presupposes personalization of trajectories [43]. Balancing the need for comparable credentials with individualized learning paths presents a fundamental challenge.
4. Conflict Between "Digital Natives" and "Digital Immigrants". Students who grew up in a digital environment have different expectations and practices than older generation teachers [35]. This generational divide affects communication patterns, pedagogical preferences, and technology adoption rates.

Effective management requires creating institutional mechanisms for coordinating interests:

- Collegial bodies with representation of different stakeholder groups;
- Transparent decision-making procedures;
- Feedback systems and opinion accounting;
- Negotiation platforms for discussing contentious issues;
- Mediation procedures for conflict resolution.

The concept of educational ecosystem is becoming key to understanding contemporary forms of smart education organization. An ecosystem represents a network of interconnected actors jointly creating educational value. Unlike a traditional educational organization with clear boundaries, an ecosystem is characterized by openness, dynamism, and multiplicity of connections.

Key Principles of Educational Ecosystems:

1. Openness. Free exchange of educational resources, data, practices across organizational boundaries enables innovation and continuous improvement.
2. Interdependence. The success of each actor depends on the effectiveness of the ecosystem as a whole, creating incentives for collaboration rather than competition.
3. Coevolution. Joint development of actors in response to environmental changes allows adaptive capacity that no single organization could achieve alone.

4. Self-Organization. Spontaneous formation of connections and practices without centralized management reflects the complexity and emergent properties of networked systems.

From a sociological perspective, educational ecosystems represent a form of network coordination alternative to both market and hierarchical mechanisms. Management in an ecosystem is carried out through orchestration—creating conditions for effective participant interaction, rather than through direct control [25, 44]. This requires new forms of governance emphasizing facilitation, convening, and platform provision rather than command and control.

The organizational culture of educational institutions plays a critical role in digital transformation processes. As E. Schein notes, culture represents a "pattern of basic assumptions" that determines perception of and reaction to changes [45].

Academic culture is characterized by a number of specific features: the value of autonomy and academic freedom, orientation toward tradition and continuity, skepticism toward "fashionable" innovations, collegiality in decision-making, and priority of content over form.

These characteristics can create cultural barriers to digital transformation [46]. Technophobia manifests as fear and distrust of digital technologies, with educators perceiving them as a threat to professional identity. For educators whose expertise is based on traditional pedagogical methods, digital tools may be seen as devaluing their accumulated knowledge. Conservatism shows itself through adherence to traditional methods and resistance to innovations with the mindset "we've always done it this way." The stability and proven effectiveness of established practices create inertia against change. Deficit of digital literacy results from insufficient competencies for effective use of digital tools. When educators lack confidence in their technological abilities, adoption becomes a source of anxiety rather than empowerment. Innovation overload creates fatigue from constant changes and reforms. In contexts where educational organizations have experienced repeated waves of reform, skepticism and exhaustion can undermine even well-designed initiatives.

Sociological studies show that resistance to change is rarely purely irrational. Behind it stand real concerns: risk of losing professional status, increased workload, uncertainty about the future [47]. Effective change management must address these legitimate concerns rather than dismissing resistance as mere obstinacy. An effective change management strategy must include involving teachers in the change planning process, training and support in mastering new technologies, demonstrating benefits through concrete examples of how digitalization improves the educational process, gradualism by avoiding radical revolutionary changes, and creating change agents—groups of enthusiasts capable of inspiring colleagues.

Smart education, while declaring the goal of ensuring equal access to quality education, paradoxically may reproduce and amplify social inequality. Sociologists speak of several levels of digital inequality.

First Level: Access Inequality. Inequality in access to digital technologies and the internet. Despite the widespread proliferation of digital devices, significant differences persist between urban and rural areas, regions, and social groups. Infrastructure limitations create fundamental barriers to participation.

Second Level: Skills Inequality. Inequality in digital skills and competencies. Having access does not guarantee the ability to effectively use technologies for educational purposes. Cultural capital and prior educational experiences shape individuals' capacity to leverage digital tools.

Third Level: Benefits Inequality. Inequality in obtaining benefits from using digital technologies. Students from privileged families are better able to convert access to technologies into educational achievements due to cultural and social capital. The "homework gap" and differential access to parental support amplify initial advantages.

Students with special educational needs present a particular problem. On one hand, digital technologies create new opportunities for inclusion: adaptive interfaces, assistive technologies, personalized trajectories. On the other hand, many digital resources are not adapted for people with disabilities [48]. Universal design principles remain more aspiration than reality in most educational platforms.

Addressing digital inequality requires targeted infrastructure investment in underserved communities, universal design standards for educational technologies, professional development for educators on inclusive digital pedagogy, monitoring and accountability for equity outcomes, and recognition that technology alone cannot overcome structural inequalities. The digital transformation of education generates multiple ethical questions requiring sociological understanding. Digital platforms collect enormous volumes of students' personal data, raising questions about data ownership, usage, access, and protection guarantees. The commodification of educational data creates risks of surveillance capitalism entering the educational sphere. Automated assessment, recommendation, and allocation systems may reproduce and amplify existing prejudices and discrimination. When algorithms are trained on historical data reflecting past inequalities, they risk perpetuating systemic biases under the guise of objective measurement. Educational technologies may use methods borrowed from the gaming industry and social networks to retain attention, raising questions about the manipulative nature of such practices where "gamification" and "engagement" strategies may prioritize addictive design over genuine learning. Monetization models of educational platforms may contradict educational goals, as platform design may prioritize data extraction over pedagogical effectiveness when user data becomes a revenue source. When using artificial intelligence systems in the educational process, the question of distributing responsibility for erroneous or biased decisions arises regarding whether the developer, the educational institution, or the user is accountable when an algorithm makes a consequential mistake.

The formation of an ethical framework for smart education requires the participation of various stakeholders—educators, students, ethics specialists, and civil society representatives. Development of codes of ethical use of educational technologies, creation of ethical audit institutions, and ensuring algorithm transparency are necessary. Effective digital transformation management requires integration of technological and social aspects through the sociotechnical approach, which assumes that technological and social systems mutually shape each other. Technologies are not neutral tools but carry certain values, power relations, and social practices, while social context simultaneously determines how technologies are used, interpreted, and adapted.

The sociotechnical approach presents implementation challenges as it requires cross-disciplinary expertise encompassing technical, pedagogical, and sociological domains, demands ongoing investment in participatory processes, may slow deployment compared to top-down technology mandates, and challenges prevailing assumptions about technological determinism. However, evidence suggests that sociotechnically informed implementations achieve greater sustainability, acceptance, and educational effectiveness than purely technical approaches.

5 Conclusion

Smart education represents a complex process of institutional transformation that extends far beyond technological modernization, affecting the fundamental foundations of the educational institution: goals, values, norms, interaction structures, and reproduction mechanisms.

This study demonstrates that a global consensus is forming regarding the key characteristics of smart education; however, their implementation varies significantly depending on sociocultural, institutional, and economic context, requiring contextualization of global principles rather than mechanical copying of "best practices." Management of educational digital transformation requires a transition from bureaucratic to adaptive, network forms of organization, where the abilities to coordinate interactions with multiple stakeholders, flexibility, and orientation toward continuous change acquire key significance. The formation of educational ecosystems as a form of network coordination plays a special role. Sociocultural factors—organizational culture, values, participants' attitudes—are critically important for the success of digital transformation, as ignoring these factors and focusing exclusively on technological aspects leads to the failure of reforms or their formalization.

Smart education carries risks of reproducing and amplifying social inequality, requiring targeted policy to ensure inclusivity and equity along with careful attention to multiple levels of digital inequality. Digitalization of education generates multiple ethical questions related to privacy, algorithmic fairness, commercialization, and manipulation, necessitating the formation of ethical frameworks and institutions of ethical control. Effective digital transformation management requires a sociotechnical approach integrating technological and social aspects, ensuring participation of various stakeholder groups, and reflection on the social consequences of technologies.

This study advances sociological understanding of educational transformation by applying neo-institutional theory to analyze smart education as an emerging institution, integrating organizational change theory with management sociology to examine digital transformation processes, developing a stakeholder-based framework for analyzing conflicts and coordination mechanisms, and demonstrating the utility of sociotechnical approaches for educational digitalization. For educational policymakers and administrators, this research suggests that digital transformation requires attention to cultural and social dimensions beyond technical infrastructure, that participatory approaches engaging diverse stakeholders produce more sustainable outcomes, that ethical frameworks and equity monitoring should be integrated from the beginning rather than added as afterthoughts, and that network governance models and ecosystem thinking offer alternatives to traditional hierarchical management.

Sociological analysis of smart education remains at an initial stage, with further research required in several directions: comparative analysis of national models of educational digital transformation management and their long-term social effects, longitudinal studies tracking the transformation of teachers' professional identity in the digital era, empirical investigation of new forms of social inequality in the digital educational environment, research on educational ecosystems as a new form of educational organization and governance, and analysis of ethical dilemmas and practices of their resolution in digitalized educational settings. Smart education represents not only a technological but also a profound social challenge requiring rethinking of fundamental principles of educational organization and management. Sociological analysis plays a critical role in understanding these processes and developing effective change management strategies. The success of digital transformation depends not on the sophistication of technologies deployed, but on the wisdom with which they are integrated into the complex social fabric of educational institutions and systems.

References

1. Avdeeva EV, Vasilenko LA, Deryabina GG et al (2024) Smart education in digital society. Dashkov and K Publishing and Trading Corporation, Moscow. ISBN 978-5-394-05927-8
2. Huang R, Liu D, Kanwar AS, Zhan T, Yang J, Zhuang R, Liu M, Li Z, Adarkwah MA (2024) Global understanding of smart education in the context of digital transformation. Open Praxis 16(4):761–789
3. Vasilenko LA, Kechinova O, Skorokhod O (2025) Innovative trends in the development of digital society. In: Digital transformation of socio-economic and technical systems: theory and practice. Springer, Cham, pp 3–13. https://doi.org/10.1007/978-3-031-85608-2_1
4. Chun S (2018) Birth and major strategies of smart education initiative in South Korea and its challenges. In: Uskov V, Howlett R, Jain L (eds) Smart education and e-Learning 2017. SEEL 2017. Smart innovation, systems and technologies, vol 75. Springer, Cham, pp 3–13
5. UNESCO IITE & Global Smart Education Network (2024) Global understanding of smart education: research report. Beijing Normal University, Beijing
6. Minina VN (2020) Digitalization of higher education and its social results. Vestnik of Saint Petersburg University. Sociology 13(1):84–101 (in Russian)
7. Block M, Khvatova T (2017) University transformation: explaining policy-making and trends in higher education in Russia. J Manag Dev 36(6):761–779

8. Bodyako AV, Rogulenko TM, Ponomareva SV (2022) Perspectives and commitments of smart education in Russia. Vestnik Universiteta 6:41–48
9. Dneprovskaya NV, Bayaskalanova TA, Zaika AA (2018) Study of digitization of Russian higher education as basis for smart education. In: 2018 IEEE conference on quality management, transport and information security. IEEE, Saint Petersburg, pp 390–394
10. Dneprovskaya NV, Komleva NV, Urintsov AI (2018) The knowledge management approach to digitalization of smart education. In: Knowledge, service, tourism & hospitality. CRC Press, London, pp 634–639
11. Uvarov AY, Gable E, Dvoretskaya IV (2019) Difficulties and prospects of digital transformation of education. In: Uvarov AY, Frumin ID (eds) HSE Publishing House, Moscow (in Russian)
12. Burnes B (2004) Managing change: a strategic approach to organisational dynamics, 4th edn. Prentice Hall, Harlow
13. Chauhan N, Karthikeyan A, Kumar NR (2024) Transforming education: the intersection of education management, technology, smart universities, and the impact of Covid-19. In: Artificial Intelligence (AI) and finance. Springer, Singapore, pp 1071–1085
14. Oliver C (1991) Strategic responses to institutional processes. Acad Manag Rev 16(1):145–179
15. Suchman MC (1995) Managing legitimacy: strategic and institutional approaches. Acad Manag Rev 20(3):571–610
16. Davenport TH, Harris JG (2007) Competing on analytics: the new science of winning. Harvard Business Press, Boston
17. Lewin K (1947) Frontiers in group dynamics: concept, method and reality in social science. Hum Relat 1(1):5–41
18. Zotov VV, Vasilenko LA (2023) Digital transformation of public administration: unity of service digital and social network aspects. Public Admin Issues (3):26–47. https://doi.org/10.17323/1999-5431-2023-0-3-26-47 (in Russian)
19. Litvintseva EA (2018) Moral as a regulator of official behavior of state civil servants. Central Russ J Soc Sci 13(6):43–56. https://doi.org/10.22394/2071-2367-2018-13-6-43-56 (in Russian)
20. Litvintseva EA, Afanasyeva NV et al (2020) Managing changes in the interaction between authorities and society: textbook for the system of training public administration personnel. Prospect Publishing House, Moscow. https://doi.org/10.31085/9785392314799-2020-232 (in Russian)
21. Litvintseva EA, Afanasyeva NV et al (2023) Social management: modern trends and technologies: textbook for the system of training public administration personnel. Prospect LLC, Moscow (in Russian)
22. Panarin VI, Puchkov OE (2020) Digital education in information society: features of transformations of management mechanisms. Profess Educ Mod World 10(2):3700–3710 (in Russian)
23. Powell WW (1990) Neither market nor hierarchy: network forms of organization. Res Org Behav 12:295–336
24. Lepsky VE (eds) (2006) Reflexive processes and management. Kogito-Center, Moscow (in Russian)
25. Beer D (2016) Metric power. Palgrave Macmillan, London
26. Neborsky EV (2017) Reconstructing the university model: transition to format 4.0. World Sci 4:25–32. (in Russian)
27. Frumin ID, Dobryakova MS, Barannikov KA, Remorenko IM (2018) Universal competencies and new literacy: what to teach today for tomorrow's success. HSE Publishing House, Moscow (in Russian)
28. Rezaev AV, Starikov VS (2017) The transformation of higher education systems in six post-soviet countries: causes and consequences of organizational change. Comp Sociol 16(1):127–153
29. Weber M (1978) Economy and society, vol 1. University of California Press, Berkeley
30. Castells M (2000) Materials for an exploratory theory of the network society. Br J Sociol 51(1):5–24

31. Sydow J, Lindkvist L, DeFillippi R (2004) Project-based organizations, embeddedness and repositories of knowledge. Org Stud 25(9):1475–1489
32. Spillane JP (2006) Distributed leadership. Jossey-Bass, San Francisco
33. Siemens G, Long P (2011) Penetrating the Fog: analytics in learning and education. EDUCAUSE Rev 46(5):30–40
34. Williamson B (2018) The hidden architecture of higher education: building a big data infrastructure for the 'Smarter University.' Int J Educ Technol High Educ 15:12
35. Prensky M (2001) Digital natives, digital immigrants. Horizon 9(5):1–6. Prokazina NV (2024) EdTech in sociological education: challenges and opportunities, risks and solutions. RUDN J Sociol 24(1):123–139 (in Russian)
36. Freeman RE (1984) Strategic management: a stakeholder approach. Pitman, Boston
37. Frooman J (1999) Stakeholder influence strategies. Acad Manag Rev 24(2):191–205. Mitchell RK, Agle BR, Wood DJ (1997) Toward a theory of stakeholder identification and salience. Acad Manag Rev 22(4):853–886
38. Sharonova S, Avdeeva E (2024) Smart education: social risks and challenges. In: IoT, AI, and ICT for educational applications. Springer, Singapore, pp 105–125
39. Campbell DT (1979) Assessing the impact of planned social change. Eval Progr Plann 2(1):67–90
40. Foucault M (1977) Discipline and punish: the birth of the prison. Pantheon Books, New York
41. Neave G (2012) The evaluative state, institutional autonomy and re-engineering higher education in Western Europe. Palgrave Macmillan, London
42. Slaughter S, Leslie LL (1997) Academic capitalism: politics, policies, and the entrepreneurial university. Johns Hopkins University Press, Baltimore
43. Karseth B, Solbrekke TD (2016) Curriculum trends in European higher education. Eur J Educ 51(3):298–307
44. Dhanaraj C, Parkhe A (2006) Orchestrating innovation networks. Acad Manag Rev 31(3):659–669
45. Schein EH (2010) Organizational culture and leadership, 4th edn. Jossey-Bass, San Francisco
46. Bergquist WH, Pawlak K (2008) Engaging the six cultures of the academy. Jossey-Bass, San Francisco
47. Ford JD, Ford LW, D'Amelio A (2008) Resistance to change: the rest of the story. Acad Manag Rev 33(2):362–377
48. Prokazina NV (2024) EdTech in sociological education: challenges and opportunities, risks and solutions. RUDN J Sociol 24(1):123–139 (in Russian)

Digital Well-Being in Smart-Vitagen Education: Psychological Resilience, Social Interaction Patterns, and Economic Costs of Screen-Intensive Learning

Nodira Rustamova, Sardorkhon Sunnatillaevich Gaibnazarov, Mukhtor Nematovich Sunnatov, and Feruza Axtamovna Ibragimova

Abstract Smart-Vitagen systems undoubtedly provide major benefits. They facilitate the accessibility to learning materials, lower geographic and temporal hurdles and allow for individualised learning pathways which are otherwise inaccessible. But with these benefits come unseen, but perhaps expensive, costs. Furthermore, continued exposure to screen-dominated modes of learning can influence psychological resilience and change the dynamics of social interaction as well as create financial loss (direct costs) or be construed as an opportunity cost to learners (and institutions). Notwithstanding concern from the public, there is uncertainty about the size and persistence of these disadvantages. It is this less-studied aspect of Smart-Vitagen education that the study will attempt to investigate systematically. In particular, it looks at the impact of sustained digital engagement on learners' psychological resilience; how social interaction is re-shaped in largely mediated learning environments; and how economic considerations-explicit and tacit-determine participation within smart educational systems. Based on theoretical framings provided by, the study employs an integrative analytical perspective for making sense of these findings. In so doing, it aims not only at identifying possible risks or risk factors but also to provide evidence-based recommendations for the optimal implementation of Smart-Vitagen towards sustainable, human-centered education.

N. Rustamova
ISFT Institute, Tashkent, Uzbekistan

S. S. Gaibnazarov (✉)
Tashkent State Medical University, Tashkent, Uzbekistan
e-mail: nasazusu@gmail.com

M. N. Sunnatov
Tashkent State University of Economics, Tashkent, Uzbekistan
e-mail: muxtorsunnatov7@gmail.com

F. A. Ibragimova
University of science and technologies, Tashkent, Uzbekistan
e-mail: f.ibragimova@usat.uz

© The Author(s), under exclusive license to Springer Nature Switzerland AG 2026
A. Kulachinskaya et al. (eds.), *Digital Transformation: What Kind of Smart Life do we have today?*, Lecture Notes in Networks and Systems 1895,
https://doi.org/10.1007/978-3-032-20582-7_21

Keywords Digital well-being · Psychological resilience · Smart-vitagen education · Social interaction · Educational psychology

1 Introduction

The proliferation of screen display devices is radically transforming education. The rapid emergence of smart-learning technologies creates huge opportunities for deliverable wisdom to learners; while, at the same time, exponentially expanding screen exposure and use-time at all levels. Accordingly, this paper aims to build a comprehensive framework for the digital well-being and education under Smart-Vitagen paradigm. It addresses three questions: 1. What are the resilient mechanisms of learners in screen-intensive learning scenarios? 2. How do learning and educational influencing factors shape and community-oriented connectivity in Smart-Vitagen education? 3. What are the economic costs and expenses associated with those educating under screen-intensive learning environments?

The considerable increase in screen time and usage could lead to various negative psychological, social and economic concerns; it is crucial to protect and promote the digital well-being of learners under the Smart-Vitagen education even including school-based learning, hybrid learning and industrial skills-training episodes [1–4]. Different from traditional education, the Smart-Vitagen education can fully cover the lifetime from toddlers to elderly. During screen-intensive situations, educations still involve parallels of physical integrity.

Conceptually, the digital well-being can thus be defined within schooling, hybrid and human training in Smart-Vitagen era; accordingly, the core rigors are psychological-resilience, social-interaction and economic-costs; when and where those influencing parties occur or accumulate [5].

Digital well-being in smart-Vitagen education encompasses psychological resilience, social interaction patterns, and economic costs linked to screen-intensive learning. Theory and research on digital resilience inform the notion of well-being as a student's adaptive capacity in response to digital challenges. To illustrate this construct, a comprehensive model of psychological resilience for adolescents is conveyed, identifying multiple dimensions, indicators, and influencing factors [6]. Connection and collaboration sustain students' engagement and connectedness across learning environments; online socialization also influences educational achievement. The shift from in-person to hybrid or online interaction modifies oral and written communication dynamics, social network structures, and peer relationships, with corresponding consequences for learning outcomes and skills development. Equity of access to digital devices and stable internet connectivity influences students' ability to interact and collaborate in screen-intensive learning. The economic costs of learning significantly affect students' well-being and are therefore included in the conceptual framework. Direct expenditures encompass devices, internet access, and the digital infrastructure of individual households and educational institutions. Opportunity costs arise in the form of lost productivity within formal education systems, which remains a crucial criterion for adopting distance education. Cost–benefit analyses of educational technology investments further contribute to the economic dimension of digital well-being.

The concept of psychological resilience, widely studied in psychology, specifically in the fields of child protection, education, and social services, also has relevance in digital learning. Defining it as the ability to successfully adapt to difficult or challenging life experiences [7], resilience incorporates various personal traits, behavior patterns, and environmental factors. Disruptive events such as family breakdown, death, illness, or violence can threaten well-being, safety, and healthy development, and transitional events like moving to a different country, changing schools, or entering adulthood at the end of compulsory education, are often accompanied by considerable stress. Digital learning environments are challenging situations for students, requiring them to adapt to numerous transitional or transformational changes associated with distance learning. Pedagogical and social frameworks, learning requirements, regulation of interactions, choice of tools, the digitalization of study material and resources, and completion of assignments and technology-related logistics are just a few of the important characteristics which differ significantly from traditional face-to-face environments. As a result, students' educational psychological resilience is critical in determining success in these new environments. Prior studies competently demonstrate the overall development of the educational psychological resilience of students learning through either a digital hybrid or distance environment. A large proportion of student respondents were found to retain three important characteristics-even before the start of distance learning; maintaining calmness in difficult situations; being optimistic, even when others are disappointed; and generating solutions when problems arose. Such factors strongly favor achieving personal learning objectives in the digital environment and constitute important considerations for reducing the cognitive load on students, thereby fostering the overall goal of learning [8].

Theoretical foundations of resilience in educational settings stem from the wider literature on child and adolescent psychology and from the model of dual process resilience, which proposes that multiple protective factors act on different processes to foster resilience under various adversities. Four main categories of protective and promotive factors apply to educational situations: individual characteristics (intelligence, temperament, personality traits); family characteristics (support, connectedness, bonding); school characteristics (care, climate, academics); and peer relationships (support, popularity, acceptance) [6].

Many learners around the world were placed in a distance learning scenario following the COVID-19 outbreak. The online teaching–learning transition raised immense concerns regarding student adaptation. During 2019–2020, a significant portion of a university's students experienced profound adaptation challenges in terms of emotional state, performance levels, and study habits when using online instruction; levels were substantially lower than pre-COVID experiences, even after several months of adjustment. These observations indicate the crucial role of psychological resilience in screen-intensive learning environments and form the upper pillars of a well-being framework [8]. Resilience encompasses the capacity to rebound from adversity and adjust to challenging circumstances. Resilient individuals adversely adapt less to shocks or become less perturbed, enabling them to function normally and maintain a relatively good emotional state. Psychological resilience is a valuable resource, especially in adversity-laden settings such as online teaching–learning transitions during pandemics. Empirical studies investigating psychological resilience in academic and business contexts emphasize the positive influence of resilience on emotional state and performance; resilience enhances adjustment, and individuals equipped with adequate resilience levels generate less negative effect and can concentrate better.

Educational paradigms have been shifting towards innovative approaches that improve knowledge retention, collaboration, motivation, and cognitive load management. Developing types and levels of interaction, either in physical or digital forms, aims to enhance student engagement and satisfaction. Blended learning, flipped learning, and smart education have been designed and proposed to address limited student interaction. An interactive learning environment focusing on resilience was developed, aiming to augment psychological resilience while reducing symptoms of obsessive–compulsive disorder [8]. Interactive technologies can deliver effective resilience interventions in an accessible, cost-effective, and flexible manner. The environment presents concepts, materials, and exercises on resilience and encompasses three cycles-introduction, exercise, and externalization. Interactive learning environments offer resilience interventions within formal education settings in an accessible and user-friendly way. The development of the educational paradigm helps in the enhancement of key areas of interaction and has observed the emergence of additional impact areas for technological training.

These interactive exercises can be incorporated into various courses, subjects, and lessons where resilience is an educational target or topic; they may also be used independently or integrated into psychological counselling. Class time and other requirements for implementing the exercises can be readily estimated, enabling their effective embedding into courses. Interactive, web-based exercises that align with both educational and psychological objectives are therefore considered valuable assets. The framework acts as a guideline for selecting technologies and pedagogy using screen-mediated instructional delivery, and aligns with specific objectives of smart learning environments.

2 Social Interaction Patterns in Smart-Vitagen Education

Digital learning is based on enriched curricular and connected learning, which are particularly relevant in the context of education. Connected learning seamlessly merges peer, family, and institutional settings, whereas enriched curricular learning broadens the educational experience to promote engagement and ultimately contributes to student well-being. Implemented on an extensive scale since 2020 and adaptive to various contexts, hybrid learning demands awareness of student behaviors and a multi-faceted approach to securing students' sense of agency and engagement [9–11].

In dense urban centers and megacities, extensive digital learning assessment campaigns continue to take place; pre- and post-COVID comparisons and trend analysis focus on environmental quality, well-being, and education [12]. The availability of smartphones and relevant applications has been subject to widespread and thorough investigation. By contrast, studies revolving around hybrid connected education, Smart-Vitagen curricula, the impact of these on student dynamics across different sectors, and the resultant implications, remain relatively rare [8].

Digital education employs diverse media and content. In flexible learning, course delivery is mainly face to face (i.e., classroom), complemented by online (e.g., recorded video, reading material) or activities. In- and out-of-classroom interaction occurs through social media. Educational technologies enhance resilience against remote learning pressures [9]. During blending approaches, instructor feedback, social networks, and group projects improve psychological resilience [10]. Providing learning opportunities consistent with pre-lockdown cycles fosters adaptation [11].

The impact of digital technology on social connections has implications for educational settings in addition to the emotional and cognitive effects. Learning today transcends physical boundaries, irrelevant of geographic proximity or time [8]. Moore's transactional distance theory posits that communication, dialogue, and social interaction represent a vital dimension of distance education models. Likewise, the Community of Inquiry framework illustrates the significance of social presence, in addition to teaching and cognitive presence, within collaborative online learning environments. Digital participants also possess an extensive array of collaborative tools, ranging from text and image-based chat to real-time audio and video [13].

The strength of peer relationships positively predicts engagement across educational contexts, including face-to-face and online environments; lower peer respect predicts disengagement. Digital education networks extend beyond the learning institution, parents, older siblings, and friends serve as models and sources of support in the higher education sector and influence full-time versus part-time continuation of studies. Understanding how peer relationships and collaboration develop in digital settings therefore remains crucial.

Assignment-related collaboration appears to diminish in digital learning environments, while social connectivity via online text chat, micro-blogging, voice, or video remains common. Teacher-led interactions dominate many digital engagement channels, suggesting that instructional approaches unintentionally facilitate top-down forms of communication. Social connectivity within higher educational networks diminishes in hybrid learning settings, whereas overall academic performance remains associated with contact.

Core peer relationships influence overall life activity and well-being but lack satisfactory treatment in digital contexts. Digital technologies can extend networks and serve as enablers for new relationships, yet pre-existing bonds remain significant.

Opportunities for peer communications offered within educational contexts strongly depend on accessibility levels. Thus, equity and access considerations are paramount. Various studies have documented how usage of those channels is shaped by family income, resources and support [13, 14]. For example, many low- and middle-income countries, family supervision is sorely lacking, posing risks of excessive digital independence. On the same note, materials, telecommunication tools and financial support for digital learning exhibits an unequal distribution in many regions [15]. Economic factors have been shown to influence the acquirement of devices that permit family access to digital content and remote schooling; additionally, they condition the availability of budget to cover connectivity costs. Without tailored solutions, inequalities in education, mental health and future opportunities may widen.

Consequently, precaution must be exercised to ensure that these effects do not worsen pre-existing inequalities. Those countries and geographic sub-regions where equitable access is lacking may face major challenges in embarking on Smart-Vitagen education schemes without risking further detriment to the more vulnerable portions of the fresh-out-of-school population. In nations exhibiting wide disparities in homes and household arrangements, sprawling family set-ups tied down by land tenure systems still exert high pressures in rapidly urbanizing settings, where individual mechanisms of resilience fail to perform as desired [16]. Adjustments to arrange suitable modes of smart-Vitagen education must therefore be undertaken, ranging from localized adjustments of individual learning plans to the different distribution of varied learning opportunities in neighborhoods, where offline public services can provide bridging support.

3 Results

Economic costs exacerbated by screen-intensive learning practices include direct expenditures for devices and software, lost opportunity due to reduced productivity, and long-term ramifications for the broader economy. Students obliged to resort to makeshift or non-ergonomic equipment report decreased comfort and increased fatigue, in turn impairing academic performance and resulting in foregone economic output. Consideration of the socio-economic impact of the COVID-19 pandemic on educational engagement is therefore essential for determining the cost–benefit ratio of investments in digital infrastructure and curricular reform to alleviate pedagogical disruption [12, 16].

3.1 Economic Costs of Screen-Intensive Learning

Smart-Vitagen education refers to creative learning which is screen-intensive. Estimated additional indirect costs (including educational costs, productivity loss of family members, and additional educational materials, etc.) relate to residential set-up developments during the outbreak etc. To illustrate the extent of economic costs further, recent increases in the use of mobile devices (tablet computers, smart phones and laptop computers) etc., among school and university students in specific countries provide an indicator of the rise in demand for digital devices [12, 17].

Aggregated measures of well-being underscore the need to evaluate the opportunity costs of digital learning. Opportunity costs appear to arise as learning processes and social engagement are channeled away from non-digital alternatives and/or family and friend networks. Notably, higher average hours of online instruction reduce, and increased hours of time on educational, child-focused and entertainment media further abate, teenagers' frequency of socializing with friends in [16, 18]. The ongoing digitalization of schooling and rise in screen-based media consumption among children and adolescents suggests both have tangible implications for social well-being (Table 1).

A diverse set of inputs continues to underlie the importance and prioritization of university education, the signal of which inevitably conveys considerable opportunity costs alongside the direct monetary value of a tertiary qualification. Modelling of adult opportunity costs positioned the equivalent hourly wage (the rate at which time could be productively transformed into income) at the center of analysis, alongside the schooling period required to obtain a qualification. Consequently, the modelling framework considers learning productivity as a real determinants of opportunity costs-where net human capital accumulation is explicitly accounted for and resonates strongly across economic education [19]. The pre- and post-schooling choices that determine remaining time-sleep, unpaid work, travel and leisure-are taken as exogenous inputs that allow communication of the entire framework in a single, well-structured equation. Below is the opportunity cost modelling for emerging adults, presenting résumé-like detail commensurate with other highlight models [1].

Estimates of adult opportunity costs appear to more than double and, as would be expected, the average required learning productivity factor likewise increases

Table 1 Economic cost dimensions of smart-vitagen education

Cost category	Description	Examples/indicators	Educational implications
Direct educational costs	Expenses directly related to digital learning participation	Tuition fees, digital platforms, paid applications	Increased financial burden on households
Indirect household costs	Costs transferred to families due to digital learning arrangements	Productivity loss of family members, residential adaptations	Redistribution of household resources
Digital device expenditure	Rising demand for personal digital devices	Tablets, smartphones, laptops among students	Widening digital divide across income groups
Supplementary learning materials	Additional digital and hybrid resources	Online subscriptions, child-focused media	Escalation of cumulative learning costs
Infrastructure adjustment costs	Adaptation of home learning environments	Internet upgrades, study space redesign	Long-term household investment pressures

to values that remain substantially below unity. Under the scenario prior to the pandemic, university graduates in the OECD context were anticipated to pursue well over ten additional years openly spent on a mix of unpaid and non-education-enriching activities-well summarized through the post-schooling choice variables of the full communication. For adolescents, presentation of further models relating to policy choice needs internalization of the monitoring of pre- and post-schooling times, yet that relates to extensive other work carried out-future analysis might involve reworking of the framework across broader age-differentiated school premises to further gauge their correspondence with the education-now-digital model (Tables 2, 3, 4).

Investing in digital education for adolescents requires a careful examination of costs and benefits, including economic implications and the impact on students'

Table 2 Opportunity costs associated with screen-intensive digital learning

Opportunity cost dimension	Observed trend	Supporting evidence	Impact on well-being
Reduced face-to-face socialization	Decline in peer interaction frequency	Increased hours of online instruction [16, 18]	Lower social connectedness
Displacement of non-digital activities	Shift away from offline leisure and family time	Growth in screen-based educational and entertainment media	Weakened family and friendship networks
Learning productivity trade-off	Efficiency gains not proportional to time invested	Modelling of net human capital accumulation [19]	Risk of diminished educational returns
Psychological opportunity costs	Increased technology dependence	Prolonged screen exposure studies [8]	Reduced overall well-being
Time reallocation effects	Rebalancing of sleep, unpaid work, travel, leisure	Exogenous time-use assumptions in models [1]	Long-term lifestyle restructuring

Table 3 Modelling framework for opportunity costs in digital education

Model component	Operational definition	Role in the framework
Equivalent hourly wage	Potential income per hour if time were monetized	Baseline valuation of time
Learning productivity factor	Efficiency of converting time into human capital	Core determinant of opportunity costs
Schooling duration	Time required to obtain qualification	Accumulated cost driver
Pre-schooling time allocation	Sleep, unpaid work, leisure before schooling	Exogenous constraint
Post-schooling time allocation	Non-education activities after schooling	Captures long-term opportunity effects

Table 4 Risks and benefits of digital education investment for adolescents

Aspect	Potential benefits	Associated risks
Educational outcomes	Increased engagement, flexible access	Declining learning efficiency
Social development	Expanded online interaction	Reduced close peer relationships
Psychological well-being	Personalized learning support	Technology dependence
Safety and ethics	Digital literacy development	Cyberbullying, online exposure risks
Economic efficiency	Scalable education delivery	Unaccounted well-being costs

psychological well-being. Digital learning environments offer the potential for increased engagement and improved educational outcomes but are also linked to greater technology dependence, reduced close peer relationships, declining educational efficiency, and social media-related risks such as cyberbullying. The detrimental influence of excessive screen exposure has been the subject of widespread debate, particularly with respect to diminished overall well-being [8]. In Finland, the increase in digital education investment owing to the COVID-19 pandemic warrants analysis of the corresponding economic costs and educational benefits. Economic assessments of education technology have sought to capture the impact of planned interventions on student learning, but they seldom account for unintended consequences affecting well-being [18].

3.2 *Methodological Approaches to Studying Digital Well-Being*

The determination of digital well-being in smart-Vitagen education encompasses a range of methodological approaches for investigating, analyzing, and monitoring three interrelated components: psychological resilience, social interaction patterns, and economic costs related to screen-intensive learning. Each component can be further understood through specific sub-elements, as illustrated in the diagram below [20].

1. Measurement of Psychological Resilience and Well-Being. Psychological resilience can be approached as a second-order construct comprising four core dimensions: composure, adaptability, acceptance, and optimism. Several related theories describe how learning and education influence such college-age constructs as perceived stress, life satisfaction, and self-control. Broader conceptualizations of well-being include both positive and negative aspects [12] and have also been validated in an educational context.
2. Assessing Social Interaction in Digital Learning. The complexity of social interaction in the digital era has led researchers to probe user involvement across multiple contexts and platforms. Interaction in online formats can be considered

at three levels: interface (e.g., words and emojis), content (e.g., course materials), and structure (e.g., Web 2.0 platforms) [4, 8, 21, 22]. The distinction between two fundamental types of online interaction- "interpersonal" and "instrumental"-has been made, as well as an investigation of digital participation by educational phase.
3. Economic Evaluation Methods in Educational Technology. Several economic evaluation methods can be utilized to assess innovative educational technologies in terms of supply-side (cost analysis) and demand-side (cost–benefit analysis) methods. Approaches to measuring direct costs such as devices, connectivity, and infrastructure follow widely accepted international guidelines. Measures of opportunity cost and productivity impact can be derived from publicly available data sources.

Large-scale investments in digital learning have heightened concerns about negative impacts on psychological well-being during the COVID-19 pandemic recovery [12]. Measuring psychological resilience in screen-intensive environments is crucial for evaluating well-being and mitigating economic losses, particularly as education shifts online, hybrid, or blended following the smart-Vitagen model [8].

4 Discussion: Policy and Practice Implications for Stakeholders

Screen-intensive learning environments are associated with indirect psychological, educational, and economic costs to both students and educational institutions. At the educational level, cost–benefit calculations typically demonstrate significant benefits from digital intervention (e.g., from e-learning) that far exceed investment (or operating) costs [12]. Yet, externalities remain unaddressed, especially concerning screen-intensive digital learning, thereby illuminating the importance of ascertaining such externalities' forms, extents, and interrelationships. Motivated by these factors, this proposal investigates higher-education digital-well-being remnants as necessarily precursory to the corresponding Smart-Vitagen Implementation Reference Framework. Understanding smart-Vitagen education's externalized digital-well-being influences provides insight toward establishing relevant internal well-being supports and surrounding both smart-Vitagen and broader education sustainability.

Given Smart-Vitagen education's high-investment and acute-risk profiles, stakeholders across levels-students, educators, institutions, policy-makers, and family members-need a realistic picture of incoming and impinging digital-well-being states. Frame-typical interventions thus illuminate stakeholder-tailored estimates on screen-intensive-learning conditions, allowing planners to suffuse the proposed Smart-Vitagen Implementation Reference Framework and associated national-scale cost-e.g., device, booking time, physical venue, student attender, learning-room type, and content-type [5].

Educational institutions must play an active role in promoting well-being among students, youth, and children exposed to screen-intensive education and remote learning environments. Educational institutions, especially from early childhood to adolescence, are vulnerable to digital pressures through technological exposure. Curricula, educational settings, and innovative digital techniques are vital to help students develop psychological resilience and foster education within a supportive framework that promotes well-being through resilience and social interaction [12, 20, 21].

Curricula must pivot from knowledge transfer, rote-learning, or instruction to nurture youth resilience or cope-up with digital pressures. Institution should use Education 4.0 technologies for interactive teaching that can capture student engagement and visual likeness to inform curriculum design [22].

Educational institutions must identify and elaborate invisible influencing-factors counter-productive to student resilience; review depth, breadth, pace, or coherence and adopt self-adaptive curtail or remedial policies to deepen learning experience. Media literacy or smart-device competency should be integrated to equip students with the ability to select and reject media. Planning of niche topics such as time-management, choice-selection, or addiction-prevention is essential to elevate students' competence to resist techno-lustration.

From a broad perspective, families and learner support systems represent a cornerstone of students' psychological well-being-one that is especially crucial in times of crisis and emergency, such as during the COVID-19 pandemic [2]. Family backgrounds exert significant influence on young populations' social self-concept, self-esteem, social identity, and mental well-being. In the digital age, the impact of screen time and digital media on family dynamics has become a subject of increased investigation [23]. Studies are being pursued across digital well-being [3], media literacy, and ethical domains to nurture students toward balanced media habits [4]. The quality of mobile communication and social platforms is linked with subjective measures of relationship quality and well-being. Conversely, excessive smartphone usage is correlated with sleep problems, depression, and anxiety levels. Internet use, notification frequencies, and social media interaction form a pathway through which daily activities and mental health are interconnected. Accordingly, family support and comprehensive learner support systems boost motivation and well-being in relation to both general and screen time–related media use.

Educational institutions are urged to strengthen family engagement and offer ongoing encouragement to ensure students apply previously acquired knowledge in home learning settings. Insufficient task feedback likewise invites additional reliance on supplementary educational resources and commercial services. While families are primarily responsible for this arrangement, the value of educational institutions remains strong; enrichment, re-teaching, and refinement, aimed at the same subject goals, contribute considerably to students' learning experiences and outcomes.

5 Conclusion

Smart dealing with uncertainty is crucial to psychological well-being under screen-intensive learning conditions. Social interaction patterns warrant further investigation not only to gauge learner engagement and social presence in digital contexts but also to establish equity in access and capacity for connectivity. Economic costs raise questions beyond merely estimating monetary requirements for devices or infrastructure. Understanding opportunity costs could yield valuable cost–benefit analyses of deploying digital education technologies. More systematically addressing the psychological and societal dimensions of digital well-being would safeguard and promote the interests of both students and institutions as learning adapts further to integrate technology and cope with changing circumstances.

Several further avenues for study remain open. A straightforward case can be made for measuring the magnitude of shifts in psychological resilience before and after pervasive adoption of technology or during periods of mandated remote learning. Such analyses would increase understanding of the specific roles played by diverse resilience factors across different educational contexts and the relative weights of those factors in affecting overall psychological well-being at the individual or cohort level. Social learning processes requiring joint interaction with and reflection on shared artefacts present interesting opportunities for epistemic-affective engagement and learner socialization. These processes exhibit unusual complexity yet seem deserving of concerted scrutiny, especially in connection with the possibility of third-space formation or the questioning of excessive individualism in the field of e-learning. Estimating educational recovery or catch-up needs following extended periods out of formal schooling can also yield insights into the balance of transaction costs, opportunity costs, and effectiveness ratios across diverse intervention approaches.

References

1. Rustamova NR (2025) Vitagenic chemistry: unveiling life-enhancing energies in chemical reactions. In: AIP conference proceedings, vol 3304(1). AIP Publishing LLC., p 040056. https://doi.org/10.1063/5.0271016
2. Rustamova NR (2025) The role of vitagenic technologies in revolutionizing machine design and functionality. In: AIP conference proceedings, vol 3304(1). AIP Publishing LLC, p 030095. https://doi.org/10.1063/5.0269690
3. Rustamova NR, Rustamov KJ (2025) Vitagen information and hydraulic drive systems in multi-purpose machinery: Enhancing performance and innovation. In: AIP conference proceedings, vol 3304(1). AIP Publishing LLC, p 030097. https://doi.org/10.1063/5.0269692
4. Stankevich D, Gaibnazarova Z, Naumovich O, Tuychieva M, Shadieva G (2025) The role of universities in digital transformation: challenges and prospects. In: Skhvediani A, Kulachinskaya A, Mateeva Petrova M (eds) Digital transformation of socio-economic and technical systems: theory and practice. Lecture notes in networks and systems, vol 1309. (2025). Springer, Cham, pp 179–193. https://doi.org/10.1007/978-3-031-85608-2_13

5. Thomas M, Choudhari SG, Gaidhane A, Syed ZQ (2022) Digital wellbeing: the need of the hour in today's digitalised and technology-driven world! NCBI
6. Sun H, Yuan C, Qian Q, He S, Luo Q (2022) Digital resilience among individuals in school education settings: a concept analysis based on a scoping review. NCBI
7. Nicolaidou I, Stavrou E, Leonidou G (2021) Building primary-school children's resilience through a web-based interactive learning environment: Quasi-experimental pre-post study. NCBI
8. Hietajärvi L, Salmela-Aro K, Tuominen H, Hakkarainen K, Lonka K (2019) Beyond screen time: Multidimensionality of socio-digital participation and relations to academic well-being in three educational phases
9. Kovacs C, Jadin T, Ortner C (2022) Austrian college students' experiences with digital media learning during the first COVID-19 lockdown. NCBI
10. Roque-Hernández RV, López-Mendoza A, Salazar-Hernandez R (2024) Perceived instructor presence, interactive tools, student engagement, and satisfaction in hybrid education post-COVID-19 lockdown in Mexico. NCBI
11. Agrawal S, Krishna SM (2021) Communication apprehension and psychological well-being of students in online learning. NCBI
12. Gui M, Gerosa T, Argentin G, Losi L (2021) Mobile connectivity and adolescent well-being. Evidence from a randomised control trial on a media education training programme in high schools. OSF Preprints
13. Odgers CL, Jensen MR (2020) Adolescent development and growing divides in the digital age. NCBI
14. Magis-Weinberg L, Ballonoff Suleiman A, Dahl RE (2021) Context, development, and digital media: implications for very young adolescents in LMICs. NCBI
15. Castillo de Mesa J, Gómez-Jacinto L, López Peláez A, Erro-Garcés A (2020) Social networking sites and youth transition: The use of Facebook and personal well-being of social work young graduates. NCBI
16. Bohnert M, Gracia P (2020) Does digital engagement influence child mental and socioemotional well-being? Evidence across two Irish cohorts, 2007–2018. OSF Preprints
17. Hollands FM, Pan Y (2018) Evaluating digital math tools in the field
18. Istenič A, Rosanda V, Volk M, Gačnik M (2023) Parental perceptions of child's play in the post-digital era: Parents' dilemma with digital formats informing the kindergarten curriculum. NCBI
19. Luigi Marconi P, Scognamiglio R, Marchiori E, Angeloni D, Mascia ML, Penna MP (2023) Impact of coding educational programs (CEP) on digital media problematic use (DMPU) and on its relationship with psychological dependence and emotional dysregulation. NCBI
20. Gaibnazarova ZT (2023) Prospects for improving the forecasting of the gross domestic product of regions based on mathematical models. In: 9th IEEE International conference on information technology and nanotechnology (ITNT), Samara, Russian Federation, 1–4, April 17–21. https://doi.org/10.1109/ITNT57377.2023.10139074
21. Gaibnazarova ZT, Solieva BT, Iminova NA (2024) Forecasting the risk of bankruptcy in poorly formalized processes. Sci Tech 23(1):80–86. https://doi.org/10.21122/2227-1031-2024-23-1-80-86
22. Gaibnazarova ZT (2024) Artificial intelligence and smart railway transport as a tool to accelerate the transformation process in Uzbekistan. In: Sari M, Kulachinskaya A (eds) Digital transformation: what are the smart cities today? Lecture notes in networks and systems, vol 846. Springer, Cham. https://doi.org/10.1007/978-3-031-49390-4_19
23. Ziankova I, Gaibnazarova Z, Malei A, Sharafanova E, Gasheva O (2025) Digital technologies for comprehensive promotion of a startup. In: Skhvediani A, Kulachinskaya A, Mateeva Petrova M (eds) Digital transformation of socio-economic and technical systems: theory and practice. Lecture notes in networks and systems, vol 1309. Springer, Cham, p 39–55. https://doi.org/10.1007/978-3-031-85608-2_4

A Diagnostic Method for Assessing Students' Ability to understand Cultural Chronotopes Based on transmedia Projections Construction in a digital Milieu

Andrey A. Bogatyrev, Elena G. Milyugina, and Olga P. Bogatyreva

Abstract The research problem is based on the gap between the lack of methodological tools and solutions on the one side and on the other side—the actual need to develop arts and culture history students' ability to attribute the chronotopes. Such an ability necessitates the development of their sensitivity towards the primary (authenticity) and the secondary (stylization, variation, allusion, parody) chronotopes, presented in works of music and arts. The study aims to theoretically substantiate and to develop a diagnostic program, allowing for eliciting and tracking the way the art and culture history students perceive, understand, and frame a musical piece or faction (treated as a media-text) within a specific chronotope. The research methodology is preconditioned by Mikhail Bakhtin's chronotope theory. The innovative technological component is based on the application of digital solutions, multimedia tools and the learners' construction of personal transmedia projections. Empirical research methods include: the pilot experimentations and a questionnaire-based survey of the learners' transcoding and interpretations. Within frame of the interdisciplinary approach, the following semiotic and pedagogical diagnostic techniques were applied: associative experimentation, learners' uncertainty and critical reflection of generic complexity of the musical media-text; the cultural provocation method (concerning semantic shifting), the method of well-grounded differentiation of the primary and the secondary genres and texts, the functional interpretation of the of secondary musical media texts within the individual system of author's artistic solutions. The elicited flaws in arts and culture history learners' level of preparedness require further elaboration of a special program for developing the ability of the learners to identify, attribute, and evaluate the cultural chronotopes. The research results provide evidence for informativity and productivity of the proposed multimodal, complex, interdisciplinary and integrative chronotopological didactic model,

A. A. Bogatyrev
University of Science and Technology MISIS, Moscow, Russia

E. G. Milyugina (✉) · O. P. Bogatyreva
Tver State University, Tver, Russia
e-mail: elena.milyugina@rambler.ru

© The Author(s), under exclusive license to Springer Nature Switzerland AG 2026
A. Kulachinskaya et al. (eds.), *Digital Transformation: What Kind of Smart Life do we have today?*, Lecture Notes in Networks and Systems 1895,
https://doi.org/10.1007/978-3-032-20582-7_22

targeted at arts and culture history learners' vocational skills formation within a smart learning environment.

Keywords Smart learning model · Smart learning environment · Chronotope · Meaning reconstruction · Associative experiment · Multimodal semiotic analysis

1 Introduction

The relevance of the study is predetermined by necessity of rethinking the system of art and lore training for Art and Culture majors at Russian Pedagogy university department. The transformative process drivers are preconditioned by methodological shifts in conception of arts, born in the second half of XX century. No less the context of teaching Arts is predetermined by ubiquitous digitization and digital transformation of Arts and Culture milieu, on the one hand expanding their functional scopes by way of creating new meaningful projections and preconditioning new challenges and educational risks on the other hand. One of important premises of our research is the educational idea that "interacting across modes frees students to think more creatively and integrate knowledge faster" [1, p. 43]. In this connection we need to take a look at the methods of modern Art Criticism, allowing to face the modern flow of expanding space of Arts and arts reception, to distinguish the key trends and tendencies.

The turning point for finding ways of authentic historical interpretation was the discovery (or birth) of chronotope as an explanative principle and an interpretive method. The methodological shift was initiated by ideas of Ukhtomsky [2], elaborated and fruitfully practices by Mikhail Bakhtin [3] within the scope of literary studies mainly, and then by Zinchenko [4] in psychology. Later on, the extension of chronotopological approach [5] had led to comprehension of its much wider, interdisciplinary nature [6], followed by a kind of a chain reaction of applying it in different other fields of artistic and aesthetic expertise, such as visual arts [7–9], architecture [10], music [11–13], theatre stage [14], cinema screen [15], media [16], pedagogics [17] social sciences and humanities in general [18, 19]. The chronotopological approach opened new opportunities for eliciting authentic meaningful connections of a work of (any) art with its contemporary cultural, historical and social background. As a result, application of chronotopological method paved the way to reflection of people's worldview, represented in the work of art, scrutinizing the axiological impact of it and estimating its performative function within capsule of time and space of its first social reception. However, the idea of Ukhtomsky [2], highlighting the major trajectory ('world lines') in historical succession of chronotopes, reflecting a macrolevel of chronotopological studies, still remain underestimated.

Introduction of discourse concept and discourse analytics significantly reshaped XX Art studies toolset and standards of analysis. This aspect of a larger methodological shift in Humanities unsealed the limitations of Formal method of work of art analytics. Eliciting and scrutinizing the underlying social constructs laid the ground for meaningful interpretation of works of art [20–22]. In Russia this trend was echoed by framing work of art as an information carrier and communication tool [23–25]. The subject of artistic information exchange was viewed as a suite of artistic text per se, the context, the author's and the recipients' intentions, social and cultural setting etc., [26, 27]. The derived multiplicity of interpretations added to relevance of the problem of identifying and the exegesis of chronotopes of culture and arts [28]. In arts exhibition industry the sensitivity to plurality has caused a methodological turn in conception of exhibition organization and guiding the visiting tour to art gallery [29]. The immediate perception of art masterpieces by the spectators has been marginalized. Instead more emphasis is laid on either social or professional art criticism discourse.

The so-named Visual Turn was enacted as a generic reaction to previously mentioned discourse studies advent, so rich with multiplicity of interpretations and plurality of aesthetic judgments [30–32]. The champions of the visual approach proclaimed the big comeback to the artifact as is, recognized as a key element of cultural heritage. This was a move to reappreciation of visual representation in world perception and art reception [33, 34]. This turn-back to visuals enriched the methodological toolkit for arts and culture (in tandem) chronotopes analysis. The "visual" methodological turn also forwarded a question on coordination of visual and discursive elements in art criticism and art education.

The ongoing digitalization of arts and culture space is tightly associated with the rise of plethora of arts media communication practices. The communications are being enriched (and possibly partly impoverished) by incoming new media text production, based on different sorts of secondary use of socium-arts-and-culture chronotopes, based on different techniques—such as, for example, parody, pastiche, developing sequels, prequels etc. Another one challenging feature is still rising level of information Noise of diverse nature (kitsch, link-baiting fakes and humbug critical reviews, Ai-generated good-for-nothing 'authentic evidences' and 'authentic reconstructions', supporting false stereotypes). New media along with (continuously transforming) classical media compete in making both information and noise viral [35]. Under the pressure of the gush of secondary media-texts and pseudo-artistic artefacts production there arises a social turn [36], affirming priority of secondary texts over authentic works of arts and true values in mass reception. So, Digital turn brings viruses into information flow (or gush). What about "vaccination" against chronotopological misconceptions and chronotopological deafness then? The remedy needs to be found in the milieu of infection. The Digital vaccine brings about the full-fledged opportunity of sharing and installing a multimedia supported chronotope reconstruction point just any time anywhere. At the same time, it allows for methodological and pedagogical interventions.

The mentioned methodological shifts, resource findings and educational risks have not yet penetrated fully the present-day paradigm in vocational training of arts and culture history learners as well as on the list of target competences [37]. The framework of university teaching arts and art criticism is still essentially based on classical discourse on the undeniable Genius and the Masterpiece, viewed from the point of Eternity, that is beyond any limits of space and time locations, as well as specific suggestions of historical processes context. The predominant method of analysis remains traditional formal [38, 39]. Elaboration and application of variety of art-criticism research tools still remains in one's infancy [40–42]. At the same time the actual scope of considering digital milieu is mostly limited to admitting electronic carrier copies for documents on paper analog. Restricting the educational practices within the pattern described may not fit the range of actual challenges, faced by modern learners and the coming demands of efficiency for educational practices.

The Methodology of Musical Chronotope Analysis. So far as our pilot experiment is dedicated to the search for authenticity in musical experience, we developed a methodology for analyzing the musical chronotope. This method is based on contemporary concepts of music theory and the theory of musical content [40, 41], interpreted from a chronotopological perspective: Melodic is considered as an organization principle of a musical text performance within the "space" of human hearing; Rhythmic is considered as the organization principle of a musical text performance over time; Texture is considered as the organization principle of a musical fabric as a polyphonic composition that shapes an artistic representation of the world; Thematicism is considered as the composer's conceptualization of the musical model of the world created and the listener's comprehension of this model. We used phenomenological tools for analyzing musical chronotopic experience [see also: 11, 12].

The formal method falls short when confronted with the XXI century view on the world and arts' life and history as dynamically changing polyphonic and multipolar process. The methodology of vocational training students at arts and culture history university department needs re-conception and reshaping, allowing for a more active and creative application of digital materials, interfaces and tools and innovative changes in the content of the targeted learners' ensemble of competences.

The described set of contradictions between the methodological and technological innovation drivers, the learners' actual educational needs and the factual theory on the one hand and teaching practice on the other hand forms the core of the overall problem of the study.

The strategic goal of the study is substantivation of the theoretical model and fulfilling an applied project, dedicated to elaboration of a pedagogic program for developing the arts and culture history department university students' ability for ideation of chronotopological cultural phenomena and their authentic meanings derivation.

The concrete research problem is elaboration of a special testing methodology for detecting and assessing the learners' level of preparedness to perceive, identify and qualify the embedded (in mediatext) sociocultural/historic and aesthetic chronotopes, allowing for decoding the meaningful messages, artistic ideas and values.

The purpose of the study is to theoretically substantiate, develop and test a diagnostic program, allowing for eliciting the university arts and culture history department students' aesthetic perception, meaningful reception and critical evaluation of an artistic media-text as a concrete historical and cultural chronotopological information carrier and the concomitant semantic procedures and processes.

The research hypothesis. The elaboration and application of a special diagnostic program, dedicated to eliciting learners' aesthetic perception, meaningful reception and critical evaluation of an artistic media-text as a concrete historical and cultural chronotopological phenomenon will allow to test and assess their current level of accomplishment of learners' pre-vocational preparedness. These include chronotopological, discourse and "visual" criticism skills along with their creative ability to construct and reconstruct meaningful associations in different arts media. The set of targeted skills will be useful for elaboration of didactic program, focused on achievement of learners' vocationally important abilities by learners'.

The range of tasks of the study:

- to determine the key indicators of testing module and the corresponding measurements criteria (presently—chronotopological, discursive, 'visual thinking', learners' creative ability to construct meaningful associations in different arts media);
- to apply the testing module, process and interpret the results acquired and draw conclusions on the actual levels of preparedness according to the four chosen indicators;
- to specify the key directions of upgrading the didactic program, based on the assessment results and the corresponding conclusions concerning arts, arts and culture history learners' vocational (un)preparedness in respect to their abilities of identifying and qualifying cultural phenomena within frame of chronotopological models and transmedia coding ("transcoding") the artistic meanings and values of works of art.

2 Research Methods

The research design is done within frame of a complex approach, essentially funded in Mikhail Bakhtin's conception of chronotope as an explanative method for historic phenomena in humanities, especially in art and culture, elaborated by Ukhtomsky [2], Bakhtin [3] and Zinchenko [4]. One can distinguish the two predominant meanings of the term in the present study: (a) time–space continuum as existential attributes of work of art; (b) time–space continuum as a format of recipient's aesthetic contemplation, aesthetic reception and—finally—evaluative qualification of a concrete work of art as a specific phenomenon of history and culture of a definite epoch.

The general research conception was framed within Bakhtin's conception [3] chronotope methodology. The gist of the conception: (a) attribution and distinction of the primary (prototypical), that is authentic chronotopes on the one hand and the secondary (gameplayed) chronotopes on the other hand—such as parody, stylization, citing, pastiche, sequels et cetera of the kind; (b^1) elicitation and specification of a system of (genre) markers and specific meanings of primary chronotopes, allowing for them to be identified as a specific historical and cultural phenomenon; (b^2) eliciting the system of specific markers and markemes of secondary chronotopes, allowing for meaningful reconstruction of its relatedness and interrelations with the preceding text and its primary meanings, as well as substantiating the pragmatics of intertextuality strategy in secondary text construction; (b^3) observation of a range of possible alternative genres of a secondary text construction and secondary chronotope formation and substantivation of the author's choice 'in casu proprio'; (b^4) construing a projection of the derived chronotope of a ('postmodernist') work of art in a system of different media expressive means—that is transmedia coding ("transcoding") of the chronotope of the subject matter work of art.

The chronotopological analysis is a crucial ability for arts and culture history learners, because "the field of discursive practice is social and therefore historical, and cannot be understood without a sense of the historical/social contingencies of the arrangement and configuration of practices and modes" [43, p. 43]. This point exposes the reason for us to apply chronotopological approach in testing leaners' sensitivity and reasoning skills in respect to poly-stylistic and polyphonic music pieces structures and interplay of meanings. Understanding an art masterpiece presupposes recipient's ability to grasp the universal, the particular and the unique in the semiotic arrangement of it. Concretization of reconstructed message of work of art is a challenging task, since "common principles of human social semiosis… have very different articulation at different times and in different places" [43, p. 43]. Achieving high level of learners' preparedness in chronotoplogical reconstruction of work of art message is a most demanding task within scope of vocational training.

Special accent in elaboration of a special testing module was laid on specific intertextual correlations (re)construction techniques, involving work of art recipients into a situation of text genre and meaning indeterminacy, challenging learners' ability for genre individuation, firm and final chronotopological attribution and meaning interpretation in form of a comprehensive and resolute exegesis of work of art message. Such challenges pertain to works of art (pieces of music as well), characterized either by necessity for stylization (e.g. in presenting picture of historic past or ancient myths and narratives) or composer's desire for a playful poly-stylistic artistic solution, permeated with spirit of postmodernist sensitivity.

Opting for the transmedia coding ('transcoding') here as a testing technique is due to a number of reasons. The first one is chronotopological and social in respect to 'modes' as relatively autonomous meaning-making systems: "mediums are shaped and organized into a range of meaning-making systems in order to articulate the meanings demanded by the social requirements of different communities" [44, p. 43]. Such systems are standardly not transparent ("available to understand") just for outsiders of the community. The second one reason is that "meaning … is always interwoven with the meanings made by all the other modes … in the context" [44, p. 43]. So, reception and interpretation of what is "communicated not only through the linguistic mode, but by other semiotic modes such as visual, sound or gesture" needs a holistic approach [45, p. 222]. The third reason is the openness of modes to changes in time [cf. 44, p. 43]. Consequently, transcoding can be viewed as an interpretative procedure and as a method of testing the recipients' understanding of work of art meaningful message.

The digital turn empowers multimedia and transmedia techniques of teaching and testing, based on implementation on tandem of semiotic and interdisciplinary approaches. In order to test the arts and culture history leaners' ability for meaningful analysis and interpretation of chronotopes a tracking of learners' self-reports was introduced, based on multimodal transmedia associations—audial, spatial, tactile, reading experience, visual and suchlike. These associations attributions can arise from one's nature-based sensitivity and one's own individual experience of communication with artistic texts, from erudition and exploring art criticism publications, and last, but not least—from directed learning activity in culture-based educational practices. The development of apperceptive base and the aesthetic judgement reasoning (mostly incrementally) contribute to learners' personal aesthetic taste formation. The extension of experience and critical judgment skills, fostered by expansion of the range of artistic works and genres experienced and mastered, preconditions formation of experiencers' memory, sensitivity to arts and culture chronotopes, followed by subjective discrimination of accountable chronotopological features, viewed as a net of distinctive indicators and associate personal constructs. Educationally stipulated learner's systematization of chronotopological constructs preconditions formation of chronotopological competence, vocationally important for arts and culture history department postgraduates. Chronotopological competence presupposes personal ability to individuate the chronotopes, to attribute intra-textual chronotopes and to specify their contribution to artistic text formation and textual meaningful message processing.

Works of different arts were chosen as the materials of the study: literary texts; visual and plastic images; musical pieces, landscape design, architectural styles and masterpieces and more. Holistic, interpretative and systemic approach were used to elicit and evaluate arts and culture history learners' preparedness to ideate, distinguish, match and contrast different chronotopes as primary (historic authentic donor-texts) and secondary (derivable back in history and represented as derivative and reused in other genres—such as, for example, parody, pastiche, developing sequels, prequels etc.). The recipient performs an attempt of a (creative) reconstruction of the chronotope(s), the genre and the meaning of artistic message, meanwhile the cultural background and the related artwork's rhetoric system provide meaningful drivers and triggers for decoding (and transcoding) processes. The recipient's interpretative predisposition, or interpretant (based on a set of intrinsic meaningful values—interpretative determinants) here is confronted with an embedded in work of art contextualizing net of "extrinsic" determinants as drivers of objective meaning management.

The empirical research experimentation results are based on a pilot testing, conducted in order to elicit and evaluate the quality of experienced multifarious aesthetic associations and meaning reconstruction—predominantly from the point of view of approximation to authenticity of predominant values distribution in the personal hearing a given music piece experience. The term "approximation" here is adopted in a threefold meaning: first, as a step-by-step process of decoding predictors of values involved in authentic meaning (re)construction; second, the elicitable admittable incompleteness (or missing some meaningful details) of the holistic understanding of the work of art message; third, the creative and variative nature of subjective meaning reconstruction. The subjective nature of understanding as a meaning reconstruction activity is disclosed through building multifarious personal projections of work of art as inseparable part of artistic communication.

The interpretative part of work of art (music piece) reception included modelling the chronotope and extended description of a sort of Augmented Reality of music piece reception. The AR (augmented reality) principle of adding value by adding (or "mixing") information channels and layers was implemented in learners' reporting multifarious projections, enabling interpretative transcoding of the work of art (music piece) predictors of meaningful messages. Transcoding (transmedia coding) extended to building recipients' meaningful associations with the chronotopological features, presented in literary texts, art manifestoes, visual and plastic images; musical pieces, landscape design, architectural styles and masterpieces etc. Learners' interpretative activity contributed to reconstruction of implicit meaningful context and reconstruction of work of art (music piece) chronotope as an informative and holistic meaningful entity.

Empirical Research Methods. The empirical part of the study was based on conducting a pilot experiment, application of a questionnaire design, transcoding and interpretation procedures, numerical calculation. Within frame of implementation on tandem of semiotic and interdisciplinary approaches, reinforced by application of digital facilities (apt for pellet-by-pellet cutting polyphonic music pieces and presenting different parts of musical body separately) the following set of semiotic diagnostic techniques were applied:

- the associative modeling method (used to study the respondents' abilities to perceive and comprehend musical stimuli—a series of musical factions);
- reflective thinking method, directed at cultural / chronotopological indeterminacy issues;
- the "cultural provocation" (semantic shift) method;
- the discursive modeling method (in order to structure associative reactions to a musical stimulus and to verbally formulate and interpret the stimulated musical experience);
- a method of visualizing semantic images in the mental space of the individual in order to test recipients' ability to visualize the analyzed sociocultural chronotope [46];
- the method of differentiation and demarcation of primary and secondary texts and genres;
- method of (tuning) interpretation of contribution and functionality of secondary genres as part of author's artistic solution program in music media text construction.

3 Research Results

The empirical experimentation was conducted in October 2025 at Tver state university (Tver, Russia) and at MISIS University of Science and Technology (Moscow, Russia). The overall population tested were postgraduate students (educational program 'Pedagogical education. Pedagogical activity in culture and arts') and linguistic postgraduate students at MISIS University, Institute of Basic Education.

The stimulus material was presented by music pieces. The learners' task was to disclose the predominant meaningful values of music piece by way of creative listening and further extended transcoding of their aesthetic experience and aesthetic judgment. A set of indicators was introduced to provide measurements of quality of recipients' interiorization and creative interpretation of aesthetic judgment, based on immediate experience of listening to either holistic or fragmented music masterpieces of different genres and epochs. The targeted recipients' projections set included tonality and mood identification; space, time and environment localization features and markers; cultural and social projections of different kind.

The key research results provide evidence of (a) existing gaps in a set of vocational reflection ability and skills development and completion, elicited by testing arts and culture history students; (b) the measured levels of personal sensitivity and idiosyncratic multimodal reception of music masterpieces.

The methodology of presenting music piece pellets/factions to different recipients allows to elicit and measure experientially the level of their music reception ability development. The introduced reception format included establishing meaningful associative correlations of music-generated chronotopological holistic images with specimens of other genera of arts (literary texts; visual and plastic images; musical pieces, dress codes, landscape design, architectural styles and masterpieces and more) and the artistic idea-based meanings. The analysis of music motifs (fixed in presented to recipients' music cuts)/a series of motifs allows to: (a) measure the level of learners' music reception skills formation; (b) measure the level of learners' preparedness for eliciting the dynamics of music image construction in changing "space—time" continuum; (c) measure the level of learners' preparedness to construct a holistic, comprehensive and consistent associations field; measure the level of learners' preparedness to exercise a set of specific vocationally important competences, concerning eliciting and evaluation of work of art/music piece meaningful dominants (chronotopological, discursive, 'visual thinking' along with constructing transmedia meaningful projections in different genera of arts). See Fig. 1.

The Fig. 1 presents informative testing results, eliciting different existing gaps in learners' competences and allowing for further elaboration of the innovative didactic program in compliance with our general research hypothesis. The hypothesis was not refuted.

PS Some interesting "side-effects" were acquired in testing non-musical students of MISIS University. In their independent intuition-based evaluation of separated pellets of music composition many of those students, presented with the first seconds

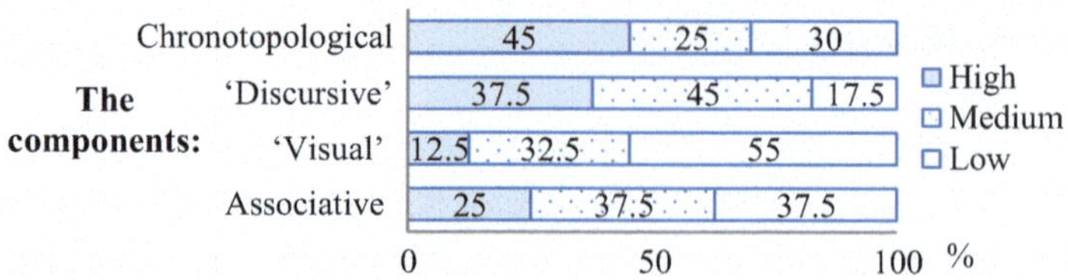

Fig. 1 The testing results, concerning measurements of learners' preparedness for meaningful interpretation, evaluation and transmedia coding the meaningful dominants of arts and culture history chronotopes, embodied in music pieces

of "Locomotive Breath" composition by Jethro Tull (1971) record marked "dark, almost not disturbed by faint or small light environment, nocturnal hours". Many also specified live playing academic piano on the ground floor—"somewhere in the open terrace of a country manor, facing a garden or a green lawn". All the students (though not studying music arts professionally) unanimously pointed to a variable retro style of the first three pellets of the composition. Melancholy notes in the music pellets stretches were also emphasized. Many questionnaire answers also mention black and white movies. Such sort of precision (or rather approximation) was not accounted on education and learned reflection, but rather attributed to sensitivity of the listeners.

Based on the conducted diagnostic results, we identified the key areas for the vocational didactic program elaboration, dedicated to forming the preparedness of arts and culture history students to comprehend and qualify cultural phenomena (within frame of chronotopological models) and transcode their artistic meanings and values.

There is a set of major methodological contradictions, pertaining to humanities as an integrate ensemble of interpretative sciences. Let us name three of those, challenging chronotopological interpretations:

- the contradiction between insularity and penetrability of meaningful worlds (historic epochs, chronotopes);
- the contradiction between informativity/redundancy of work of art|music piece and the corresponding cultural background on the one hand and its relative opaqueness and impenetrability on the other; the latter being grounded (a) on the basic asymmetry of the music author's and the music recipient's algorithms of text construction and decoding the text message; (b) on incompleteness of learners' preparedness for decoding the work of arts/music piece meaningful message authentically;
- the contradiction between the subjectively perceived self-evident self-sufficiency of recipients' aesthetic experience in the communication event and the actual gaps and distortions, detectable in its impoverished representation in their personal projections. This contradiction is also discoverable through meaningful dominants shift in transcoding the artistic message.

Application of digital tools and solutions allows for intensification of teaching learners the productive hermeneutic schemata of elicitation stylometric features of either music piece or works of arts of other kind.

4 Discussion

The targeted model for arts and culture history learners' chronotopological analysis and interpretation vocational skills formation is based at large on semiotics and multimodal semiotic text analysis methods. The suggested chronotopological approach addresses multifarious semiotic analysis and meaningful interpretation and evaluation of art masterpieces. The targeted interpretation activity is featured as multimodal, interdisciplinary and poly-artistic, covering audial and visual, plastic and scenic arts. Here we have presented a music piece analytics skills testing faction of a much larger targeted didactic model, covering audial and visual, plastic and scenic arts representations of cultural chronotopes, empowered by nets of associative ideas and their interconnections.

How much new and innovative the presented piece of empirical experience should be estimated? Literature review has elicited no similar interdisciplinary diagnostic model for teaching arts and culture history students—ideation, comprehension and interpretation of cultural chronotopes. Nevertheless, applications of different sets of semiotic analysis methods have been used productively in classical and present-day theory and history of arts, mainly not exceeding the boundaries of specific art.

Within the frame of art history studies the combination of semiotic micro- and macro-analysis and interpretation method was nourished by the ideas of American semiotician Charles S. Peirce, Swiss art historian Heinrich Wölfflin, French semiologist Roland G. Barthes, Russian semiologist and culturologist Juri Lotman [47–50]. However, the development of such poly-artistic multimodal and multi-layered method [51] has not met systemic completion in educational practices. Its concrete applications this method is limited to visual arts, mainly omitting connections to different other arts and genres [7, 52]. In teaching history of music cultures, the method is integrated into musical conservatoire educational courses [40–42], not exceeding the boundaries of music. The range of tasks for mastering semantic fields and larger spaces at the crossroads of aesthetic, ethical, scientific, religious and philosophical ideas and developing tools for analyzing the author's position and presence in a musical text have just been set recently [53, pp. 327–331, 54, p. 62]. The semiotic method of analyzing events and chronotopes in theater and film studies is standardly confined to these specific forms of art practices [55–57]. Some attempts to take a look beyond the limits of one type of art have been made, when associated with the application of fractal analysis to art history texts [58]. However, little attention to the chronotopic thinking of the author was paid.

To recapitulate with, the proposed here didactic model for arts and culture history learners' chronotopological analysis and interpretation vocational skills formation provides a unique integration of goals, multimodal poly-artistic scopes and teaching tools—such as multimodal poly-artistic semiotic and associative analysis, synthesis of artistic and cultural codes, synesthetic and chronotopological interpretation. The acquired experiential testing results provide evidence for informativity and expected productivity of the targeted chronotopological didactic model. Its multimodal, complex, interdisciplinary and integrative nature opens new educational opportunities and provides ample grounds for its fruitful implementation in teaching arts practices.

Limitations and Future Research

The conducted experiential study tested one module of a larger poly-artistic diagnostic program. The holistic diagnostic program includes similarly structured components of different arts, represented in either whole masterpiece or selected segments, exposed to learners' aesthetic perception and evaluation. The experiential exploration of art masterpieces is pre-designed as a meaningful reconstruction of embedded audial and visual, plastic and scenic chronotopes, charged with ensembles of associative ideas and their interconnections. Further development of smart digital tools will expand the precision of diagnostics of arts and culture history learners' vocational preparedness.

The tested module is a part of a poly-artistic diagnostic program, which in its own turn is a component of a larger didactic model, directed at arts and culture history learners' chronotopological analysis and interpretation vocational skills formation. The main purport of the didactic model is teaching students correct ideation, comprehension and interpretation of cultural chronotopes as a crucial part of art criticism competence. The chronotopological analysis and synthesis skills foreground deep understanding and authentic evaluation of the works of art meaningful messages. The main directions of further development of the targeted didactic model are as follows.

First and foremost a digitized collection of relevant data/corpora is needed. This needs to be presented as an accessible online interactive resource, staffed with educational, instructional and artistic materials. A most important functional component of it needs to be presented by digital analytical tools, applicable both online and offline in synchronous and asynchronous modes, the facilities of distant networking, cloud-storing, cloud-services applications being part of design. The portal hosts different artistic content, which is organized in accord with the "world map" and "palimpsest" principles, the latter one connected to a time-line, assisted by hyperlinks for flexible addressing biographies, geography, space and time of masterpieces creation. This ensures the application of the principle of historicism and contributes to authenticity of the historical scope and view of the organization of artistic fabric by means of chronotopological analysis.

The key directions in teaching applied semiotic analysis of the rhetoric / hermeneutic organization of artistic fabric by means of chronotopological analysis include:

- the study of musical scores (decomposition of a digitized performance of a score into separate parts of instruments, facilitated by application of digital instruments);

analysis of their epiphenomena: epoch, setting, people, interiors, artistic environments; application of techniques of transposition, modulation, temporal shifts, mixing based on digital facilities, including neural network; (re)construction of a color standard for the synesthetic perception of a musical piece; analysis of the artistic and emotional vocabulary (reaction words), used by a population in describing music, based on a coloristic transcoding system; exploring correlations and mismatches, provided by application of automated analysis systems results (e.g. BAAL [59]);
- study of visual arts masterpieces (such as paintings, drawings etc.) with application of digital tools of coloristic analysis in order to clarify the functions of coloristic resources in (re-)creating artistic space and time (e.g. Gradients.app [60]);
- analysis of multimodal media-texts of synthetic art forms using methods of element-wise decomposition, combinatorics, semantic shifts, etc.;
- development of musical sensitivity and providing instruction and training productive schemes of analysis and interpretation of postmodern, polystylistic, polyphonic musical/non-musical works;
- teaching integration of musical elements into a historical/cultural narrative structures and storytelling of polyphonic musical art works.

5 Conclusions

Multimodal learning and testing techniques not only "empower students to generate meaning and make sense of themselves, their experiences" [cf. 1, p. 43], but also to boost learners' vocational skills in understanding of continuity of arts and culture historical chronotopes, develop personal sensitivity to arts languages and stylistics. Multimodal interpretation of works of art bridges the gap between sensory and spiritual reception, elicits interconnectedness of different art genera masterpieces, stipulates in-depth understanding of their creation epoch and message destination. Digital age tools allow for facilitated reconstruction of art chronotopes anytime anywhere.

The elicited deficiencies of arts and culture history learners' preparedness demand thorough attention and elaboration of a special pedagogical program for developing students' multimodal abilities to perceive, identify and qualify the embedded (in media-text) sociocultural/historic and aesthetic chronotopes, allowing for decoding the meaningful messages of art pieces of different epochs. The implementation of such multimodal transcoding didactic program would improve the overall practice of teaching research methods in the field of arts history and cultural studies, as well as stipulate the development of new productive teaching and learning technologies and particular techniques. The prospects of the research are related to further development of a formative vocational training program, substantiated in the presented research, and the evaluation of its educational outcomes.

References

1. Bouchey B, Castek J, Thygeson J (2021) Multimodal learning. In: Ryoo J, Winkelmann K (eds) Innovative learning environments in STEM higher education. Springer, Cham, pp 35–54
2. Ukhtomsky AA (2002) Dominanta. Stat'i raznyh let. 1887–1939. Piter, Sankt-Peterburg
3. Bakhtin MM (1975) Voprosy literatury i estetiki. Issledovaniya raznyh let. Khudozhestvennaya literatura, Moskva
4. Zinchenko VP (2000) Mysl' i slovo Gustava Shpeta: vozvrashchenie iz izgnaniya. URAO, Moskva
5. Lukin AN (2009) Khronotop kul'tury. Vestnik CHGAKI 4(20):38–45
6. Akopov GV (2020) Iskusstvo v kategoriyah soznaniya: khronotopy, obrazy, diskursy. Soglasie, Moskva
7. Balkind EL (2013) Khronotop v izobrazitel'nom iskusstve. Uchenye zapiski KFU imeni V.I. Vernadskogo. Sociologiya. Pedagogika. Psihologiya 26(4):246–255
8. Belomoeva OG (2020) Osobennosti khronotopa v drevnerusskoj ikonopisi. Kul'tura i Civilizaciya 10(5A):322–327
9. Boguslavskaya EL (2025) Mul'tmodal'naya struktura khronotopa avtorskogo polikodovogo diskursa pejzazhnoj zhivopisi v kontekste politiki pamyati. Diskurs-Pi 22(2):126–144
10. Öztürk Z, Arıdağ L (2022) The chronotopological emergence from the spatial narrative to performance. ITU A|Z 19(3):615–630
11. Astakhova OA (2020) Muzykal'nyj khronotop. Pero, Moskva
12. Mozgot SA (2014) Konceptual'noe prostranstvo v muzyke kompozitorov XIX veka. Problemy Muzykal'noj Nauki 4(17):20–24
13. Yarmukhametova RS (2020) The chronotope in the worldview and music eras. Aspirantskiy Vestnik Povolzhiya 20(3–4):63–66
14. Mazar S (2023) Idyllic chronotope as one of the principles of organizing time and space: The example of handel's "Alcina" staged by Katia Korabelnik. Psychol Res 13(11):537–541
15. Poznin VF (2019) Khudozhestvennoe prostranstvo i vremya v ekrannom khronotope. Vestnik SPbGU. Iskusstvovedenie 9(1):93–109
16. Radionceva ES (2025) Khronotopicheskie dominanty mediaprostranstva BANI-mira. Kommunikologiya 13(1):25–38
17. Ostapenko AA (2010) Khronotop kak perekrestok proshlogo i budushchego, nastoyashchego i… «ponaroshechnogo». Kul'turno-Istoricheskaya Psihologiya 6(1):2–6
18. Chirkov VF (2005) Teoriya khronotopa kak metodologicheskaya baza ponimaniya i izucheniya lokal'nyh mestnyh kul'tur. Omskij Nauchnyj Vestnik 3(32):187–192
19. Vasil'ev VA (2021) Ispol'zovanie khronotopa v social'no-gumanitarnom poznanii. Social'no-Gumanitarnye Znaniya 2:75–85
20. Foucault M (1966) Les mots et les choses. Gallimard, Paris
21. Derrida J (1967) De la grammatologie. Les Éditions de Minuit, Paris
22. Fairclough N (1995) Critical discourse analysis: papers in the critical study of language. Longman, London and New York
23. Bulatova AP (1999) Konceptualizaciya znaniya v iskusstvovedcheskom diskurse. Vestnik Moskovskogo universiteta. Seriya 9: Filologiya 4:34–49
24. Elina EA (2002) Verbal'nye interpretacii proizvedenij izobrazitel'nogo iskusstva. Nominativno-kommunikativnyj aspekt. SGSEU, Saratov
25. Olomskaya NN, Bazhan AS (2022) Art criticism discourse: specificity and classification. Kazan Ling J 5:346–359
26. Zharkova UA (2011) Voploshchenie znakovoj prirody izobrazitel'nogo iskusstva v iskusstvovedcheskom diskurse (na materiale nemeckoyazychnyh muzejnyh katalogov) Vestnik Chelyabinskogo Gosudarstvennogo Universiteta. Filologiya. Iskusstvovedenie 33(248):49–52
27. Ivanova TE, Akhidzhakova MP (2021) Iskusstvovedcheskoe diskursivnoe prostranstvo: pragmasemanticheskie osobennosti organizacii. Gumanitarnye i Social'nye Nauki 4:125–135
28. Crossley S (2007) A chronotopic approach to genre analysis: an exploratory study. Engl Specif Purp 26(1):4–24

29. O'Neill P (2007) The curatorial turn: from practice to discourse. In: Curating contemporary art and performance. University of Chicago Press, Chicago, pp 13–28
30. Paola M (2019) The translational turn in visual culture. Rev Estud Glob Y Arte Contemp 6(1):1–13
31. Purgar K (2019) Pictorial appearing: image theory after representation. Verlag, Bielefeld
32. Reutov AS (2018) Vizual'nye issledovaniya sovremennoj kul'tury: fenomenologicheskij aspekt: dis.... kand. filos. nauk. N. Novgorod
33. Simakova SI (2018) Vizual'nyj povorot – novaya filosofiya obraza v sredstvah massovoj kommunikacii. Vestnik Chelyabinskogo gosudarstvennogo universiteta. Filologicheskie Nauki 10(114):225–232
34. Voevodina LN (2023) Vizual'nyj povorot i novye kul'turnye praktiki. Vestnik MGUKI 3(113):97–103
35. Elyamany N (2021) A chronotopic approach to identity performance in musical numbers: a choreo-musical case study of 'Rewrite the Stars' and 'This Is Me.' Vis Commun 22(2):278–296
36. Francis MA (2014) The dark side of art's social turn: towards a discourse on the impact of art's use of social forms upon the social. J Cult Res 18(1):25–43
37. Prikaz Minobrnauki Rossii ot 03.12.2015 N 1404 (s izm. ot 17.11.2023) «Ob utverzhdenii federal'nogo gosudarstvennogo obrazovatel'nogo standarta vysshego obrazovaniya po napravleniyu podgotovki 50.04.03 Istoriya iskusstv (uroven' magistratury)» (zaregistrirovano v Minyuste Rossii 31.12.2015 N 40499). Konsul'tantPlyus, https://www.consultant.ru/document/cons_doc_LAW_192455/. Accessed 1 Dec 2025
38. Yakovleva NA, Chagovec TP, Babiyak VV [i dr.] (2024) Analiz i interpretaciya proizvedeniya iskusstva. Khudozhestvennoe sotvorchestvo. 9-e izd., ster. Planeta muzyki, Sankt-Peterburg
39. Gumenyuk AN, Chujko LV (2025) Iskusstvovedenie. Morfologiya plasticheskih iskusstv. Yurajt, Moskva
40. Kholopova VN (2024) Teoriya muzyki. Melodika. Ritmika. Faktura. Tematizm. 6-e izd., ster. Planeta muzyki, Sankt-Peterburg
41. Kazanceva LP (2025) Analiz muzykal'nogo soderzhaniya. 2-e izd., ster. Planeta muzyki, Sankt-Peterburg
42. Kudryashov AY (2010) Teoriya muzykal'nogo soderzhaniya. Khudozhestvennye idei evropejskoj muzyki XVII–XX vv. 2-e izd., ster. Planeta muzyki, Sankt-Peterburg
43. Kress G, Van Leeuwen T (2001) Multimodal discourse. The modes and media of contemporary communication. Oxford University Press Inc., New York, Arnold, London
44. Kress G, Jewitt C, Ogborn J, Tsatsarelis C (2001) Multimodal teaching and learning. The Rhetorics of the science classroom. Continuum, London, New York
45. Machin D, Mayer A (2012) How to do critical discourse analysis. A multimodal introduction. Sage, London
46. Makarova EA (2010) Vizualizaciya kak introekciya smysloobrazov v mental'noe prostranstvo lichnosti. Sputnik+, Moskva
47. Peirce CS (1960) Collected papers. Harvard University Press, Harvard
48. Wölfflin H (1915) Kunstgeschichtliche grundbegriffe: das problem der stilentwickelung in der neueren Kunst. Bruckmann, München
49. Barthes R (1977) Image, music, text. Fontana Press, London
50. Lotman YM (2004) Semiosfera: Kul'tura i vzryv. Vnutri myslyashchih mirov. Iskusstvo-SPB, Sankt-Peterburg
51. Sonesson G (1988) Methods and models in pictorial semiotics. Lund University, Lund
52. Nikol'skaya TM (2017) Ispol'zovanie principov semioticheskogo analiza dlya izucheniya proizvedenij zhivopisi v vuze (na primere tvorchestva E. Munka). Vestnik TGU 6(170):85–93
53. Shnitke A (1990) Polistilisticheskie tendencii v sovremennoj muzyke. In: Kholopova VN, Chigareva EI (eds) Al'fred Shnitke. Iskusstvo, Moskva, pp 327–331
54. Kazantseva LP, Volkova PS (2021) Semantic analysis of a musical work. In: Khannanov ID, Ruditsa R (eds) Proceedings of the worldwide music conference. Springer, Cham, pp 59–64
55. Pesochinskij NV (2010) Semiotika kak teatrovedcheskij metod. Teatron. Nauchnyj al'manah 2(6):3–13

56. Anastas'eva IL (2019) Neverbal'naya semiotika scenicheskogo iskusstva. Vestnik Moskovskogo universiteta, Seriya 19. Lingvistika i Mezhkul'turnaya Kommunikaciya 4:88–96
57. Geroeva LM (2023) Semiotika teatral'nogo iskusstva kak metod refleksivnogo vozdejstviya na zritelya. Sovrem Pedagog Obraz 12:307–310
58. Anisimova E (2014) Princip sovmeshcheniya vizual'nogo, semioticheskogo i vremennogo fraktalov v «Analiticheskom iskusstve» P. Filonova. Iskusstvoznanie 4:307–319
59. BAAL http://vaal.ru/. Accessed 1 Dec 2025
60. Gradients.app https://gradients.app/en. Accessed 1 Dec 2025

Adaptive Learning Algorithms and Educational Equity: A Multi-Layered Psychosocial and Economic Evaluation Model for Migrant Students in Smart Classrooms

Khasanboy Abdusamatov, Sardorkhon Sunnatillaevich Gaibnazarov, Iroda Abdushukurovna Bakiyeva, and Sevara Marat kizi Kurbanbaeva

Abstract The rapid expansion of digital learning environments has intensified long-standing debates about equity, access, and learner well-being, particularly for students with migration backgrounds. While adaptive learning algorithms promise personalized instruction and improved academic trajectories, their psychological, social, and economic implications remain insufficiently examined. This study develops a multi-layered evaluation model to investigate how algorithm-driven personalization functions within smart classroom ecosystems serving migrant learners. The model integrates three analytical dimensions: (1) psychological resilience under technology-saturated instruction, (2) shifting patterns of social participation and belonging shaped by smart-classroom interaction structures, and (3) direct and indirect economic pressures associated with sustained engagement in digital learning systems. Drawing on contemporary research in educational psychology, migrant education, and learning analytics, the study highlights the tensions between enhanced access to high-quality digital resources and the hidden burdens that adaptive systems may impose on vulnerable learner groups. The findings emphasize the need to recalibrate smart-education policies so that personalization

K. Abdusamatov
ISFT Institute, Tashkent, Uzbekistan

S. S. Gaibnazarov (✉)
Tashkent State Medical University, Tashkent, Uzbekistan
e-mail: nasazusu@gmail.com

I. A. Bakiyeva
University of science and technologies, Tashkent, Uzbekistan
e-mail: i.bakiyeva@usat.uz

S. M. Kurbanbaeva
Tashkent State University of Economics, Tashkent, Uzbekistan
e-mail: s.kurbanbaevam@gmail.com

© The Author(s), under exclusive license to Springer Nature Switzerland AG 2026
A. Kulachinskaya et al. (eds.), *Digital Transformation: What Kind of Smart Life do we have today?*, Lecture Notes in Networks and Systems 1895,
https://doi.org/10.1007/978-3-032-20582-7_23

technologies reinforce—not weaken—educational equity. The proposed framework provides conceptual and methodological guidance for institutions seeking evidence-based approaches to designing inclusive adaptive-learning environments for migrant students.

Keywords Adaptive learning · Migrant students · Educational equity · Smart classrooms · Psychosocial factors · Economic costs

1 Introduction

Adaptive learning algorithms have demonstrated favorable potential in addressing equitable and fair educational practices for migrant students. Comprehensive statistical evidence indicates that migrant students share common underlying difficulties in adapting to changing learning environments, retrieving learning resources, and reformulating cognitive and social constructs. An organization-oriented multi-layered evaluation model for these algorithms in migrant education is needed. Available literature covering adaptive education intervention models has not addressed the perspectives of migrant learners within intelligent or smart classrooms.

The educational policy framework does not underline the importance of financial budgeting, resource allocation, or return on investment for educational technology, particularly adaptive learning platforms. Restricted funding mechanisms sharply limit the growth and development of adaptive learning. Current evaluations generally overlook psychosocial dimensions that adversely influence student learning. Motivation, identity, social belonging, and well-being are established as the prominent psychosocial barriers that many students in intimate and distant learning face. The effect of an intelligent classroom on educational equity and overall learning improvement has not yet received detailed consideration within the literature.

Adaptive learning systems implement technologies such as data mining, user modelling, recommendation systems, gamification, and collaborative learning for theory-driven pedagogical objectives like planning, authoring, presentation, sequencing, and engaging [1]. Existing frameworks focused on structuring adaptive educational actions but neglected to delineate learning analytics, personalization, feedback loops, and adaptability as core components. Learning analytics involves tracking student interactions and providing feedback with the aim of boosting engagement, achievement, and retention and relies on temporal records of specific behaviors; adaptations to action sequences reflect a mismatch between current and target competencies, teacher guidelines, or both; adaptations targeting content presentation adjust cognitive load according to learner characteristics, design principles, or other criteria; adaptations addressing access regulation, communication support, or social-awareness enhancement cater to the specific requirements of the collaboration mode selected [2]. Engaging pedagogical strategies such as gamification are used to motivate low-achieving learners. Adaptive learning technologies thus embody: (i) monitoring learner activity, (ii) analyzing contextual, behavioral, intermediate, and final results, and (iii) selecting and enacting decisions after the assessment of the state.

Educational equity encompasses the fair and just distribution of educational resources, opportunities, and treatment among all individuals and groups. The provision of equal opportunities and fair treatment ranks among the foremost social justice and fundamental human rights issues worldwide. Education equity strives to achieve equality across distinct and diverse dimensions, including gender, class, race, ethnicity, disability, language, parental background, cultural capital, and geographic location. Migrant education operates as a powerful entry point for the realization of educational equity due to the typically disadvantaged status of this population. Specific barriers encountered by migrant learners include low enrolment, high dropout rates, weak language support, and infrequent assessment of their learning conditions. Ingersoll (cited by Shankar [3]) reported that, while these pupils constitute less than 10% of total student populations, their share of programs for at-risk students generally ranges from 30 to 50%.

Policy documents and targeted interventions highlight the need for special assistance in relation to language use, additional pedagogical consultation, individual instruction, further training for instructors, peer tutoring, and special teaching materials [4]. Socioeconomic challenges and digital divides contribute to technology use disparities that further exacerbate educational inequalities among migrant students [5, 6]. Smart classrooms offer digital devices, internet access, and high-quality learning resources; leverage their superior educational quality, efficiency, and effectiveness; and follow provincial and local regulations limiting technology application in nursery and primary schools, but countries and regions advance smart education in different ways. Digital resources should not exclusively benefit those households that can afford them, and pedagogy and culture should play more significant roles than technology [7].

Adaptive learning systems facilitate personalized education by acquiring data on student behavior, iteratively updating learner profiles, and furnishing immediate individual feedback [3]. Adaptive technologies like Yixue Squirrel AI have garnered attention for their potential benefits in adapting to diverse learning needs. Assessments of its impact in middle school English and mathematics courses yielded positive outcomes relative to traditional instruction and other platforms. Embedded artificial intelligence analyzes student interactions to estimate knowledge states and craft customized learning pathways. Large-scale comparisons indicate adaptive courses enhance performance more effectively than conventional alternatives. The Adaptive Learning Guidance System (ALGS) constitutes a pedagogically principled adaptive architecture, balancing the marginalization of teacher roles with the capacity to leverage big data. Unlike many prior systems, ALGS solicits teacher input throughout the educational process via a hybrid recommendation mechanism integrated within a three-stage framework. Models governing users, content, and adaptation serve specific functions: the user model accumulates data on learners for greater personalization; the content model organizes knowledge in accordance with instructional goals; the adaptation model employs collaborative filtering, a hybrid recommendation scheme that combines machine-generated and teacher-generated suggestions, and a personalization engine to optimize recommendations for both learners and instructors. Initial collaborative filtering operates on minimal data, with system-generated recommendations identifying skills requiring reinforcement and teacher engagement remaining integral to decision-making [1].

Migrant education in the United States epitomizes the pursuit of educational equity for underserved student populations [8]. Although migrant education programs have existed since 1966, the chronic nature of the problem reflects a persistent national challenge. Educational equity encompasses the distribution of not only resources but also opportunities and treatment to ensure that each student's needs are met [9]. Equity thus emphasizes access to the appropriate curriculum and instructional strategies that consider students' diverse backgrounds, including those of refugee and asylee learners as well as migrants and displaced persons affected by circumstances such as armed conflict [8]. Migrant programs are intended to help address those basic needs of students who must frequently change schools, especially in regard to school records, placement tests, and even personal hygiene. Migrant students represent an example of how equity of opportunity—one of the fundamental goals of education in a democracy—has been systematically denied by ever-growing changes in educational policy and organization.

The central learning activities in educational psychology are motivation, cognition, and knowledge construction. Knowledge and understanding from socioeconomic, educational, cultural, and psychological perspectives are crucial. Endeavors to motivate learners to cherish learning, to deepen knowledge cognizance, and to consider how to monitor learning processes are vital in promoting learning outcomes. Motivation and caring from authorities can enhance social belonging, quell anxiety about learning, and further encourage social development. Adaptive learning algorithms could be an effective tool [10].

Education plays a key role in enabling personal growth, social satisfaction, and economic opportunity. Educational equity is a multifaceted challenge in many societies, especially regional education inequality and suburban-rural learning resource gap [11]. Another topic of concern is equity and the State of Social Justices of different ethnicities. Learning analytics, adaptive learning, and automatic assessment algorithms draw particular attention. Educational technologies should further explore how to promote and monitor equitable learning among migrant students.

Migrant students may find it difficult to access learning resources in smart classrooms due to socio-economic background, language barrier and so on. Appropriate support mechanisms have to be provided in time. Failure to do so will create a gap between students who receive adaptive mechanisms and the other. The countries get back from lockdown, the hybrid model has come to an end, smart classrooms are slowly considered as the formal learning space applied in special events, the use of educational technology, both hardware and connections by teachers are recommended to be used as a strategy, not during all classes, not also apply the same design for all subjects [12].

2 Methods

Models that apply evaluation metrics from different domains have the potential to support efficient learning designs and educational delivery while providing insights on the trade-offs among multiple criteria [13]. Nonetheless, the preservation of student privacy when collecting data is a prominent concern for education systems. A promising avenue for addressing this issue lies in the adoption of the data ecosystem framework together with established technical solutions [14].

Smart classrooms feature a range of tools designed to facilitate assessment of students' understanding and engagement. These tools include feedback, signal, annotation, and achievement plugins. Their effectiveness can be evaluated through metrics that quantify learning gain, equity, motivation, engagement, stress, and teaching cost.

Adaptive learning systems have received increasing attention as potential solutions to the multifaceted problem of educational inequality [15]. Escalating migration flows are producing large numbers of students who are constrained in their educational trajectories [16, 17]. Although systems of this nature may contribute to greater access and equity, they can also inadvertently reinforce inequity in learning opportunities. Multi-layered approaches to modelling the performance of such systems enable an enhanced assessment of their potential equity impacts and trade-offs across a variety of dimensions.

Adaptive learning systems rely on data analytics to personalize instruction in real time and reconfigure pedagogical scenarios to meet learners' evolving needs. Learning analytics infers these needs from system-generated data and educational ecosystems facilitate data collection through diverse digital procedures. In education, the data ecosystem comprises the collection, governance, sharing, and security of data, regulating system characteristics or safeguards plays a crucial role in the operation of adaptive learning systems and envelops distinct data phases and requirements. Attention to data ecosystems and privacy safeguards is vital for learning-design decision making during the incorporation of adaptive-learning technologies within educational programs. Institutional-level proportionality frameworks inform educators regarding fitted data ecosystems according to context [15, 18].

Learning analytics, procedures through which data are gathered and analyzed to inform learning-design decisions, operates within educational ecosystems. In education, the data ecosystem encapsulates data collection, governance, sharing, and security and characterizes institutional-level educational systems. Regulations for adapting learning analytics prompt measures for data ownership, sharing, security, or algorithmic fairness early in adaptive-learning system implementation. When introducing adaptive-learning technologies, awareness of institutional-level frameworks elaborating, prescribing, or regulating systemic characteristics—including data ecosystems and privacy safeguards—enhances learning-design decision making, allows educators to align adaptive-learning arrangements with institutional-level stipulations, and promotes discrimination against a broader sociocultural structure comprised of educational sectors and agencies [8].

Algorithmic fairness refers to the determination of whether an algorithm is fair or biased according to defined metrics, including demographic information [8, 19]. It plays a crucial role in educational data mining systems, where models may exhibit biases against certain groups. These biases can arise from either covariate shifts of the inputs or a lack of shared representation among protected groups [20]. In adaptive learning systems employing student information, performance prediction models adhering to algorithmic fairness throughout the modeling process are paramount.

A variety of factors contribute to bias in predictions of academic performance, including the region or area of residence, economic background, and enrolment in prerequisite courses [21, 22]. Prior to training, groups are identified based on protected attributes, and performance metrics are examined across these groups to ascertain fairness. To further enhance equity, disparity indicators such as the difference in learning gains among groups receiving similar assistance may be employed. These considerations apply to migrant students, who constitute a distinct group of learners with unique characteristics and requirements.

A comprehensive compliance process, encompassing strict adherence to educational data mining guidelines, documentation of positive and negative indicators, and systematic review by practitioners and researchers, is advisable. Since educational datasets are subject to constant updating, routine examination of fairness is necessary, accompanied by adjustment of corresponding actions.

To investigate the impacts of adaptive learning systems on migrant students and evaluate equity outcomes under a multi-layered framework, a robust methodology is essential. This section outlines the proposed data collection and instrumentation, evaluation metrics, and experimental design.

The evaluation employs a combination of quantitative and qualitative methods. To capture attitudinal dynamics, a pre-intervention survey gathers demographic, behavioral, and psychosocial information along with learning design preferences [13]. During the intervention, two logs—a digital interaction log generated by the learning platform and a perturbation log recording pauses in activity, inactivity, and cognitive disturbances—enable longitudinal tracking of academic progress and identify impediments to learning [2, 8]. Initially, comprehension assessments gauge knowledge state; the Smileyometer, administered intermittently and upon task completion, reflects subjective engagement and cognitive load; and to measure motivation, belonging, stress, and personal significance, a brief psycho-emotional check comprising 10 items and a reflection comment at cycle end provide input for analyzing student well-[8]. Finally, a qualitative post-intervention questionnaire prompts open-ended and multiple-choice feedback along with institutional attribute verification and peer influence exploration.

Four types of impact are considered through appropriate metrics: (1) academic gain assessed by learning system records; (2) equity indicators computed from participants' feedback input using the Educational Opportunity Analysis (EDOP) [19]; (3) psychosocial score variation derived from pre- and post-intervention survey responses on motivation, belonging, stress, and engagement (Mishra et al. 2022); and (4) economic effect evaluated via systematic approaches to identify input, output, and expenditure from the implemented smart classroom learning plan.

To mitigate confounding influences, an experimental design with sampling and control conditions is established. The intervention, based on the OCAC course, targets Primary 3 students; the experimental group includes three migrant learners with the same language proficiency attending other courses in the same school, while the control group consists of non-enrolled peers. Forward and backward links in the knowledge network are monitored longitudinally to ensure equivalence between migrant and non-migrant conditions. Furthermore, to strengthen data reliability, different teachers and pedagogies are employed, and robustness checks are performed to validate influence degree and direction, preconditioned by control condition significance.

3 Results

Learning occurs within contexts, and technology-enhanced educational contexts can exemplify adaptability. Adaptive systems for learning reside at the intersection of technology and pedagogy; such systems can be defined in terms of four primary constructs: learning analytics, personalization, feedback loops, and adaptability. Learning analytics embraces the gathering and analysis of observations that provide insights into learners, learning processes, and learning environments. Personalization tailors the learning experience to the needs, preferences, and understanding of individual learners, providing customized learning paths, content, and support. Feedback loops provide information, interpretations, and recommendations concerning the results of learning experiences; such feedback sustains learner effort, enhances understanding, fosters reflection, and contributes to the development of metacognition. Adaptability indicates that content, experiences, and opportunities to reflect vary in response to observations concerning learners, learning, and the context of learning. A diversity of adaptive systems exists, informed by differences in education technology and the conceptualization of learning [2]. Specific systems utilize intelligent tutoring to guide learner activity, simulation-based learning to facilitate practice through experience, and learning management systems that serve as primary gateways into the educational experience.

Smart classrooms, equipped to capture a range of data relevant to learning contexts, further enable adaptive learning and contribute to educational equity for migrant learners. Evidence suggests that equity issues, location-based challenges (such as school drop-out, academic engagement, distance to school, and language of instruction), and technology-supported learning environments are relevant to out-of-school marginalization. Access to adequate facilities, equipment, broadband, and digital competence are contributing equity factors facing these learners [23].

Adaptive learning algorithms operate on student data to assess learning progress, but learners can differ dramatically in their educational assets and needs. An effective class-wide algorithm that accommodates students' varied profiles should foster equity as well as enhance overall performance [24]. "Equity" takes on multiple meanings in learning contexts (Ivanov 2020), but equitable adaptive arrangements commonly address issues such as access to resources, engagement levels, relevance to personal goals, and opportunities to develop a diverse range of skills (Tables 1 and Table 2).

A multi-layered evaluation model for adaptive content materials depends on a data ecosystem, algorithmic fairness measures, and privacy safeguards. The model also supports smart-classroom infrastructures featuring diverse educational technologies, pedagogical methods, and classroom practices. From an educational perspective, the model has been applied to assess smart-classroom systems for migrant students in the Republic of Korea. The proposed framework clarifies how a multi-layered evaluation model can facilitate the development, deployment, and enhancement of

Table 1 Equity dimensions in adaptive learning contexts

Equity dimension	Description	Relevance to adaptive learning algorithms	Evaluation indicators
Access to resources	Equal availability of digital tools and learning materials	Algorithmic content delivery and accessibility	Device ownership, login frequency
Engagement level	Degree of active participation in learning activities	Personalized task difficulty and pacing	Session duration, interaction counts
Goal relevance	Alignment with learners' personal and academic goals	Individualized learning pathways	Self-reported relevance, completion rates
Skill diversity	Development of cognitive, social, and digital skills	Variety of adaptive tasks and activities	Competency mapping, skill progression
Social inclusion	Integration into peer and classroom communities	Collaborative and interactive AL modules	Peer interaction frequency

Table 2 Structure of the multi-layered psychosocial and economic evaluation model

Layer	Core components	Data sources	Evaluation purpose
Data ecosystem layer	Learning logs, behavioral traces, sensor data	LMS platforms, IoT devices	Monitoring learning behavior
Algorithmic layer	Adaptive algorithms, fairness constraints	Model outputs, bias metrics	Ensuring equitable personalization
Psychosocial layer	Resilience, engagement, social interaction	Surveys, classroom observations	Assessing psychological impact
Economic layer	Direct and opportunity costs	Institutional and learner records	Evaluating cost sustainability
Ethics and privacy layer	Data protection, consent protocols	Policy and compliance checks	Maintaining trust and legality

adaptive-learning systems in smart-classroom settings, where educational analytics, artificial intelligence, and digital content support personalized learning experiences (Table 2).

Educational institutions employ a plethora of tools and arrangements to instrument students' learning processes through feedback, guidance, and hints; still, instructor-assisted interventions remain paramount. Adaptive Learning (AL) has emerged in the educational technology arena as a system that records users' actions and automatically adjusts content according to each learner's characteristics [8]. AL systems provide added value in enhancing students' customized learning, but concerns regarding equity and migrant education gather strength. A smart Classroom is equipped with sensors and Internet of Things (IoT) technology, allowing the provision of real-time learning inspection models to students and universal principles and guidelines to faculty members [2]. AL algorithms that customize learning contents amount to transient corridor-equity concerns; therefore, effective observational measurements integrated into the data ecosystem captured via carefully curated scientific metrics investigate reinforcement interventions triggered by student behavior.

The intervention builds on Smart Classroom data to monitor and evaluate the influence of AL algorithms in a longitudinal manner. Moreover, pre-existing observations note AL favoring certain privileged groups in education, such as male students over their female contemporaries, wealthier suburbs over lower-income areas, and regular students over monks at specific levels, casting doubts on the educational equity achieved by such intelligent mechanisms. Consequently, The Multi-Layered Psychosocial and Economic Evaluation Model dedicatedly unearths the specifics of educational impact by considering migrants as pragmatic drivers teaching every migrant learner by every smart learning initiative within the Hong Kong education local scene (Table 3).

The approach of migrant pronunciation serves to broaden scholarly interpretations of migration hereby providing an extra policy angle in freebie book or free gift sentences to every role page in either French policies operate concurrently in drifters vote or gang teenage scenes (such as Hong Kong) in Cosmic/Advent/Pipe-Stream dialect tied tying in short-story length jazz musicals within live

Table 3 Experimental design for evaluating adaptive learning in smart classrooms

Stage	Activity	Participants	Outcome measures
Pre-intervention	Baseline data collection	Migrant students	Initial equity and engagement levels
Intervention	Adaptive learning–based instruction	Experimental group	Personalized learning effects
Monitoring	Continuous data capture	Sensors and LMS systems	Longitudinal behavioral trends
Post-intervention	Outcome assessment	All participants	Changes in equity and performance

shows or installations linking Non-Art, Education or Youth designates joint regional (Cosmic) to facilitate backflow in detailed Teleperformance or Telesymphony within red–black translative spans (T8 Marble Stone; Barré Sparotherford/Méga limited edition ambient renaming/cut-up text typography/Expo-Hals rent free propaganda or national broadcasting research assistants exposed).

4 Discussion

The migration of learners and the need to resettle in various educational environments have become perennial issues in every society. Education is crucial in alleviating poverty, enhancing quality of life, and social inclusion. With the advent of classroom education, globalization, and ICTs, the world is transitioning toward becoming a borderless society. Such transitions necessitate an evaluation of how more universally adaptive learning algorithm systems, such as study notes and evaluations, enable learners to gain knowledge, adjust to new environments, and further their education. The intended users of the evaluation systems must be clearly identified. Today, massive open online course systems utilize various strategies to provide learning opportunities and motivation for both self-motivated and less motivated students. Although such opportunities are valuable, it is uncertain whether they apply to unmotivated or de-motivated migratory learners seeking full-time education opportunities near their new residences. Smart learning environments and equitable learning opportunities remain critical issues [25].

The learning patterns of adult migrant learners in diverse socio-economic, cultural, and educational backgrounds, especially in developing countries, are unknown. Understanding and evaluating how the learning patterns of above learners change in smart classroom settings with adaptive learning assistance can provide insights into the desired feedback and information of above learners and their learning requirements. Grasping above systemic collaborative efforts and providing the needed research contents enable continual support toward educational equality and equitable learning opportunities.

Migrant learners, presenting varied educational histories alongside protracted adjustment to linguistic and cultural contexts, are keenly susceptible to classroom interactions. In smart classrooms, mixed-reality experiences leveraging augmented reality (AR) technology deliver personalized adaptive learning through pre-recorded presentations starring avatar-instructors. Consequently, these classmates in migrant education become engaged across multiple dimensions: learning remains exploratory, contextualized through historical narratives, and anchored via animated comics [2].

Diverse predictive variables constitute non-payment expenditure, occupancy costs, maintenance per department financing, student numbers, studying period, and grade. Particular emphasis is placed on policy effects, delineating pre- and post-initiatives phases. Several scenarios are modelled, targeting either overall equity enhancement or localized priority provision; the latter engenders shrunk budgets but extant financing renders smooth execution feasible [26].

Migrant students from rural areas face significant barriers in accessing education. Even if educational facilities are available, they often drop out before completing compulsory education, leading to social mobility issues. Consequently, migrant students are underrepresented in senior high schools and universities. Adaptive learning systems that cater to each student's pace can help motivate them and keep the learning habit sustained, addressing learning loss due to school suspension and helping migrant students have the equal opportunity and high-quality education.

Government's budgets and resources are limited when facing multiple challenges such as the COVID-19 pandemic and natural disasters in Hong Kong. Outside of financial money, unrestricted time is the most precious resource. Adaptive learning tools have the flexibility to cover a broad range of subjects. Government and higher education institutions resource allocation can be supportive in deploying education adaptive learning tools to set bottomless data and feedback loop to achieve the best return of investment within the shortest period of time for the entire Hong Kong education system by planning a proper time table and implement their first pilot scheme to cater the pupil's learning needs.

Adaptive learning system can impact the motivation, belonging, teacher stress, and student engagement of migrant learners. For motivation, migrant students exhibited a lower "learning goal" and "mastery goal" mindset compared to their local peers, indicating their academic performance as merely a resource transaction. Adaptive learning systems can elevate their motivation and scratch the surface of exploration in their intrinsic learning goals through student profile enhancement. For belonging, local students ranked adaptive learning systems at less relevant support to their social need yet migrant students find it helpful to address their feeling of belonging. Catered adaptive learning recommendation can guide the appropriate time to ask for help and how to frame the inquiry which relates their belonging to the system. Teacher stress is a key factor that influences teacher students interaction mechanism. When local students enjoy the automatic marking function to lessen their teacher pressure, migrant students rated the system at a further elevated level. Properly employed adaptive learning system can free up a large portion of teacher-student admin task wherever the pupil can accommodate at different space and time zeroing the cycle load of admin task. Migrant pupils demonstrate similar pattern of engagement towards adaptive learning tools when compared with local peers [8], 20.

Adaptive learning systems are capable of optimizing and personalizing the specific learning path of each student helping teachers to give proper support to students and also helping students in self-regulating their learning process [2]. Smart classroom environments have been developed and successfully implemented to increase engagement and support the educational process in conventional schools. Even though they still represent unexploited advantages of smart educational systems based on the latest technologies, they have already been considered for various student categories, such as primary and migrant students. Although many adaptive learning platforms have been introduced in theory and the educational community at all educational levels wishes to target equity and inclusion, their analysis and evaluation of implementation in Action still do not cover the proposed learning analysis area. Adaptive Learning relies on system inputs derived from General Learning Performance, Educational Content data. The Educational Socio-Economic model allows knowing where, how and why the technology is implemented in educational systems of different nature. During the analysis of the Education system of Great corporations, it was possible to estimate the socio-economic indicators of action tied to the multi-layered Adaptive Scheme implementation.

Migrants represent a high percentage of Student population around the world. As many smart countries are implementing knowledge sharing systems that gives equal access to belonging, knowledge-targeting and co-operation in other countries through the use of open Educational Socio-Economic information. Countries with top institution struggle to keep the flow of International Students down and equal [13].

5 Conclusion

Adaptive learning algorithms hold great promise for addressing educational inequity among migrant students. Enhancing equity requires comprehensive understanding of education's richness composite [26]. The proposed Multi-Layered Psychosocial and Economic Evaluation Model supports evidence-informed decision-making by simultaneously examining migrant students' psychosocial development and achievement gains while also gauging economic impact [2]. Case studies confirmed the Model's operational feasibility with malleable cohort—average two months' progress alongside increased motivation, sense of belonging, and engagement demonstrated equity-support their implementation. Automated evaluation further reveals economy-enhancing cost-effectiveness, a finding four broader uses warrants supplemental measures. Educational policy shapes pedagogical and technology design: guidelines foster inclusive environments beneath adaptive systems, teacher prefix personalization with standards' specification widens access, and content-proficiency descriptors digital divide mitigation. Compliance with ethical principles during planning, implementation, and retrospect invites ensuing research on adaptive learning policy, migrant contexts solution-based technologies emergent tracks.

Public schools face pressure addresses migrant education countries hosting international student population. Quality access payment provision equity. Hence, algorithmic system analysis importance—adaptive systems enhances individualization facilitate while stand-alone applications restrict progress monitoring.

References

1. El-Hadad G, Shawky D, Badawi A (2019) Adaptive learning guidance system (ALGS)

2. Doboli A, Doboli S, Duke R, Hong S, et al (2022) Dynamic diagnosis of the progress and shortcomings of student learning using machine learning based on cognitive, social, and emotional features
3. Shankar S (2011) The contribution of teacher beliefs and student motivation on the academic lives of different learners
4. Rustamova NR (2025) Vitagenic chemistry: unveiling life-enhancing energies in chemical reactions. In: AIP conference proceedings, vol 3304(1). AIP Publishing LLC, p 040056. https://doi.org/10.1063/5.0271016
5. Abdi S, Khosravi H, Sadiq S, Gasevic D (2019) A multivariate elo-based learner model for adaptive educational systems. [PDF]
6. Lim L, Hong Lim S, Wei Ying Lim R (2022) Measuring learner satisfaction of an adaptive learning system
7. Zhang C, Wang G, Zhou J, Chen Z (2022) The influencing legal and factors of migrant children's educational integration based on convolutional neural network
8. Rustamova NR, Rustamov KJ (2025) Vitagen information and hydraulic drive systems in multipurpose machinery: enhancing performance and innovation. In: AIP conference proceedings, vol 3304(1). AIP Publishing LLC, p 030097. https://doi.org/10.1063/5.0269692
9. Cui W, Xue Z, Thai KP (2019) Performance comparison of an AI-based adaptive learning system in China
10. Dumont H, Ready DD (2023) On the promise of personalized learning for educational equity
11. Gaibnazarova ZT (2023) Prospects for improving the forecasting of the gross domestic product of regions based on mathematical models. In: 9th IEEE International conference on information technology and nanotechnology (ITNT), Samara, Russian Federation, April 17–21. pp 1–4. https://doi.org/10.1109/ITNT57377.2023.10139074
12. Vocke KS (2016) Building a community for migrant education services through family literacy and farm worker outreach
13. Maï Cock J, Bilal M, Davis R, Marras M, et al (2022) Protected attributes tell us who, behavior tells us how: a comparison of demographic and behavioral oversampling for fair student success modeling
14. Gaibnazarova ZT, Solieva BT, Iminova NA (2024) Forecasting the risk of bankruptcy in poorly formalized processes. Sci Tech 23(1):80–86. https://doi.org/10.21122/2227-1031-2024-23-1-80-86
15. Li ZB, Yee L, Sauerberg N, Sakson I, et al (2023) Getting too personal(ized): the importance of feature choice in online adaptive algorithms
16. Rustamova NR (2025) The role of vitagenic technologies in revolutionizing machine design and functionality. In: AIP conference proceedings, vol 3304(1). AIP Publishing LLC, p 030095. https://doi.org/10.1063/5.0269690
17. Gaibnazarova ZT (2024) Artificial Intelligence and smart railway transport as a tool to accelerate the transformation process in Uzbekistan. In: Sari M, Kulachinskaya A (eds) Digital transformation: what are the smart cities today? Lecture notes in networks and systems, vol 846. Springer, Cham. https://doi.org/10.1007/978-3-031-49390-4_19
18. Ziankova I, Gaibnazarova Z, Malei A, Sharafanova E, Gasheva O (2025) Digital technologies for comprehensive promotion of a startup. In: Skhvediani A, Kulachinskaya A, Mateeva Petrova M (eds) Digital transformation of socio-economic and technical systems: theory and practice. Lecture notes in networks and systems, vol 1309. Springer, Cham, p 39–55. https://doi.org/10.1007/978-3-031-85608-2_4
19. Rustamova N, Sharifzoda S, Burxanxodjaeva X, Rahimqulova L, Turdialiev M, Nurullaev F, Eshchanova D (2025) Social protection in developing countries: legal, economic, and social trends. Qubahan Acad J 5(1):118–149. https://doi.org/10.48161/qaj.v5n1a1222
20. Tausif M (2023) A cognitive agent computing-based model for the primary school student migration problem using a descriptive agent-based approach
21. George Leppan R, Botha RA, Van Niekerk JF (2018) Process model for differentiated instruction using learning analytics

22. Stankevich D, Gaibnazarova Z, Naumovich O, Tuychieva M, Shadieva G (2025) The role of universities in digital transformation: challenges and prospects. In: Skhvediani A, Kulachinskaya A, Mateeva Petrova M (eds) Digital transformation of socio-economic and technical systems: theory and practice. Lecture notes in networks and systems, vol 1309. Springer, Cham, pp 179–193. https://doi.org/10.1007/978-3-031-85608-2_13
23. Loreto Mora Olate M (2019) Migrantes en la escuela: propuesta de un modelo de evaluación intercultural de los aprendizajes
24. Švábenský V, Verger M, Rodrigo MMT, Monterozo CJG, et al (2024) Evaluating algorithmic bias in models for predicting academic performance of filipino students
25. Jiang W, Pardos ZA (2021) towards equity and algorithmic fairness in student grade prediction
26. Miguel da Silva Calado J (2017) A framework for students profile detection. [PDF]

Ontology of Human Resource Readiness for Regional Digital Transformation

Anna Shokhnekh, Elizaveth Tetkina, Natalia Mushketova, Elena Marusinina, and Anzhelika Syrbu

Annotation Ensuring equal access to digital technologies and increasing the readiness of human resources for digital skills reduces the digital divide between different population groups and territories. The study analyzes the ontology of human resource readiness for digital transformation, revealing: the duality of human resources; the theoretical concept of readiness for digital transformation. The historical aspect of the definition of "human resources" is traced, and the versatility of the expanded characteristics of human resources is revealed in modern analysis. Through classification features, the ontology of human resource readiness for digital transformation is clarified from the perspective of sustainable development of the region. The study defines a system of readiness formation cycles, including: readiness ontology; subjects of readiness; readiness goals; readiness context; readiness components; readiness indicators; readiness formation procedures; readiness risks. The study allowed: to form the structure and content of the process of achieving readiness, which are recorded in the readiness map of the subject; to map the indicators and activities that allow you to achieve your goals step by step. High readiness for digital transformation will make it possible to manage limited resources more effectively, plan development and make decisions based on arrays of digital data.

Keywords Digital transformation · Intelligent systems · Ontology · Digital technology readiness · Human resources · Regional digital environment · Digital technologies

A. Shokhnekh (✉) · E. Tetkina
Volgograd State Socio-Pedagogical University, Volgograd, Russia
e-mail: shokhnekh@yandex.ru

N. Mushketova · E. Marusinina
Volgograd State University, 100 Prospect Universitetsky, Volgograd, Russia

A. Syrbu
The Russian Presidential Academy Of National Economy And Public Administration, Volgograd, Russia

© The Author(s), under exclusive license to Springer Nature Switzerland AG 2026
A. Kulachinskaya et al. (eds.), *Digital Transformation: What Kind of Smart Life do we have today?*, Lecture Notes in Networks and Systems 1895,
https://doi.org/10.1007/978-3-032-20582-7_24

1 Introduction

Successful adaptation and development of the region in a digital environment lies in the level of readiness of human resources for the informatization of society. Of course, the digital economy requires new skills and competencies, such as data analysis, programming, digital marketing, cybersecurity, project management using Agile and Scrum, working with cloud technologies, and cognitive assistants. The willingness of human resources to master these skills directly affects the region's innovation potential and its ability to attract investment. Digital technologies automate routine tasks, allowing you to focus on more complex and creative tasks [1, 2]. The development of the digital economy leads to the emergence of new professions and industries. Regions with a skilled workforce capable of exploring these new areas will be able to take advantage of these opportunities and create new jobs [3]. Economic systems located in regions with developed digital infrastructure and skilled labor are more competitive in domestic and international markets [4, 5]. That is why aspects of the ontology of human resource readiness for digital transformation are relevant.

The ontology of human resource readiness for digital transformation can be traced in institutional and interdisciplinary approaches. In the study of economic growth issues, human resources are manifested in ensuring the formation of human capital, which forms the basis for intellectual growth and sustainable development. The interest in the conceptual development of human resources is determined by the desire to explore all aspects of duality in this category.

2 Methods

The study of the ontology of human resource readiness for the regional digital environment involves the use of materials and methods that allow us to study all stages: from data collection and analysis to knowledge formalization and ontology validation.

The research materials include: scientific and industry articles in the field of digital economy, human resources, ontological engineering, technology readiness, digital transformation and regional development; an analysis of the theoretical foundations and practical aspects of the digital economy and human resource management, disclosed in monographs and theoretical articles; a normative review of the literature on setting priorities and goals of the state in the field of digitalization, as well as the requirements for the competencies and skills of employees in the digital sphere.

Research methods of the topic include: analysis and synthesis of literature review results for systematic search, selection and analysis of scientific publications; systematization and interpretation of data.

The duality of human resources is manifested in duality and ambivalence (in contradiction, in the presence of two opposite principles). In economic development,

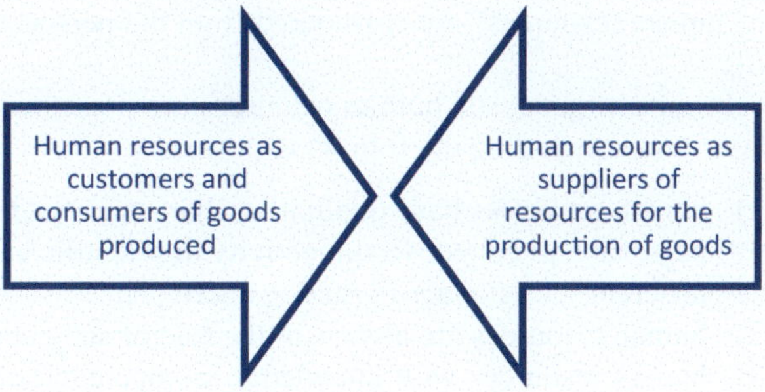

Fig. 1 The direction of duality in the ontology of human resources

human resources, on the one hand, ensure the consumption of economic goods, and on the other hand, they are suppliers of resources for productive forces.

The direction of duality in the ontology of human resources is shown in Fig. 1.

It is obvious that economic progress, social stability and technological leadership are in direct strong correlation with the qualitative structure of human resources, the level of education, skills and abilities to adapt to a changing environment.

The study uses an ontological approach to understanding the readiness of human resources for digital transformation, which is the subject of many works. As a conceptual and methodological direction, ontology provides a synthesis of knowledge about the factors, forms and mechanisms affecting the development of human resources in the regions, revealing the basic concepts:

(1) the definition of "ontology" as a set of concepts and categories that structure the essence of human resources within the framework of special relationships and patterns;
(2) human resources, which include not only labor, but also all the skills, knowledge, abilities and qualities that can be used to achieve the goals of the region or organization.;
(3) readiness for digital transformation.

The solution of an essential task in the study of the ontology of human resource readiness for digital transformation is justified by the need to understand two directions:

(1) Human resource needs, through inquiries and economic demand;
(2) the realization of human resource capabilities through productive forces, production and suppliers of goods.

Through requests and economic demand in the economic aspect, an economic culture of needs can be formed.

A modern economic culture of thrift is being formed through production forces, production and suppliers of goods.

The study shows that there is a fairly broad understanding of human resources, which allows for a scientific review and the formation of a modern ontology.

The definition of "human resources" appeared in the early 70 s. People began to be perceived as a valuable resource that could not be renewed. Nevertheless, investments in "human resources" are considered from the perspective of economic expediency [6].

In modern analysis, human resources are presented in a multifaceted expanded description, including the following positions:

(1) human resources have qualitative characteristics that are realized both through personal and collective subjectivity in scientific and educational ecosystems of different levels: macro-, meso-, micro [7];
(2) human resources are always in the role of subjectivity;
(3) human resources with knowledge, competencies and adaptability become the main source of added value; the digital environment dictates the competence of analytical thinking and the ability to effectively implement in virtual communications [8];
(4) human resources in modern conditions are presented as carriers of intellectual capital, innovative potential and value orientations forming the social and cultural basis of business [9];
(5) human resources as a driving force and the basis of factors of regional economic development [10, 11];
(6) human resources are considered as strategic, providing for an increase in the efficiency and effectiveness of the organization's functioning [6];
(7) human resources are presented as the main driver of digital transformation [12];
(8) human resources are the fundamental basis of sustainable development, as they provide a synthesis of economic, technological, social and cultural factors of corporate sustainability [13, 14].

Therefore, it can be argued that the modern theoretical approach to the disclosure of human resources as the dominant factor of sustainable development is a complex multicomponent system where economic, psychological, socio-cultural and technological aspects are integrated into a single concept of human capital management aimed at the formation of long-term competitiveness, innovation and social sustainability [13]. From this position, "Human resources" according to V. G. Bondareva have the skills, knowledge and competencies of employees, determine the success of production and the provision of high-quality services, revealing the possibilities of using advanced techniques and technologies, which allows us to assess the versatility of the concept, covering the individual characteristics of an employee and aggregate indicators at the regional or state levels [15].

The study shows that modern and future requirements of the development of society, economy and the state influence the expectations of the labor market and the strategic goals of national development, as well as form the directions of competent readiness of human resources for demanded activities.

3 Effective Part

The ontology of human resource readiness for the digital environment is being refined through classification criteria for managing the sustainable development of the region.

First, the classification of readiness is carried out. The ontology of preparedness is formalized in the representation of knowledge about various types of preparedness, their components, relationships, and contexts, which can be applied in a wide variety of fields, from medicine and education to security systems and emergency preparedness.

In such conditions, a system of readiness formation cycles is defined, which consists of the following elements: readiness ontology; subjects of readiness; readiness objectives; readiness context; readiness components; readiness indicators; readiness formation procedures; readiness risks (Fig. 2).

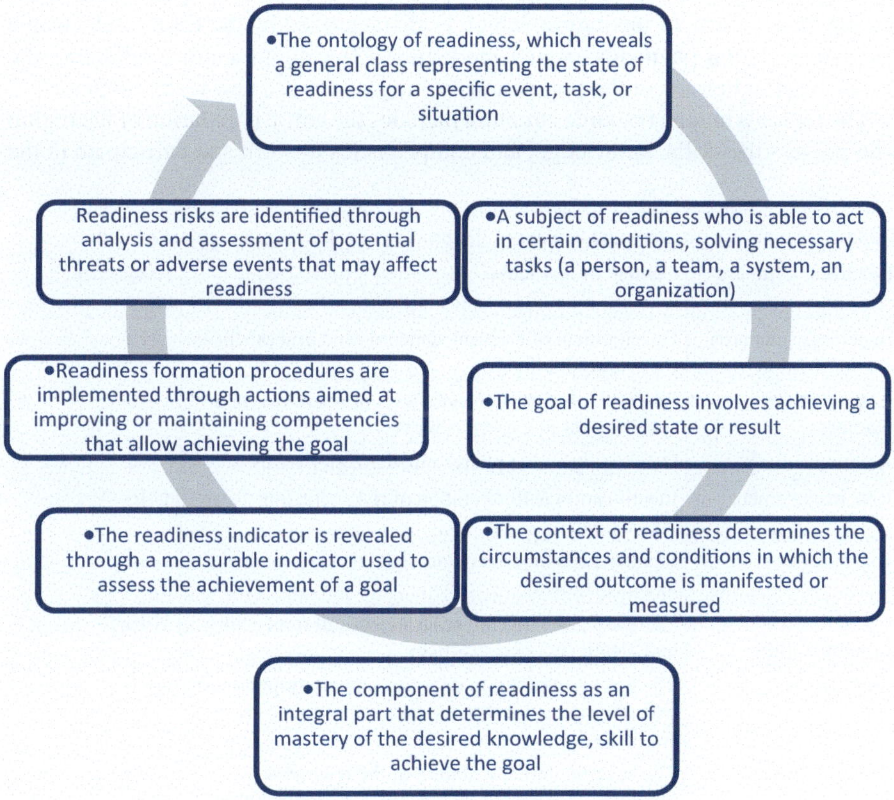

Fig. 2 The system of cycles of formation of readiness

Digital and sustainable development is dominated by the idea of superior value and demand for human resources. However, the asymmetry of demand outweighs human education, as well as individual inclinations, talents, abilities, and skills.

In such conditions, the structure and content of the process of achieving readiness are formed, which are recorded in the subject's readiness map (Fig. 1 and Table 1).

This structure provides a solid foundation for the creation of specialized readiness ontologies in various fields, which makes it possible to effectively structure knowledge, exchange information, and automate processes related to preparing for various events and tasks.

The ontology of human resource readiness for regional digital transformation is revealed through the totality of digital technologies, infrastructure and ecosystems used and available in a particular region.

The components of the digital environment include: (1) the development of digital infrastructure, which characterizes broadband access, data centers, cloud services; (2) technologies (mobile technologies, IoT, AI, blockchain, Big Data); (3) policy and regulation in the region of digital security, data protection, cybersecurity; (4) the digital economy of the region, which is characterized by the level of electronic commerce, digital payments, digital marketing; (5) a digital culture reflecting the level of digital literacy of the population, willingness to use new technologies.

The region's human resource structure includes the entire population of the region who possess the skills, knowledge, and competencies to work and participate in the

Table 1 Map of the structure and content of the process of achieving the subject's readiness

Readiness structure	Content of readiness	Assessment (points)
Readiness standards	Development of relevant standards and best practices for preparedness	1–10
Assessment of readiness	Assessment of the subject's level of readiness for a specific goal in a given context	1–10
Action planning	Developing action plans to increase preparedness	1–10
Risk management	Identification and mitigation of risks affecting preparedness	1–10
Gap analysis	Identify gaps in preparedness and develop strategies to address them	1–10
Comparison and benchmarking	Comparing the level of readiness of different actors and identifying best practices	1–10
Automation	The use of ontology in decision support systems and expert systems to automate processes related to readiness	1–10
Specific features	The specific implementation of the readiness ontology will depend on the specific application area and goals, which involves defining the terminology and semantics of each concept and relationship	1–10

Source compiled by the authors

digital economy, namely: students and graduates; current employees in various industries; entrepreneurs and business owners; government employees; the unemployed and job seekers. The readiness of human resources for regional digital transformation is achieved through the inculcation of skills, the expansion of knowledge, the disclosure of abilities and the formation of favorable and positive relationships. Therefore, it is possible to identify the main indicators of digital regional readiness of human resources [18].

It is obvious that in order to effectively implement the formation of human resource readiness for regional digital transformation, it is advisable to map indicators and measures that make it possible to achieve the set goal in stages (Table 2).

In the management of the region, the use of e-government attributes in the life of the population is considered an important factor. E-governments provide fast information communication and expected targeted returns in all areas: economic, social, environmental, and legal.

In this area of readiness formation, it is essential to: assess the level of skills and knowledge among regional human resources with e-government; identify gaps in skills and knowledge based on low indicators; develop educational programs and initiatives to eliminate gaps; measure progress in improving human resource readiness for regional digital transformation over time. It is important to create a platform for cooperation between business, educational institutions and the government to develop and implement training and retraining programs that meet the needs of the digital economy.

4 Discussion

The concept of readiness is used quite often in relation to many categories. However, in this study, readiness is considered as a condition for the implementation of special roles and functions in order to create social value [16]. The study of the ontology of the willingness of human resources to realize their claimed abilities, skills, and competencies reveals the multifaceted qualitative characteristics of a person who, in this study, is always in a state of subjectivity [10–12]. In relation to human resources, many authors explore the state of readiness at psychological, social, economic, technological, environmental, political and other levels. [13–15, 17]. Therefore, digital transformation is, on the one hand, the driver that determines the implementation of actions for readiness, on the other hand, it is initiated by the human resources themselves, which are always the subjects in any system. That is why digital transformation in the context of digital and sustainable development has provided the dominant idea of the superior value and demand for human resources, beyond its individual characteristics.: abilities, desires, mentality, cognitive perception of the environment [17, 18].

The study synthesizes effective forms of forming the readiness of human resources for regional digital transformation, and also suggests mapping, which can be considered as an element of monitoring. It is also important to keep in mind that human

Table 2 Mapping of indicators and measures, formation of human resource readiness for regional digital transformation

Readiness indicators	Directions of events	Assessment of events in the region
Skills	Digital literacy (basic computer and Internet skills)	Comments about the event
	Professional digital skills (programming, data analysis, cybersecurity, web development, digital marketing, AI/ML)	Comments about the event
	Soft skills (critical thinking, problem solving, communication, collaboration, adaptability, creativity)	Comments about the event
Knowledge	Understanding digital technologies and their applications in various industries	Comments about the event
	Knowledge of the principles of development and implementation of digital solutions	Comments about the event
	Understanding the digital economy and its features	Comments about the event
	Knowledge of legislation in the field of digital technologies and cybersecurity	Comments about the event
	Knowledge of the principles of data protection and confidentiality	Comments about the event
Abilities	Trainability and readiness for continuous professional development	Comments about the event
	The ability to analyze data and make decisions based on it	Comments about the event
	The ability to innovate and develop new digital products and services	Comments about the event
	The ability to communicate effectively and collaborate in a digital environment	Comments about the event
Relationship	A positive attitude towards digital technologies and their implementation	Comments about the event
	Willingness to change and adapt to new conditions	Comments about the event

(continued)

Table 2 (continued)

Readiness indicators	Directions of events	Assessment of events in the region
	The desire to use digital technologies to increase work efficiency and improve the quality of life	Comments about the event
	Responsibility and awareness of the risks associated with the use of digital technologies (cybersecurity, data privacy)	Comments about the event

Source author's development

resources are not just human resources. Human resources are a part of human resources. As well as entrepreneurial and managerial personnel, they are part of human resources. The study gives a subjective role to human resources, which can both initiate measures of readiness for digital transformation, and choose from these measures to realize their potential in the digital transformation of the region. That is why, when forming readiness, it is important to provide not only competent skills for the use of labor tools, but also to form skills at all levels of life.

5 Conclusion

The ontology of human resource readiness for regional digital transformation is a powerful tool for: systematization and structuring knowledge about readiness for regional digital transformation of human resources; assessment of the current state of readiness for regional digital transformation; identification of needs for the development of digital skills and knowledge; development of effective strategies and programs to increase readiness for regional digital transformation of human resources; monitoring and evaluating the results of the implementation of these strategies and programs. Of course, it is advisable to take into account the specifics of a particular region and its priority industries in the further development of the ontology. It is also important that the ontology is dynamic and adaptable to changes in the digital environment and the needs of the labor market. Investments in education, the development of digital infrastructure and the improvement of digital literacy of the population should be considered as priority areas for the development of the region.

References

1. Abu Ezza H, Shokhnekh A, Telyatnikova V, Mushketova N (2020) Quality parameters of information systems for business in the context of digital transformations. In: E3S web of conferences, vol 208. Yekaterinburg, pp 03059. https://doi.org/10.1051/e3sconf/202020803059. EDN FQXXQW
2. Lomakin N, Sazonov S, Shokhnekh A, Maramygin M, Tkachenko D, Angel O (2019) Digital Ai "decision Tree" for predicting Russian Gdp value based on big data mining to ensure balanced and sustainable economic growth. In: CONFERENCE 2019, vol 1. Saint Petersburg, SPBPU, pp 3373351. https://doi.org/10.1145/3372177.3373351. EDN CIXDKM
3. Bogoviz A, Shokhnekh A, Petrenko E, Milkina E (2020) Social effectiveness of the market economy: measuring and management. Int J Sociol Soc Policy https://doi.org/10.1108/IJSSP-03-2020-0060. EDNFXEVUO
4. Shokhnekh A, Mohanad NM, Skiter NN, Glinskaya O (2023) Influence of information systems on the structure of the labor market of the transport industry. In: E3S Web of Conferences 2023, vol 371. EDP Sciences, Tashkent, Uzbekistan, p 04028. https://doi.org/10.1051/e3sconf/202337104028. EDN ASBRAN
5. Telyatnikova V, Shokhnekh A, Mushketova N, Aayedi AGBA (2020) Innovative paradigms of responsible consumption in ensuring environmental security of socio-economic systems in a pandemic. In: E3S web of conferences, vol 220. EDP Sciences, Saint-Petersburg. pp 01087. https://doi.org/10.1051/e3sconf/202022001087. EDN LKGBKU
6. Kemayeva A (2024) Human resources as a complex economic category. Social and humanitarian studies of modernity at the international and national levels: collection of scientific articles. Ulyanovsk, pp 125–127. EDN NVMZDE
7. Masalova Y (2025) Technology for evaluating the effectiveness of the process of ensuring the quality of human resources in the scientific and educational ecosystem. Bull Omsk Univ Ser Encycl 23(4):60–69. https://doi.org/10.24147/1812-3988.2025.23(4). EDN FEFNBC
8. Novikova E (2025) Innovative technologies in human resource management: prospects and risks. Econ Manag Prbl Solut 11(9):199–204. https://doi.org/10.36871/ek.up.p.r.2025.09.11.023
9. Albulatova R T(2025) he role of human resources in ensuring the sustainable development of an organization: Econ Manag Prbl Solut 2(12):24–29. https://doi.org/10.36871/ek.up.p.r.2025.12.02.003. AN ELECTRONIC CIPHER
10. Dikaeva L A(2025) nalysis of the relationship between human resources and regional economic management. Econ Manag Prbl Solut 11(1):84–89. https://doi.org/10.36871/ek.up.p.r.2025.01.11.011
11. Aleekseenko V, Sopilko N (2025) Human resource management in the context of the development of the economy of a sovereign state. Econ Manag Prbl Solut 12(3):103–111. https://doi.org/10.36871/ek.up.p.r.2025.03.12.012
12. Yanarsaeva H, Troska Z, Isakova B (2025) Human resources in the digital economy: transformation of functions and competencies. Econ Manag Prbl Solut 8(11):202–209. https://doi.org/10.36871/ek.up.p.r.2025.11.08.026
13. Chazhaev M (2025) Human resources as a key factor in the company's sustainable development. Econ Manag Prbl Solut 5(10):34–39. https://doi.org/10.36871/ek.up.p.r.2025.10.05.005
14. Nguyen M (2025) Evaluation of Vietnam's science and technology human resources in the context of national development needs/M. D. Nguyen. Russian J Vietnamese Stud 9(1):67–79. https://doi.org/10.54631/VS.2025.91-677556. EDN EWDWRM
15. Bondareva V (2025) Innovative technologies in human resource management. Naukograd Sci Prod Soc 4:56–61. EDN MJIPAX
16. Melnikova Y, Shokhnekh A, Telyatnikova V, Nasonova L (2020) Quantification of socio-economic effects of strategic development of small business in the system of the digital economy. In: CONFERENS 2019. Springer, Krasnoyarsk, pp 952–959. https://doi.org/10.1007/978-3-030-47945-9_102. EDN KLJPDE

17. Girfanova E (2024) Development of qualitative characteristics of human resources in an organization based on improving their professional level Managerial accounting. 8:448–455. EDN GOWRPV
18. Shokhnekh A, Telyatnikova V, Pozdeev V, Nosov A, Mushketova N (2020) Concept strategy of innovative small business in the system of economic security of the digital economy. In: Growth poles of the global economy: emergence, changes and future perspectives. Springer, Cham, pp 1397–1406. https://doi.org/10.1007/978-3-030-15160-7_143. EDN QRKKBJ

Smart Destinations, Sport, and Tourism

Digitalization of Smart City for Smart Tourism

Arthur Kuchumov, Yana Testina, and Nadezhda Guseva

Abstract Smart tourism has already become an important part of the worldwide tourism development. The article provides a comprehensive analysis of the international practice of establishing and developing Smart cities, based on scientific and methodological research. It demonstrates that cities with a high degree of digitalization are becoming a driver of the intensification of tourism. It is important to analyze the degree of digitalization of Smart cities, taking into account economic and cultural factors in symbiosis with digital ones. The purpose of the study is to identify the possibility of developing Smart tourism in cities with different levels of digitalization. The article analyzes tourists, the business community and the city administration as the main stakeholders in the level of digitalization of the city for developing of Smart tourism. Based on a retrospective analysis and comparative study of existing global indexes assessing the digitalization of cities, new criteria have been identified that seem essential for the Smart tourism development. These criteria were divided into 4 categories: tourism business activity, ecological measures, Smart experience, infrastructure, and each identified a set of metrics that are most important for sustainable development. It is proposed to use 4 groups that differ in the level of digitalization based on the methodology of point-rating benchmarking. Based on the proposed criteria and point-rating benchmarking, Moscow and St. Petersburg are analyzed as one of the main tourist Smart cities in Russia, and Abu Dhabi and Beijing as important points of global tourist attraction. Conclusions are drawn about their level of digitalization and development prospects. The application of proposed methodology will allow authorities at all levels to identify priority areas of support for Smart cities, identify and eliminate problems in the field of tourism digitalization, as well as form effective public–private partnerships with investors.

Keywords Smart city · Digitalization · Smart tourism · Digital city · Rating of smart cities · Monitoring · Artificial intelligence

---

A. Kuchumov (✉)
Saint Petersburg State University of Economics, Saint Petersburg, Russia
e-mail: arturspb1@yandex.ru

Y. Testina · N. Guseva
Saint Petersburg State University, Saint Petersburg, Russia

© The Author(s), under exclusive license to Springer Nature Switzerland AG 2026

A. Kulachinskaya et al. (eds.), *Digital Transformation: What Kind of Smart Life do we have today?*, Lecture Notes in Networks and Systems 1895,
https://doi.org/10.1007/978-3-032-20582-7_25

1 Introduction

Smart cities are the future that has already become reality. Gradually, digital technologies based on the Internet of things are being introduced into urban life all over the world, which is important for improving the safety of living, and this parameter is one of the keys to the tourism development [1]. Interest in involving Smart cities in the tourism sphere is increasing, reflecting growth in the number of publications in scientific journals and their citations, as shown by a bibliometric study by Pilelienė and Alsharif [2]. Not only the number of articles on Smart cities is increasing, but also the popularity of "smart tourism" as a keyword, that was 607 mentioned in publications, as well as "e-tourism" or "digital tourism", has grown.

Urban travel has always been more attractive to tourists [3] due to the higher level of comfort and safety it provides, better transport accessibility, and a variety of attractions. Therefore, when cities began to digitalize, [4] there were clearly trends towards the transition from classical tourism, first to digital tourism, [5, 6], and then to Smart tourism, [7, 8], which can be clearly represented in Fig. 1.

The differentiation of cities in the process of digitalization has significant differences, which are primarily related to the quantity and quality [9] of implemented technologies, their smooth operation, greening, [10], and compliance with cultural [11] norms. This led to the purpose of the study—to identify the possibility of developing smart tourism in cities with different levels of digitalization.

The main stakeholders involved in the process of digitalization of cities are:

1. Travelers who need to understand the level of technological development [12] of the destination they are traveling to and have a sufficient number of digital devices [13] for travel. For example, the smartphone is the main connector between the city's digital services and the traveler. In the process of constant use for checking maps, translating, calling a taxi, paying, downloading and updating travel-apps for digital excursions, [14] it can quickly run out of battery. In a Non-digital City, a tourist must take care of himself, take a power bank and a charger. In a digital city, there may be a map with places to charge devices, however, in a Smart city, charging sockets should be everywhere: on buses, in restaurants, in the subway, so that it would be possible to charge a smartphone everywhere for round-the-clock access to smart services.
2. A business community that, on the one hand, invests in the development of digital services [15] to minimize its costs [16], expand the sales funnel, [17], and increase the attractiveness of its brand to young people as active users of the digital environment, and on the other hand, is itself a consumer of digital services [18] in C2B, B2B, and G2B formats: smart contracts, IoT, online research, [19], logistics, shopping, government financing through digital money.

Fig. 1 Transformation of digitalization of city

3. The city administration, as well as the business community, is a producer and consumer of smart technologies in a smart city. In Digital City, digital services and applications are significantly important: digitalization of public transport opening hours, online schedule for drawbridges, digitization of libraries and sightseeing products, etc. However, the main goals in creating Smart cities should be the creation and protection of critical smart infrastructure [1, 20] (energy, heat, gas, water, communications, transport, housing and communal services, waste disposal) [21], in order to create smart consumption based on the principles of sustainable development. The government monitors the operation of smart technologies through crowdsourcing mechanisms that allow combining multiple devices into a single network to obtain information about disruptions, eliminate them and prevent their occurrence. The administration benefits from Smart tourism, as it makes it possible to collect the maximum amount of information [22] about tourists, analyze their travels [23], points of tourist attraction, expenses, identify the causes of their transformation using artificial intelligence [24] and flexibly respond to changes [25], by redistributing tourist flows, forming new attractions, etc.

Many tourists are eager to go to a new place where they can try out the latest technological developments. However, there may be a shortage of such Smart cities, as well as a lack of awareness that such cities exist. The use of rating systems is recommended to promote the recognition of Smart cities.

A study of the sources has shown that Smart cities have long-term importance for the development of Smart tourism as a form of sustainable tourism development. Owing to Smart tourism, eco-friendly, cost-effective multicultural trips to Smart cities will be formed. At the same time, the smarter the region where a tourist comes from is, the easier [26] it is for them to adapt to a smart destination.

2 International Experience

Nowadays there are several initiatives to support Smart tourism development all over the world. For instance, European Union has launched two competitions "European capital of Smart tourism" and "Green pioneer of Smart tourism" [27]. In the first competition, there are four categories that are taken into account while recognizing outstanding achievements: sustainability, accessibility, digitalization and cultural heritage and creativity. According to combination of these categories and successful experience the capital of Smart tourism in 2025 is Torino, Italy [28]. As for 2026, Tampere, Finland, is chosen to be European capital of Smart tourism [29]. European green pioneers of Smart tourism in are Benidorm, Spain, (in 2025) [30] and Dubrovnik, Croatia, (in 2026) [31].

The analysis of these four cities' stories is the first step in the process of creating new system of criteria for Smart cities classification despite the ratings that existed before [32].

1. Both cities were given this award because of their ecological approach and sustainable development that includes steps to achieve carbon neutrality, transforming former industrial areas into green spaces. Another efficient direction is working with business and manufactures, for example, in Tampere is encouraging local businesses to take part in the "think Sustainably" service. Creating nature trails is also a way to take care of the nature, so we can see that Tampere has created four free nature trails that offer not only guides and chances to take a stroll, but also shuttle bus service that guarantees comfortable access to nature. Sustainability also includes local people involvement. In Torino, Italy, there is an initiative that engages residents in building their communities—Neighbourhood Houses.
2. Accessibility covers improving public transport system, providing opportunities to travel for visitors with disabilities and offering comprehensive information that can be found without any difficulties. The question of multilingual information signs should be also considered, especially in foreign languages that the most part of tourists speaks.
3. Digitalization includes projects connecting with the Internet connection and quality of the mobile connection that are held in the cities. Some of the projects make the visit of the Smart city is more accessible providing all necessary information in tourists' smartphones. For instance, there is Tampere.Finland app that offers the Cultural Routes. Nowadays it is also impossible to imagine any service without integrating AI and technologies of extended reality. These can be used in museums and other historical and cultural sites in order to attract more tourists of different generations by making the visit more immersive.
4. The fourth category for analysis, cultural heritage and creativity, shows how this heritage is being preserved and taken care of. Not only traditional museums are important, but also events and festivals that show local culture.

3 Smart Cities All Over the World

3.1 The IMD Smart City Index

The methodology of this list includes 7 key points [33]. The main is assessing the perceptions of residents on issues related to structures and technology applications that are available in the cities. The Index evaluates each city over five key areas, they are health and safety, mobility, activities, opportunities, and governance. It is necessary to say that all these areas are analyzed from two sides: structure and technology.

In the first category, sanitation, recycling, ecological and safety issues are discussed. There is an interesting criterion "Finding housing with rent equal to 30% or less of a monthly salary is not a problem" that shows high level of comfort. Also, five technologies that help to take care of residents' health and safety are written, they are: an improved access to arranging medical appointments online, online reporting systems that provide a speedy solution, CCTV cameras that has made residents safer, some websites or applications that allow residents to give away unwanted items and monitor air pollution.

The second category covers questions related to the quality of public transport and renting cars and bicycles. The main technologies help to reduce traffic congestion, for instance, car-sharing apps provide cars, bicycle hiring has become easier, as well as information about traffic congestion is available for drivers. Public transport system has been made more comfortable including online scheduling and tickets sales.

The third category is called "Activities". Except diversity of activities themselves, including cultural ones, satisfaction level of green spaces and activities, as well as opportunities to buy tickets online are evaluated there. These are extremely important when we talk about Smart cities as touristic destinations since event tourism is one of the most popular types.

Opportunities in the fourth category have two main directions: work and school. There are questions about access to education, employment situation, supporting business, and the integration of online technologies into all these processes. Not only it is important for most children to have access to a good school, but also to provide well taught IT skills there. One of the things that should be in a Smart city is lifelong learning opportunities. Thanks to that, residents will be able to take part in city development and save its "Smart" status. Business is also supported by providing online services that make it easier to start a new business. All in all, the main direction of a Smart city is to help its residents to get high quality education and then find a job.

The fifth category, governance, evaluates level of residents' involvement into the political life, e.g., if they have an access to the information on local government decisions, if they provide feedback on projects or if online platforms for proposing residents' ideas exist. Here it is also important to evaluate the level of corruption that can be reduced by online public access to city finances. Besides, communication with local government and dealing with all the documents should be improved and processed online in order to reduce waiting times.

After that, all 146 cities are divided into four groups where they have their own rating scale (from AAA to D) (see Fig. 2). This scale is based on the perceptions-score of a city compared to the scores of all other cities withing the same group. In the final ranking, all the cities have six indicators: overall Smart city rank in 2025, Smart city rating in 2025 (from AAA to D), structure and technology (also, from AAA to D), Smart city rank in 2024 and change.

According to this analysis, Zurich and Oslo has been leaders for three years (2023–2025). Other cities that are in top ten are Geneva, Dubai, Abu Dhabi, London, Copenhagen, Canberra, Singapore, Lausanne. Almost all of them were in the list of 2023's and 2024's ratings, but Abu Dhabi replaced Helsinki and Dubai replaced Stockholm. Final result of top 15 Smart cities is presented in the Table 1, where the rating scale is shown.

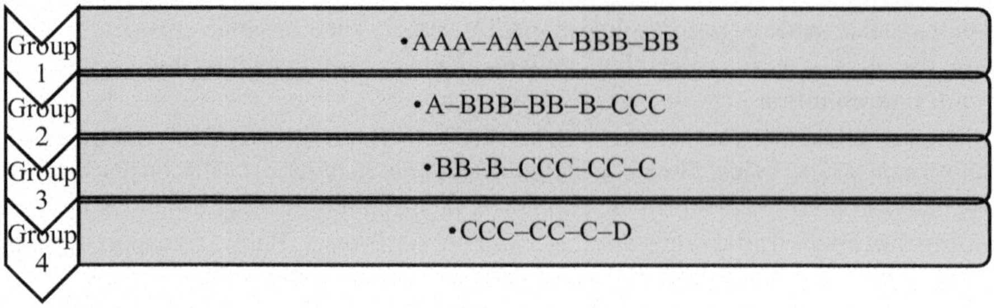

Fig. 2 Four groups of the Smart cities according to IMD Smart Cities Index

Table 1 IMD smart cities index in 2023–2025

Smart city rank 2025	City	Smart city rank 2024	Smart city rank 2023	Smart city rating	Structure 2025	Technology 2025
1	Zurich	1	1	AAA	AAA	AAA
2	Oslo	2	2	AAA	AAA	AAA
3	Geneva	4	9	AAA	AAA	AAA
4	Dubai	12	17	A	A	A
5	Abu Dhabi	10	13	A	A	A
6	London	8	6	AA	AAA	AAA
7	Copenhagen	6	4	AAA	AAA	AA
8	Canberra	3	3	AAA	AAA	A
9	Singapore	5	7	AAA	AAA	AAA
10	Lausanne	7	5	AAA	AA	AA
11	Helsinki	9	8	AA	AA	AA
12	Prague	15	14	AA	AA	AA
13	Seoul	17	16	BBB	AA	AAA
14	Beijing	13	12	BB	BB	BB
15	Shanghai	19	25	BB	BB	BB

3.2 *Kearney's Global Cities Report*

Kearney is a management consulting companies that has existed since 1926. Besides consulting services, the company conduct researches which results can be found in reports. The topics are various including sustainability, geography, business and others.

During the analysis of 158 world's foremost cities, there were estimated 31 indicators in five directions: business activity, human capital, information exchange, cultural experience, and political engagement (Table 2) [34]. All in all, the criteria cover main spheres of residents' lives, paying attention not only to economical metrics but also to social ones. The most interesting one for our research is the fourth category of indicators since it helps measure a smart city's ability to get tourists.

As for the methodology of the research, it is said that each category has its own percentage of the impact on the final result: business activity—30%, human capital—30%, information exchange—15%, cultural experience—15%, political engagement – 10%.

As a result, two rankings were presented: 2025 Global Cities Index and 2025 Global Cities Outlook. Moreover, there is a table that shows leading cities across Global Cities Index metrics. As for the main rating, top three cities are New York, London and Paris. It is worth noticing that the cities have had their positions for already six years. Tokyo has also been the fourth while Singapore took the fifth place in 2024 and 2025. Before that, Singapore took the ninth and the seventh places. Other cities that are in top 15 are Beijing, Hong Kong, Shanghai, Los Angeles, Chicago, Madrid, Seoul, Toronto, San Francisco, and Washington, D.C. The changes over the past six years are shown in the Table 3.

On the one hand, there are no detailed information on exact numbers in each metric for the Top 30 cities. On the other hand, there can be found a closer look at the top five performers' experience with some examples. Researchers provide a very broad description of a forecast, especially for some particular countries, for example, China and India. Besides, a thought-provoking part about the influence of AI has been added to this report which makes the paper useful for further researches.

Table 2 Global cities index metrics

Business activity	Human capital	Information exchange	Cultural experience	Political engagement
• Fortune 500 • Top global services firms • Capital markets • Air freight • ICCA conferences • Unicorn companies	• Foreign-born population • Top universities • Population with tertiary degree • International student population • Number of international schools • Ease of entry	• Broadband subscribers • News agency bureaus • Freedom of expression • Online presence • Internet speed • Data center presence	• Museums • Visual and performing arts • Sporting events • International travelers • Culinary offerings • Sister cities	• Embassies and consulates • Think tanks • International organizations • Political conferences • Local institutions with global reach

Table 3 The top 15 cities in the Global Cities Index

Rank 2025	City	Rank 2024	Rank 2023	Rank 2022	Rank 2021	Rank 2020
1	New York	1	1	1	1	1
2	London	2	2	2	2	2
3	Paris	3	3	3	3	3
4	Tokyo	4	4	4	4	4
5	Singapore	5	7	9	9	9
6	Beijing	6	5	5	6	5
7	Hong Kong	9	10	10	7	6
8	Shanghai	8	13	16	101	12
9	Los Angeles	7	8	6	5	7
10	Chicago	10	11	7	8	8
11	Madrid	13	12	19	19	16
12	Seoul	11	14	13	17	17
13	Toronto	12	15	18	20	19
14	San Francisco	14	17	15	11	13
15	Washington, D.C	15	19	12	14	10

4 Results: New Methodology for Rating Smart Cities as Smart Destinations

The analysis of the previous two systems of ranking Smart cities has shown that usually Smart cities are considered only as cities where it is comfortable to live. There is a lack of "touristic" criteria and metrics that explains the need for creating them. Thus, the list of Smart cities can be different since not all of the previous leaders can follow all the criteria.

4.1 New Criteria and Metrics

Tourism business plays vital role in tourism development since it studies trends and tendencies, analyzes the statistics and creates new offers for a market. Here it is important to have a look at the available information about B2B and B2c communication channels, as well as NPS data that is given in close to real-time mode.

In our opinion, a Smart destination follows the "sustainable path" that should take into account social, economic and ecological issues. That is why, criteria in this direction will give the highest weight on the final score. Ecological measures will be the second criteria where the experience of low carbon footprint is evaluated, for example, ecological transport and hospitality services, as well as, ways to support protected natural areas and prevent over tourism.

Besides, the usage of smart technologies will be estimated, especially official apps or website that provide all the necessary information. It is hard to imagine any website or social media without AI integration [35], so there is such metric in the category "Smart experience".

In addition, level of touristic infrastructure [36] is necessary to be estimated since the touristic flow in the future depends on the satisfaction level [37] of current tourists. Transportation is a huge part of any trip so the transport system and its integration are one of the main metrics in this category. Electric transportation is one of the main directions for sustainable development, so it is important for a Smart destination to provide electric cars rent as well as to have charging places for such cars.

All in all, there are four main categories of criteria for Smart Destination Index that can be seen in the Table 4: tourism business activity, ecological measures, Smart experience and infrastructure. The first category has 3 metrics, the second one—4, the third one—5, the fourth one—6, in total there are 18 metrics.

Each category has its own weight depending on the importance. Thus, ecological measures and Smart experience have 1,0, infrastructure 0, 9 and tourism business activity 0, 8. Metrics also have their own scores but for each one the score is the same—1 point. The highest result for the first category is 2, 4, for the second one—4, for the third one—5, for the fourth one—5,4.

All the cities that are analyzed can be divided into 4 groups depending on the final result and their scores. The groups and description are written in the Table 5. Cities that have less than 6 points are not Smart destinations but potential contenders to be a smart destination.

Table 4 New criteria for smart destinations index

Tourism business activity	Ecological measures	Smart experience	Infrastructure
• NPS data is got in close to real-time mode • Online portal for tourism business is available • Level of usage of social media and messengers for B2C communication	• Number of electric buses and trams • Percentage of hotels and restaurants offering low-carbon footprint services • Implement technologies to monitor the monuments condition • Systems to manage tourist flows in protected natural areas	• Official apps for tourists are available • AI chat bots are used as a 24/7 support • Self-check-in/check-out hotels exist • Non-cash and NFC payment systems • Number of sites that offer AR, VR, XR technologies	• Public transport development • Car sharing services are available • Integration of transport systems • Number of the charging places for electric vehicles • Interactive and offline maps are available • Quantity and quality of QR-codes for getting information

Table 5 Classification of smart destinations by digitalization level

Group	Number of points	Description
Group 1 Future pioneer	14–16,8	Such city does not follow a trend but creates trends itself. Future Pioneer uses the latest technology, including AI, Internet of Things, Big Data to form unique touristic experience
Group 2 Confident navigator	12–13,9	City uses data and technology rather wisely to manage tourist flows, make services more personalized, as well as solve current issues more efficient
Group 3 Active explorer	8,8–11,9	City is in the process of smart solutions integration into the tourism sphere. It has found successful cases and support them but still is on the stage of optimization all the smart things usage
Group 4 Digital pathfinder	6–8,7	A city that is only in the beginning of the "smart path". It has defined the goals and development directions, has started new projects but does not have a huge number of results and cases

4.2 *Application of the New Criteria*

In order to prove the efficiency of the new criteria, it is necessary to analyze several cities. The choice was based on the rating of the most visited cities in Russia and two foreign cities were chosen to compare. As for Russian cities, Moscow and Saint Petersburg are traditionally most popular cities among tourists. Foreign cities for comparison are Abu Dhabi and Beijing since these cities draw more and more Russian tourists' attention thanks to visa policy and increasing interest to the oriental culture. Final results with the scores for each category are shown in the Table 6.

1. ***Moscow*** is the capital and the biggest city of Russia. Recent times, there are more and more pieces of news about technological development in the city. Despite the fact that it does not have any information about NPS data in close to real-time mode, the B2B and B2C communication is well developed. There is no a particular online platform as a website but there is a Telegram channel and VKgroup that is called "Mostourism" where useful information for both business and tourist is published.

Ecological measures are also highly estimated since electric transport is well developed, for instance, public transport system includes river electric boats, electric buses.

Table 6 Target cities analysis on each category of criteria

City	Tourism business activity	Ecological measures	Smart experience	Infrastructure	Total
Moscow	1,6	2,5	5	5,4	14,5
Saint Petersburg	1,6	2	4	4,5	12,1
Abu Dhabi	1,2	2	3	4,5	10,7
Beijing	1,6	3,5	5	5,4	15,5

The percentage of electric public transport all the trips on public transport is 80%. There is a lack of information and statistics on monitoring monuments condition and systems to manage tourist flows in protected natural areas that can mean there are not so many successful cases of using technologies in these directions. However, there is one example, Mosnature, that takes care of natural areas including reducing negative impact on them.

The category "Smart experience" has the highest score because there are a lot of examples of smart things. For instance, an app "Know Moscow" is available for any device, there is a lot of information about history and culture, as well as famous people in both Russian and English languages. Hospitality industry is also well-developed including examples of hotels with self-check-in/check-out services.

Infrastructure quality is also rather high, including charging places for electric vehicle that are more than 350 around the city. Car sharing services are available and affordable, as well as public transport system is well-organized, including interactive and offline maps. Each year more and more QR-codes appear that helps tourists pay and get new information.

2. ***Saint Petersburg*** is called "Cultural capital of Russia" and gets more and more tourists with different purposes every year. There is also a platform for both tourists and business that is called "visit-petersburg". Business activity is supported by holding different tours and events for tour operators' and hospitality industry's specialists. Besides the website, there are social media pages that post useful content that attracts tourists.

Ecological measures are taken gradually, including public transport system transformation adding more electric vehicles. More trams and trolleybuses have appeared since the beginning of 2025. There are six ecological trails on the territory of the city's protected natural area, so they are under the Direction of protected natural areas of Leningrad region.

Smart experience is a bit poorer than in Moscow but still is available. There is no a particular app about the city but there are others useful apps to make holiday more comfortable. It is possible to use non-cash payment system almost everywhere in the city, as well as NFC system is wide spread, including in public transport. AR technology is used in different sites and websites of the places to make the visit of a place more interactive.

Public transport system is well developed but the metro does not have the same number of stations as Moscow has which is explained with the geographical features. There are also charging places for electric cars but their number is much lower in comparison with Moscow. All the services can be seen on the maps that continue working in offline mode.

3. ***Abu Dhabi*** is a city in UAE and the safest destination for travelers in 2025 which is the main criteria for a Smart city. The city has its own touristic website that tells why it is necessary to visit and gives useful information for visitors, including MICE tourists. However, there is no information or a part for business. Level of usage of social media is high since there are links to five world famous media that attract tourists all over the world, including China.

Nature is under the control of the Environment Agency of Abu Dhabi which has claimed that the transportation system is going to be reformed including using renewable energy technologies. As for protected natural areas, there are not data on how they are controlled and managed in the direction of preventing over tourism. Also, there is no information about monitoring the monuments condition.

Official app for tourists is available, there are guides on where to go and what to do, and interactive map. Some other examples of Smart experience are also provided, including non-cash payment systems, as well as contactless payment in transport.

As for the transportation, this system is well developed and includes buses, trams and ferries. Car sharing, including electric cars, and bicycle renting services are also available. With the growth of electric vehicle, there is an increasing number of charging stations in the city that are located not only near the block of flats but close to main destinations, hotels and restaurant.

4. ***Beijing***, the capital city of China, is in the top Smart cities in different ratings. The official website for travelers is a part of the main Government website. There is information about seasons, things to do and places to visit. The main difference from the previous cities' websites is the number of available foreign languages—there are 10 of them. As for social media, it is hard to estimate since most popular European social media are not used or banned in China, that is why the main channel for B2C communication is an official website, but if a tourist has an access to Chinese social media, then they might find necessary information there.

Electric public transport is a part of sustainable development program, that is why the underground system is well-developed, and some of the buses were replaces with the ones equipped with hydrogen fuel cell systems. As for the social part of sustainability, it is important to highlight the practice of taking care of historical heritage with the help of both technology and people. This cooperation makes it possible to monitor the condition of the Beijing Central Axis.

Despite the fact that there is no exact app about Beijing, there are a lot of useful apps for tourists. All of them are available and provide smart experience, for instance apps for getting taxi or online shopping. As for the AI chat bots that are popular in China, there is no a particular on for tourists but DeepSeek can solve any problem and tourist can trust it. Non-cash and NFc payment systems in the capital of China are also developed, so the level of comfort is high.

Infrastructure also helps to have various Smart experience, for instance car sharing services are available, including renting an electric car. Using the las one is comfortable as there are more than 50 charging places. Maps and navigation is also clear in both online and offline modes.

To sum up, there are two cities that can be called "Future Pioneer"—Moscow and Beijing, "Confident Navigator" is Saint Petersburg and Active Explorer is Abu Dhabi. It important to stress that all of the cities are the Smart ones and continue their sustainable development.

5 Conclusion

All in all, analysis of the Smart cities ratings has shown the importance of classifying all the cities using criteria. At the same time, the criteria do not cover the touristic direction of a city development, that is why the need of a new system of criteria has been proved.

As a result, 4 groups of criteria were created including the scores each metric gives. These criteria were checked during the analysis of four cities. It led to dividing these cities into four groups depending on the final result, which can mean the level of digitalization and Smart tourism destination. This kind of rating can be broadened in the further research, as well as the system of criteria can be expanded for different regions and types of destinations.

To sum up, it is worth mentioning that Smart tourism development is possible in cities with different level of digitalization. The main idea here is to adapt smart technologies as much as it possible and integrate them into residents' everyday life that will attract more tourists.

References

1. Testina Y, Kuchumov A, Boykova J, Voloshinova M SMART CITIES as drivers of digitalization and environmentalization of the arctic. In: SPBPU IDE '21: proceedings of the 3rd international scientific conference on innovations in digital economy. pp 220–227. https://doi.org/10.1145/3527049.3527133
2. Pilelienė L, Alsharif AH (2025) Going smart: a Bibliometric review of Smart tourism, Smart destination, and Smart hospitality research. Baltic J Econ Stud 11(5):92–112. https://doi.org/10.30525/2256-0742/2025-11-5-92-112
3. Wu D (2024) City walk in a gap day: potential and opportunities for tourism and leisure. Tour Rev 79(9):1576–1581. https://doi.org/10.1108/TR-09-2023-0614
4. Karpova GA, Kuchumov AV, Voloshinova MV, Testina YS (2019) Digitalization of a tourist destination. In: Proceedings of the 2019 International SPBPU scientific conference on innovations in digital economy (SPBPU IDE '19). p 3373342. https://doi.org/10.1145/3372177.3373342
5. Zheng M (2025) Impact of the digital economy on cultural -tourism integration. Highlights Bus Econ Manag 64:170–189. https://doi.org/10.54097/bs9xy091
6. Gong XY, Wang HF (2019) Research on the development and effects of contemporary digital economy. E-Government 8(51–62):7
7. Muharis M, Taufik M, Citra Kencana N (2025) Implicature of smart tourism in enhancing the reputation of tourism destinations. ARMADA J Penelitian Multidisipl 3(7):240–248. https://doi.org/10.55681/armada.v3i7.168
8. Kuchumov A, Karpova G, Testina Y (2024) Smart destinations for advanced development of tourism. In: Sari M, Kulachinskaya A (eds) Digital transformation: what are the smart cities today?. Lecture notes in networks and systems, vol 846. Springer, Cham. https://doi.org/10.1007/978-3-031-49390-4_4
9. Sustacha I, José Francisco B, Eduardo Del V (2023) The role of technology in enhancing the tourism experience in smart destinations: a meta-analysis. J Destin Mark Manag 30. https://doi.org/10.1016/j.jdmm.2023.100817

10. Kuchumov A, Testina Y, Chaikovskaya A, Maslova E (2020) Influence of digital innovations on environmentalization of tourist destinations. In: Proceedings—International scientific conference on innovations in digital economy, SPBPU IDE 2020., 3444466, pervasivehealth: pervasive computing technologies for healthcare, association for computing machinery, 2020 International scientific conference on innovations in digital economy, SPBPU IDE 2020. Saint–Petersburg, Российская Федерация. https://doi.org/10.1145/3444465.3444466
11. Buhalis D, Aditya A (2015) Smart tourism destinations enhancing tourism experience through personalisation of services. Inform Commun Technol Tour 2015:377–389. https://doi.org/10.1007/978-3-319-14343-9_28
12. Acuna R (2025) Digitalization and its effect on customer satisfaction. Res J Bus Econ 3(3):190–198. https://doi.org/10.61424/rjbe.v3i3.572
13. Bai S, Li Z, He H, Fan W (2025) Natural versus urban: unveiling the effect of visual complexity on user engagement with destination marketing content. J Hosp Tour Technol ahead-of-print No. ahead-of-print. https://doi.org/10.1108/JHTT-04-2025-0302
14. Saltik Z, Uludag O, Andrlić B (2025) Digital pathways to sustainability: eco-travel apps and gen Z's eco-friendly travel behaviors. Tour Hosp 6(5):247. https://doi.org/10.3390/tourhosp6050247
15. Xue J, Xia Y (2025) Research status on enterprise digitalization, green transformation, and their collaborative development. Sci J Econ Manag Res 7(11):107–117. https://doi.org/10.54691/k96ar718
16. Marcinkevicius G, Vilkas M (2023) The affordances of digital technologies for business processes integration. J Syst Inf Technol 25(1):74–90. https://doi.org/10.1108/JSIT-06-2022-0149
17. Zaman SAA, Vilkas M, Rauleckas R, Zaman SI (2025) Uncovering the digitalization paradox: how investment in digitalization affects revenue enhancement and firm performance. Benchmark Int J ahead-of-print No. ahead-of-print. https://doi.org/10.1108/BIJ-06-2024-0531
18. Leung D (2019) Artificial intelligence in tourism and hospitality: a review of theories and applications. J Hosp Tour Technol 10(2):289–305. https://doi.org/10.1108/JHTT-12-2018-0118
19. Yin Q, Guo C, Yan Z (2025) From intervention to prevention! Effects of peer progress information on user engagement. Internet Res ahead-of-print No. ahead-of-print. https://doi.org/10.1108/INTR-05-2024-0817
20. Zeng S, Tanveer A, Fu X, Gu Y, Irfan M (2022) Modeling the influence of critical factors on the adoption of green energy technologies. Renew Sustain Energy Rev 168:112817
21. Kucherov YN, Bushuev VV, Ivanov AV, Korev DA, Utz SA, Sheikhina AV (2019) Development of new technologies energy life support smart cities. J Environ Earth Energy Study (JEEES) (3):49–69
22. Gu M, Xie J, Zhang Y, Yang L, Huo B (2025) Supply chain digitalization and transparency: the privacy calculus perspective. Indus Manag Data Syst ahead-of-print No. ahead-of-print. https://doi.org/10.1108/IMDS-06-2025-0834
23. Podsakoff PM, MacKenzie SB, Lee JY, Podsakoff NP (2003) Common method biases in behavioral research: a critical review of the literature and recommended remedies. J Appl Psychol 88(5):879–903
24. Heeks R (2006) Implementing and managing eGovernment: an international text. SAGE Publications Ltd. https://doi.org/10.4135/9781446220191
25. Tingli J, Aw EC, Huam H, Tan GW (2025) Artificial intelligence at the wheel: exploring tourists' trust in autonomous vehicles through a time-lagged survey. J Hosp Tour Technol ahead-of-print No. ahead-of-print. https://doi.org/10.1108/JHTT-04-2025-0334
26. Shilvy AS (2025) Research trend: SMART people and digital literacy in the SMART city Era, vol 1(1). In: Proceeding of The 5th ASPIKOM International communication conference (AICCON 2025). https://doi.org/10.31947/aiccon2025.v1i1.47794
27. European capital and green pioneer of smart tourism. https://smart-tourism-capital.ec.europa.eu/index_en. Accessed 21 Nov 2025

28. Torino (2025) European capital of smart tourism. https://smart-tourism-capital.ec.europa.eu/torino-2025-european-capital-smart-tourism_en. Accessed 21 Nov 2025
29. Tampere (2026) European capital of smart tourism. https://smart-tourism-capital.ec.europa.eu/winners-and-finalists/cities-2026/tampere-2026-european-capital-smart-tourism_en. Accessed 21 Nov 2025
30. Benidorm (2025) European green pioneer of smart tourism. https://smart-tourism-capital.ec.europa.eu/benidorm-2025-european-green-pioneer-smart-tourism_en. Accessed 21 Nov 2025
31. Dubrovnik (2026) European green pioneer of smart tourism. https://smart-tourism-capital.ec.europa.eu/winners-and-finalists/cities-2026/dubrovnik-2026-european-green-pioneer-smart-tourism_en. Accessed 22 Nov 2025
32. Nakamura H (2025) World smart cities ranking for doing business in climate change. In: Lackner M, Sajjadi B, Chen WY (eds) Handbook of climate change mitigation and adaptation. Springer, Cham. https://doi.org/10.1007/978-3-031-84483-6_16
33. Smart City Index (2025) https://www.imd.org/smart-city-observatory/home/. Accessed 22 Nov 2025
34. Global cities report: accelerating through volatility (2025) https://www.kearney.com/service/national-transformations-institute/gcr. Accessed 22 Nov 2025
35. Yong X, Xinxin T, Su Z et al (2020) Construction and application of digital creative platform for digital creative industry based on smart city concept. Comput Electr Eng 87:106748. https://doi.org/10.1016/j.compeleceng.2020.106748
36. Lee SJ, Hunter WC, Chung N (2020) Smart tourism city: developments and transformations. Sustainability 1203958
37. Veenhoven R (1996) Developments in satisfaction research. Soc Indic Res 37:1–45

The manufacturer's authorised representative in the EU is Springer Nature Customer Service Centre GmbH, Europaplatz 3, 69115 Heidelberg, Germany. If you have any concerns regarding our products, please contact ProductSafety@springernature.com

Printed and bound by CPI Group (UK) Ltd, Croydon, CR0 4YY
16/09/2026
02217781-0005